S. Northshield   10/1/86

Grundlehren der mathematischen Wissenschaften 249

A Series of Comprehensive Studies in Mathematics

Kai Lai Chung

Lectures from Markov Processes to Brownian Motion

With 3 Figures

Springer-Verlag
New York   Heidelberg   Berlin

Kai Lai Chung
Department of Mathematics
Stanford University
Stanford, CA 94305

AMS Subject Classifications (1980): 60Jxx

Library of Congress Cataloging in Publication Data
Chung, Kai Lai, 1917–
    Lectures from Markov processes to Brownian
motion.

    (Grundlehren der mathematischen Wissenschaften;
249)
    Bibliography: p.
    Includes index.
    1. Markov processes.   2. Brownian motion pro-
cesses.   I. Title.   II. Series.
QA274.7.C48   519.2′33   81-14413
                                       AACR2

Printed in the United States of America.

9 8 7 6 5 4 3 2 1

ISBN 0-387-90618-5 Springer-Verlag   New York   Heidelberg   Berlin
ISBN 3-540-90618-5 Springer-Verlag   Berlin   Heidelberg   New York

Contents

Chapter 5
Potential Developments

Preface

This book evolved from several stacks of lecture notes written over a decade and given in classes at slightly varying levels. In transforming the overlapping material into a book, I aimed at presenting some of the best features of the subject with a minimum of prerequisities and technicalities. (Needless to say, one man's technicality is another's professionalism.) But a text frozen in print does not allow for the latitude of the classroom; and the tendency to expand becomes harder to curb without the constraints of time and audience. The result is that this volume contains more topics and details than I had intended, but I hope the forest is still visible with the trees.

The book begins at the beginning with the Markov property, followed quickly by the introduction of optional times and martingales. These three topics in the discrete parameter setting are fully discussed in my book *A Course In Probability Theory* (second edition, Academic Press, 1974). The latter will be referred to throughout this book as the *Course*, and may be considered as a general background; its specific use is limited to the material on discrete parameter martingale theory cited in §1.4. Apart from this and some dispensable references to Markov chains as examples, the book is self-contained. However, there are a very few results which are explained and used, but not proved here, the first instance being the theorem on projection in §1.6. The fundamental regularity properties of a Markov process having a Feller transition semigroup are established in Chapter 2, together with certain measurability questions which must be faced. Chapter 3 contains the basic theory as formulated by Hunt, including some special topics in the last three sections. Elements of a potential theory accompany the development, but a proper treatment would require the setting up of dual structures. Instead, the relevant circle of ideas is given a new departure in Chapter 5. Chapter 4 grew out of a short compendium as a particularly telling example, and Chapter 5 is a splinter from unincorporated sections of Chapter 4. The venerable theory of Brownian motion is so well embellished and ramified that once begun it is hard to know where to stop. In the end I have let my own propensity and capability make the choice. Thus the last three sections of the book treat several recent developments which have engaged me lately. They are included here with the hope of inducing further work in such fascinating old-and-new themes as equilibrium, energy, and reversibility.

I used both the Notes and Exercises as proper non-trivial extensions of the text. In the Notes a number of regretably omitted topics are mentioned, and related to the text as a sort of guide to supplementary reading. In the Exercises there are many alternative proofs, important corollaries and examples that the reader will do well not to overlook.

The manuscript was prepared over a span of time apparently too long for me to maintain a uniform style and consistent notation. For instance, who knows whether "semipolar" should be spelled with or without a hyphen? And if both $|x|$ and $\|x\|$ are used to denote the same thing, does it really matter? Certain casual remarks and repetitions are also left in place, as they are permissible, indeed desirable, in lectures. Despite considerable pains on the part of several readers, it is perhaps too much to hope that no blunders remain undetected, especially among the exercises. I have often made a point, when assigning homework problems in class, to say that the correction of any inaccurate statement should be regarded as part of the exercise. This is of course not a defense for mistakes but merely offered as prior consolation.

Many people helped me with the task. To begin with, my first formal set of notes, contained in five folio-size, lined, students' copybooks, was prepared for a semester course given at the Eidgenössiche Technische Hochschule in the spring of 1970. My family has kept fond memories of a pleasant sojourn in a Swiss house in the great city of Zürich, and I should like to take this belated occasion to thank our hospitable hosts. Another set of notes (including the lectures given by Doob mentioned in §4.5) was taken during 1971–2 by Harry Guess, who was kind enough to send me a copy. Wu Rong, a visiting scholar from China, read the draft and the galley proofs, and checked out many exercises. The comments by R. Getoor, N. Falkner, and Liao Ming led to some final alterations. Most of the manuscript was typed by Mrs. Gail Stein, who also typed some of my other books. Mrs. Charlotte Crabtree, Mrs. Priscilla Feigen, and my daughter Marilda did some of the revisions. I am grateful to the National Science Foundation for its support of my research, some of which went into this book.

August 1981   Kai Lai Chung

Chapter 1

Markov Process

1.1. Markov Property

We begin by describing a general Markov process running on continuous time and living in a topological space. The time parameter is the set of positive numbers, considered at first as just a linearly ordered set of indices. In the discrete case this is the set of positive integers and the corresponding discussion is given in Chapter 9 of the *Course*. Thus some of the proofs below are the same as for the discrete case. Only later when properties of sample functions are introduced will the continuity of time play an essential role. As for the living space we deal with a general one because topological properties of sets such as "open" and "compact" will be much used while specific Euclidean notions such as "interval" and "sphere" do not come into question until much later.

We must introduce some new terminology and notation, but we will do this gradually as the need arises. Mathematical terms which have been defined in the *Course* will be taken for granted, together with the usual symbols to denote them. The reader can locate these through the Index of the *Course*. But we will repeat certain basic definitions with perhaps slight modifications.

Let $(\Omega, \mathscr{F}, P)$ be a probability space. Let

$$\mathbf{T} = [0, \infty).$$

Let $\mathbf{E}$ be a locally compact separable metric space; and let $\mathscr{E}$ be the minimal Borel field in $\mathbf{E}$ containing all the open sets. The reader is referred to any standard text on real analysis for simple topological notions. Since the Euclidean space $\mathbf{R}^d$ of any dimension d is a well known particular case of an $\mathbf{E}$, the reader may content himself with thinking of $\mathbf{R}^d$ while reading about $\mathbf{E}$, which is not a bad practice in the learning process.

For each $t \in \mathbf{T}$, let

$$X_t(\omega) = X(t, \omega)$$

be a function from Ω to $\mathbf{E}$ such that

$$X_t^{-1}(\mathscr{E}) \subset \mathscr{F}.$$

This will be written as

$$X_t \in \mathscr{F}/\mathscr{E};$$

and we say that X_t is a random variable taking values in $(\mathbf{E}, \mathscr{E})$. For $\mathbf{E} = \mathbf{R}^1$, $\mathscr{E} = \mathscr{B}^1$, this reduces to the familiar notion of a real random variable. Now any family $\{X_t, t \in \mathbf{T}\}$ is called a *stochastic process*. In this generality the notion is of course not very interesting. Special classes of stochastic processes are defined by imposing certain conditions on the random variables X_t, through their joint or conditional distributions. Such conditions have been formulated by pure and applied mathematicians on a variety of grounds. By far the most important and developed is the class of Markov processes that we are going to study.

Borel field is also called σ-field or σ-algebra. As a general notation, for any family of random variables $\{Z_\alpha, \alpha \in A\}$, we will denote the σ-field generated by it by $\sigma(Z_\alpha, \alpha \in A)$. Now we put specifically

$$\mathscr{F}_t^0 = \sigma(X_s, s \in [0, t]); \qquad \mathscr{F}_t' = \sigma(X_s, s \in [t, \infty)).$$

Intuitively, an event in $\mathscr{F}_t^0$ is determined by the behavior of the process $\{X_s\}$ up to the time t; an event in $\mathscr{F}_t'$ by its behavior after t. Thus they represent respectively the "past" and "future" relative to the "present" instant t. For technical reasons, it is convenient to enlarge the past, as follows.

Let $\{\mathscr{F}_t, t \in \mathbf{T}\}$ be a family of σ-fields of sets in $\mathscr{F}$, such that

(a)   if $s < t$, then $\mathscr{F}_s \subset \mathscr{F}_t$;
(b)   for each t, $X_t \in \mathscr{F}_t$.

Property (a) is expressed by saying that "$\{\mathscr{F}_t\}$ is increasing"; property (b) by saying that "$\{X_t\}$ is adapted to $\{\mathscr{F}_t\}$". Clearly the family $\{\mathscr{F}_t^0\}$ satisfies both conditions and is the minimal such family in the obvious sense. Other instances of $\{\mathscr{F}_t\}$ will appear soon. The general definition of a Markov process involves $\{\mathscr{F}_t\}$ as well as $\{X_t\}$.

Definition. $\{X_t, \mathscr{F}_t, t \in \mathbf{T}\}$ is a Markov process iff one of the following equivalent conditions is satisfied:

(i) $\forall t \in \mathbf{T}, A \in \mathscr{F}_t, B \in \mathscr{F}_t'$:
$$P(A \cap B | X_t) = P(A | X_t) P(B | X_t).$$
(ii) $\forall t \in \mathbf{T}, B \in \mathscr{F}_t'$:
$$P(B | \mathscr{F}_t) = P(B | X_t).$$
(iii) $\forall t \in \mathbf{T}, A \in \mathscr{F}_t$:
$$P(A | \mathscr{F}_t') = P(A | X_t).$$

The reader is reminded that a conditional probability or expectation is an equivalence class of random variables with respect to the measure P. The equations above are all to be taken in this sense.

We shall use the two basic properties of conditional expectations, for arbitrary σ-fields $\mathscr{G}$, $\mathscr{G}_1$, $\mathscr{G}_2$ and integrable random variables Y and Z:

(a) If $Y \in \mathscr{G}$, then $E\{YZ|\mathscr{G}\} = YE\{Z|\mathscr{G}\}$;

(b) If $\mathscr{G}_1 \subset \mathscr{G}_2$, then

$$E\{E(Y|\mathscr{G}_1)|\mathscr{G}_2\} = E\{E(Y|\mathscr{G}_2)|\mathscr{G}_1\} = E\{Y|\mathscr{G}_1\}.$$

See Chapter 9 of *Course*.

Let us prove the equivalence of (i), (ii) and (iii). Assume that (i) holds, we will deduce (ii) in the following form. For each $A \in \mathscr{F}_t$ and $B \in \mathscr{F}'_t$ we have

$$E\{1_A P(B|X_t)\} = P(A \cap B). \tag{1}$$

Now the left member of (1) is equal to

$$E\{E[1_A P(B|X_t)]|X_t\} = E\{P(A|X_t)P(B|X_t)\}$$
$$= E\{P(A \cap B|X_t)\} = P(A \cap B).$$

Symmetrically, we have

$$E\{1_B P(A|X_t)\} = P(A \cap B),$$

which implies (iii).

Conversely, to show for instance that (ii) implies (i), we have

$$P(A \cap B|X_t) = E\{E(1_A \cdot 1_B|\mathscr{F}_t)|X_t\}$$
$$= E\{1_A P(B|\mathscr{F}_t)|X_t\} = E\{1_A P(B|X_t)|X_t\}$$
$$= P(B|X_t)E\{1_A|X_t\} = P(B|X_t)P(A|X_t).$$

From here on we shall often omit such qualifying phrases as "$\forall t \in \mathbf{T}$". As a general notation, we denote by $b\mathscr{G}$ the class of bounded real-valued $\mathscr{G}$-measurable functions; by C_c the class of continuous functions on $\mathbf{E}$ with compact supports.

Form (ii) of the Markov property is the most useful one and it is equivalent to any of the following:

(iia) $\forall Y \in b\mathscr{F}'_t$:

$$E\{Y|\mathscr{F}_t\} = E\{Y|X_t\}.$$

(iib) $\forall u \geq t, f \in b\mathscr{E}$:

$$E\{f(X_u)|\mathscr{F}_t\} = E\{f(X_u)|X_t\}.$$

(iic) $\forall u \geq t, f \in C_c(E)$:

$$E\{f(X_u)|\mathscr{F}_t\} = E\{f(X_u)|X_t\}.$$

It is obvious that each of these conditions is weaker than the preceding one. To show the reverse implications we state two lemmas. As a rule, a self-evident qualifier "nonempty" for a set will be omitted as in the following proposition.

Lemma 1. *For each open set G, there exists a sequence of functions $\{f_n\}$ in C_c such that*

$$\lim_n \uparrow f_n = 1_G.$$

This is an easy consequence of our topological assumption on **E**, and gives the reader a good opportunity to review his knowledge of such things.

Lemma 2. *Let S be an arbitrary space and $\mathbb{D}$ a class of subsets of S. $\mathbb{D}$ is closed under finite intersections. Let $\mathbb{C}$ be a class of subsets of S such that $S \in \mathbb{C}$ and $\mathbb{D} \subset \mathbb{C}$. Furthermore suppose that $\mathbb{C}$ has the following closure properties:*

(a) *if $A_n \in \mathbb{C}$ and $A_n \subset A_{n+1}$ for $n \geq 1$, then $\bigcup_{n=1}^{\infty} A_n \in \mathbb{C}$;*
(b) *if $A \subset B$ and $A \in \mathbb{C}, B \in \mathbb{C}$, then $B - A \in \mathbb{C}$.*

Then $\mathbb{C} \supset \sigma(\mathbb{D})$.

Here as a general notation $\sigma(\mathbb{D})$ is the σ-field generated by the class of sets $\mathbb{D}$. Lemma 2 is Dynkin's form of the "monotone class theorem"; the proof is similar to Theorem 2.1.2 of *Course*, and is given as an exercise there. The reader should also figure out why the cited theorem cannot be applied directly in what follows.

Let us now prove that (iic) implies (iib). Using the notation of Lemma 1, we have by (iic)

$$E\{f_n(X_u)|\mathscr{F}_t\} = E\{f_n(X_u)|X_t\}.$$

Letting $n \to \infty$ we obtain by monotone convergence

$$P\{X_u \in G|\mathscr{F}_t\} = P\{X_u \in G|X_t\}. \tag{2}$$

Now apply Lemma 2 to the space **E**. Let $\mathbb{D}$ be the class of open sets, $\mathbb{C}$ the class of sets A satisfying

$$P\{X_u \in A|\mathscr{F}_t\} = P\{X_u \in A|X_t\}. \tag{3}$$

Of course $\mathbb{D}$ is closed under finite intersections, and $\mathbb{D} \subset \mathbb{C}$ by (2). The other properties required of $\mathbb{C}$ are simple consequences of the fact that each member of (3), as function of A, acts like a probability measure; see p. 301 of *Course* for a discussion. Hence we have $\mathbb{C} \supset \mathscr{E}$ by Lemma 2, which means that (3) is true for each A in $\mathscr{E}$, or again that (iib) is true for $f = 1_A, A \in \mathscr{E}$. The class of f for which (iib) is true is closed under addition, multiplication by a constant, and monotone convergence. Hence it includes $b\mathscr{E}$ by a standard approximation. [We may invoke here Problem 11 in §2.1 of *Course*.]

To prove that (iib) implies (iia), we consider first

$$Y = f_1(X_{u_1}) \cdots f_n(X_{u_n})$$

where $t \leq u_1 < \cdots < u_n$, and $f_j \in b\mathscr{E}$ for $1 \leq j \leq n$. For such a Y with $n = 1$, (iia) is just (iib). To make induction from $n - 1$ to n, we write

$$E\left\{\prod_{j=1}^{n} f_j(X_{u_j}) \Big| \mathscr{F}_t\right\} = E\left\{E\left[\prod_{j=1}^{n} f_j(X_{u_j}) \Big| \mathscr{F}_{u_{n-1}}\right] \Big| \mathscr{F}_t\right\}$$

$$= E\left\{\prod_{j=1}^{n-1} f_j(X_{u_j}) E[f_n(X_{u_n}) | \mathscr{F}_{u_{n-1}}] \Big| \mathscr{F}_t\right\}. \tag{4}$$

Now we have by (iib)

$$E[f_n(X_{u_n}) | \mathscr{F}_{u_{n-1}}] = E[f_n(X_{u_n}) | X_{u_{n-1}}] = g(X_{u_{n-1}})$$

for some $g \in b\mathscr{E}$. Substituting this into the above and using the induction hypothesis with $f_{n-1} \cdot g$ taking the place of f_{n-1}, we see that the last term in (4) is equal to

$$E\left\{\prod_{j=1}^{n-1} f_j(X_{u_j}) E[f_n(X_{u_n}) | \mathscr{F}_{u_{n-1}}] \Big| X_t\right\} = E\left\{E\left[\prod_{j=1}^{n} f_j(X_{u_j}) \Big| \mathscr{F}_{u_{n-1}}\right] \Big| X_t\right\}$$

$$= E\left\{\prod_{j=1}^{n} f_j(X_{u_j}) \Big| X_t\right\}$$

since $X_t \in \mathscr{F}_{u_{n-1}}$. This completes the induction.

Now let $\mathbb{D}$ be the class of subsets of Ω of the form $\bigcap_{j=1}^{n} \{X_{u_j} \in B_j\}$ with the u_j's as before and $B_j \in \mathscr{E}$. Then $\mathbb{D}$ is closed under finite intersections. Let $\mathbb{C}$ be the class of subsets Λ of Ω such that

$$P\{\Lambda | \mathscr{F}_t\} = P\{\Lambda | X_t\}.$$

Then $\Omega \in \mathbb{C}$, $\mathbb{D} \subset \mathbb{C}$ and $\mathbb{C}$ has the properties (a) and (b) in Lemma 2. Hence by Lemma 2, $\mathbb{C} \supset \sigma(\mathbb{D})$ which is just $\mathscr{F}'_t$. Thus (iia) is true for any indicator $Y \in \mathscr{F}'_t$ [that is (ii)], and so also for any $Y \in b\mathscr{F}'_t$ by approximations. The equivalence of (iia), (iib), (iic), and (ii), is completely proved.

Finally, (iic) is equivalent to the following: for arbitrary integers $n \geq 1$ and $0 \leq t_1 < \cdots t_n < t < u$, and $f \in C_c(\mathbf{E})$ we have

$$E\{f(X_u) | X_t, X_{t_n}, \ldots, X_{t_1}\} = E\{f(X_u) | X_t\}. \tag{5}$$

This is the oldest form of the Markov property. It will not be needed below and its proof is left as an exercise.

1.2. Transition Function

The probability structure of a Markov process will be specified.

Definition. The collection $\{P_{s,t}(\cdot,\cdot), 0 \le s < t < \infty\}$ is a *Markov transition function* on $(\mathbf{E}, \mathscr{E})$ iff $\forall s < t < u$ we have

(a) $\forall x \in \mathbf{E}: A \to P_{s,t}(x, A)$ is a probability measure on $\mathscr{E}$,
(b) $\forall A \in \mathscr{E}: x \to P_{s,t}(x, A)$ is $\mathscr{E}$-measurable;
(c) $\forall x \in \mathbf{E}, \forall A \in \mathscr{E}$:

$$P_{s,u}(x, A) = \int_{\mathbf{E}} P_{s,t}(x, dy) P_{t,u}(y, A).$$

This function is called [*temporally*] *homogeneous* iff there exists a collection $\{P_t(\cdot,\cdot), 0 < t\}$ such that $\forall s < t, x \in \mathbf{E}, A \in \mathscr{E}$ we have

$$P_{s,t}(x, A) = P_{t-s}(x, A).$$

In this case (a) and (b) hold with $P_{s,t}$ replaced by P_t, and (c) may be rewritten as follows (*Chapman-Kolmogorov equation*):

$$P_{s+t}(x, A) = \int_{\mathbf{E}} P_s(x, dy) P_t(y, A). \tag{1}$$

For $f \in b\mathscr{E}$, we shall write

$$P_t f(x) = P_t(x, f) = \int_{\mathbf{E}} P_t(x, dy) f(y).$$

Then (b) implies that $P_t f \in b\mathscr{E}$. For each t, the operator P_t maps $b\mathscr{E}$ into $b\mathscr{E}$, also $\mathscr{E}_+$ into $\mathscr{E}_+$, where $\mathscr{E}_+$ denotes the class of positive (extended-valued) $\mathscr{E}$-measurable functions. The family $\{P_t, t > 0\}$ forms a semigroup by (1) which is expressed symbolically by

$$P_{s+t} = P_s P_t.$$

As a function of x and A, $P_t(x, A)$ is also cailed a "kernel" on $(\mathbf{E}, \mathscr{E})$.

Definition. $\{X_t, \mathscr{F}_t, t \in \mathbf{T}\}$ is a homogeneous Markov process with (P_t) as its transition function (or semigroup) iff for $t \ge 0, s > 0$ and $f \in b\mathscr{E}$ we have

$$E\{f(X_{t+s}) | \mathscr{F}_t\} = P_s(X_t, f). \tag{2}$$

Observe that the left side in (2) is defined as a function of ω (not shown!) only up to a set of P-measure zero, whereas the right side is a completely determined function of ω since X_t is such a function. Such a relation should be understood to mean that one version of the conditional expectation on the left side is given by the right side.

Henceforth a homogeneous Markov process will simply be called a Markov process.

The distribution μ of X_0 is called the *initial distribution* of the process. If $0 \le t_1 < \cdots < t_n$ and $f \in b\mathscr{E}^n$, we have

$$E\{f(X_{t_1}, \ldots, X_{t_n})\}$$
$$= \int \mu(dx_0) \int P_{t_1}(x_0, dx_1) \cdots \int P_{t_n - t_{n-1}}(x_{n-1}, dx_n) f(x_1, \ldots, x_n) \quad (3)$$

where the integrations are over $\mathbf{E}$. In particular if f is the indicator of $A_1 \times \cdots \times A_n$, where $n \ge 1$ and each $A_j \in \mathscr{E}$, this gives the finite-dimensional joint distributions of the process.

If Ω is the space of all functions from $\mathbf{T}$ to $\mathbf{E}$, it is possible by Kolmogorov's extension theorem (see e.g. Doob [1]) to construct a process with the joint distributions given in (3). *We shall always assume that such a process exists in our probability space.* If x is any point in $\mathbf{E}$, and $\mu = \varepsilon_x$ (the point mass at x), the corresponding process is said to start at x. The probability measure on the σ-field $\mathscr{F}^0$ generated by this process will be denoted by P^x, and the corresponding expectation by E^x. For example if $y \in b\mathscr{F}^0$:

$$E^x(Y) = \int_\Omega Y(\omega) P^x(d\omega),$$

and if $Y = 1_A(X_t)$, where $A \in \mathscr{E}$, then the quantity above reduces to

$$P^x(X_t \in A) = P_t(x, A). \quad (4)$$

Furthermore for each $\Lambda \in \mathscr{F}^0$, the function $x \to P^x(\Lambda)$ is $\mathscr{E}$-measurable. For $\Lambda = X_t^{-1}(A)$ this follows from (4) and property (b) of the transition function. The general case then follows by a monotone class argument as in the proof that (iib) implies (iia) in §1.1.

The Markov property (2) can now be written as

$$P\{X_{s+t} \in A | \mathscr{F}_t\} = P^{X_t}\{X_s \in A\} = P_s(X_t, A) \quad (5)$$

where $t \ge 0$, $s > 0$, $A \in \mathscr{E}$. Beware of the peculiar symbolism which allows the substitution of X_t for the generic x in $P^x(\Lambda)$. For instance, if $s = t$ in the second member of (5), the two occurrences of X_t do not have the same significance. [There is, of course, no such confusion in the third member of (5).] Nevertheless the system of notation using the superscript will be found workable and efficient.

We want to extend the equation to sets more general than $\{X_{s+t} \in A\} = X_{s+t}^{-1}(A)$. This can be done expeditiously by introducing a "shift" $\{\theta_t, t \geq 0\}$ in the following manner. For each t, let θ_t map Ω into Ω such that

$$\forall t: (X_s \circ \theta_t)(\omega) = X_s(\theta_t\omega) = X_{s+t}(\omega). \tag{6}$$

With this notation we have

$$X_{s+t}^{-1} = \theta_t^{-1} X_s^{-1},$$

so that (5) becomes

$$P\{\theta_t^{-1}(X_s^{-1}(A))|\mathcal{F}_t\} = P^{X_t}\{X_s^{-1}(A)\}. \tag{7}$$

In general if $\Lambda \in \mathcal{F}^0$, then $\theta_t^{-1}\Lambda \in \mathcal{F}'_t$ (proof?), and we have

$$P\{\theta_t^{-1}\Lambda|\mathcal{F}_t\} = P^{X_t}\{\Lambda\}. \tag{8}$$

More generally, if $Y \in b\mathcal{F}^0$, we have

$$E\{Y \circ \theta_t|\mathcal{F}_t\} = E^{X_t}\{Y\}. \tag{9}$$

The relations (8) and (9) follow from (7) by Lemma 2 of §1.1.

Does a shift exist as defined by (6)? If Ω is the space of all functions on $\mathbf{T}$ to $E: \Omega = E^{\mathbf{T}}$, as in the construction by Kolmogorov's theorem mentioned above, then an obvious shift exists. In fact, in this case each ω in Ω is just the sample function $X(\cdot, \omega)$ with domain $\mathbf{T}$, and we may set

$$\theta_t\omega = X(t + \cdot, \omega),$$

which is another such function. Since $X_s(\omega) = X(s, \omega)$ the equation (6) is a triviality. The same is true if Ω is the space of all right continuous (or continuous) functions, and such a space will serve for our later developments. For an arbitrary Ω, a shift need not exist but it is always possible to construct a shift by enlarging Ω without affecting the probability structure. We will not detail this but rather postulate the existence of a shift as part of our basic machinery for a Markov process.

For an arbitrary probability measure μ on $\mathcal{E}$, we put

$$P^\mu(\Lambda) = \int P^x(\Lambda)\mu(dx), \qquad \Lambda \in \mathcal{F}^0. \tag{10}$$

This is the probability measure determined by the process with initial distribution μ. For instance, equation (9) remains true if the E there is replaced by E^μ. Note that P^μ (in particular P^x) is defined so far only on $\mathcal{F}^0$, in contrast to P which is given on $\mathscr{F} \supset \mathscr{F}^0$. Later we shall extend P^μ to a larger σ-field by completion.

The transition function $P_t(\cdot, \cdot)$ has been assumed to be a strict probability kernel, namely $P_t(x, E) = 1$ for every $t \in \mathbf{T}$ and $x \in \mathbf{E}$. We will extend this by allowing

$$P_t(x, E) \le 1, \qquad \forall t \in \mathbf{T}, x \in \mathbf{E}. \tag{11}$$

Such a transition function is called *submarkovian*, and the case where equality holds in (11) [*strictly*] *Markovian*. A simple device converts the former to the latter as follows. We introduce a new $\partial \notin \mathbf{E}$ and put

$$\mathbf{E}_\partial = \mathbf{E} \cup \{\partial\}, \qquad \mathscr{E}_\partial = \sigma\{\mathscr{E}, \{\partial\}\}.$$

The new point ∂ may be considered as the "point at infinity" in the one-point compactification of $\mathbf{E}$. If $\mathbf{E}$ is itself compact, ∂ is nevertheless adjoined as an isolated point. We now define P'_t as follows for $t > 0$ and $A \in \mathscr{E}$:

$$
\begin{aligned}
P'_t(x, A) &= P_t(x, A), \\
P'_t(x, \partial) &= 1 - P_t(x, E), &&\text{if } x \ne \partial; \\
P'_t(\partial, \mathbf{E}) &= 0, \qquad P'_t(\partial, \{\partial\}) = 1.
\end{aligned}
\tag{12}
$$

It is clear that $P'_t(\cdot, \cdot)$ on $(\mathbf{E}_\partial, \mathscr{E}_\partial)$ is Markovian. Let $(X_t, \mathscr{F}_t)$ be a Markov process on $(\mathbf{E}_\partial, \mathscr{E}_\partial)$ with (P'_t) as transition function. Notwithstanding the last two relations in (12), it does not follow that ∂ will behave like an "absorbing state" (or "trap"). However it can be shown that this may be arranged by an unessential modification of the probability space. We shall assume that this has already been done, so that ∂ is absorbing in the sense that

$$\forall \omega, \forall s \ge 0: \{X_s(\omega) = \partial\} \subset \{X_t(\omega) = \partial \text{ for all } t \ge s\}. \tag{13}$$

Now we define the function ζ from Ω to $[0, \infty]$ as follows:

$$\zeta(\omega) = \inf\{t \in \mathbf{T} : X_t(\omega) = \partial\} \tag{14}$$

where, as a standard convention, $\inf \varnothing = \infty$ for the empty set $\varnothing$. Thus $\zeta(\omega) = \infty$ if and only if $X_t(\omega) \ne \partial$ for all $t \in \mathbf{T}$, in other words $X_t(\omega) \in \mathbf{E}$ for all $t \in \mathbf{T}$. The random variable ζ is called the *lifetime* of the process X.

The observant reader may remark that so far we have not defined $P_0(\cdot, \cdot)$. There are interesting cases where $P_0(x, \cdot)$ need not be the point mass $\varepsilon_x(\cdot)$, then x is called a "branching point". There are also cases where $P_0(\cdot, \cdot)$ should be left undefined. However, we shall assume until further notice that we are in the "normal" case where

$$\forall x \in \mathbf{E}_\partial : P_0(x, \cdot) = \varepsilon_x(\cdot),$$

namely P_0 is the identity operator. Equivalently, we assume

$$\forall x \in \mathbf{E}_\partial: P^x\{X_0 = x\} = 1. \tag{15}$$

Before proceeding further let us give a few simple examples of (homogeneous) Markov processes.

EXAMPLE 1 (Markov chain).

$\mathbf{E}$ = any countable set, for example the set of positive integers.

$\mathscr{E}$ = the σ-field of all subsets of $\mathbf{E}$.

We may write $p_{ij}(t) = P_t(i, \{j\})$ for $i \in \mathbf{E}$, $j \in \mathbf{E}$. Then for any $A \in \mathbf{E}$, we have

$$P_t(i, A) = \sum_{j \in A} p_{ij}(t).$$

The conditions (a) and (c) in the definition of transition function become in this case:

(a) $\forall i \in \mathbf{E}: \sum_{j \in \mathbf{E}} p_{ij}(t) = 1$;

(c) $\forall i \in \mathbf{E}, k \in \mathbf{E}: p_{ik}(s + t) = \sum_{j \in \mathbf{E}} p_{ij}(s)p_{jk}(t)$;

while (b) is trivially true. For the submarkovian case, the "=" in (a) is replaced by "$\leq$". If we add the condition

(d) $\forall i \in \mathbf{E}, j \in \mathbf{E}: \lim_{t \downarrow 0} p_{ij}(t) = \delta_{ij}$;

then the matrix of transition function

$$\Pi(t) = (p_{ij}(t))$$

is called a "standard transition matrix". In this case each $p_{ij}(\cdot)$ is a continuous function on $\mathbf{T}$. See Chung [2] for the theory of this special case of Markov process.

EXAMPLE 2 (uniform motion).

$\mathbf{E} = R^1 = (-\infty, +\infty)$; $\mathscr{E}$ = the classical Borel field on R^1.

For $x \in R^1$, $t \geq 0$, we put

$$P_t(x, \cdot) = \varepsilon_{x+t}(\cdot).$$

Starting from any point x, the process moves deterministically to the right with uniform speed. This trivial example turns out to be the source of many counterexamples to facile generalities. A slight modification yields an example for which (15) is false. Let $\mathbf{E} = \{0\} \cup (-\infty, -1] \cup [1, \infty)$,

$$P_0(0, \{1\}) = \tfrac{1}{2}, \qquad P_0(0, \{-1\}) = \tfrac{1}{2};$$
$$P_t(x, \cdot) = \varepsilon_{x+t}(\cdot), \quad \text{if } x \geq 1; t \geq 0;$$
$$P_t(x, \cdot) = \varepsilon_{x-t}(\cdot), \quad \text{if } x \leq -1; t \geq 0;$$
$$P_t(0, \cdot) = \tfrac{1}{2}\{\varepsilon_{1+t}(\cdot) + \varepsilon_{-1-t}(\cdot)\}.$$

Note that although P_0 is not the identity, we have $P_0 P_t = P_t P_0 = P_t$ for $t \geq 0$.

EXAMPLE 3 (Poisson process). $\mathbf{E} = \mathbf{N} = $ the set of positive (≥ 0) integers or the set of all integers. For $n \in \mathbf{N}$, $m \in \mathbf{N}$, $t \geq 0$:

$$P_t(n, \{m\}) = \begin{cases} 0 & \text{if } m < n, \\ \dfrac{e^{-\lambda t}(\lambda t)^{m-n}}{(m-n)!}, & \text{if } m \geq n. \end{cases}$$

Note that in this case there is spatial homogeneity, namely: the function of the pair (n, m) exhibited above is a function of $m - n$ only.

EXAMPLE 4 (Brownian motion in R^1).

$$\mathbf{E} = R^1, \ \mathscr{E} = \mathscr{B}^1.$$

For real x and y and $t > 0$, put

$$p_t(x, y) = \frac{1}{\sqrt{2\pi t}} \exp\left[-\frac{(y-x)^2}{2t} \right]$$

and define the transition function as follows:

$$P_t(x, A) = \int_A p_t(x, y) \, dy, \qquad t > 0;$$
$$P_0(x, A) = \varepsilon_x(A).$$

The function $p_t(\cdot, \cdot)$ is a *transition probability density*. In this case it is the Gaussian density function with mean zero and variance t. As in Example 3 there is again spatial homogeneity, indeed p_t is a function of $|x - y|$ only. This example, and its extention to R^d, will be the subject matter of Chapter 4.

1.3. Optional Times

A major device to "tame" the continuum of time is the use of random times. The idea is very useful even for discrete time problems and has its origin in considering "the first time when a given event occurs". In continuous time it becomes necessary to formalize the intuitive notions in terms of various σ-fields.

We complete the increasing family $\{\mathscr{F}_t, t \in \mathbf{T}\}$ by setting

$$\mathscr{F}_\infty = \bigvee_{t \in \mathbf{T}} \mathscr{F}_t.$$

Recall that the notation on the right side above means the minimal σ-field including all $\mathscr{F}_t$, which is not the same as $\bigcup_{t \in \mathbf{T}} \mathscr{F}_t$. Although we shall apply the considerations below to a Markov process, we need not specify the family $\{\mathscr{F}_t\}$ to begin with.

Definition. The function $T : \Omega \to [0, \infty]$ is called optional relative to $\{\mathscr{F}_t\}$ iff

$$\forall t \in [0, \infty): \{T \le t\} \in \mathscr{F}_t.$$

The preceding relation then holds also for $t = \infty$ by letting $t \uparrow \infty$ through a sequence. Define

$$\forall t \in (0, \infty): \mathscr{F}_{t-} = \bigvee_{s \in [0,t)} \mathscr{F}_s;$$

$$\forall t \in [0, \infty): \mathscr{F}_{t+} = \bigwedge_{s \in (t,\infty)} \mathscr{F}_s.$$

We have clearly for each $t : \mathscr{F}_{t-} \subset \mathscr{F}_t \subset \mathscr{F}_{t+}$.

Definition. The family $\{\mathscr{F}_t\}$ is called right continuous iff

$$\forall t \in [0, \infty): \mathscr{F}_t = \mathscr{F}_{t+}.$$

It follows from the definition that the family $\{\mathscr{F}_{t+}\}$ is right continuous. Note the analogy with a real-valued increasing function $t \to f(t)$.

Proposition 1. *T is optional relative to* $\{\mathscr{F}_{t+}\}$ *if and only if*

$$\forall t \in [0, \infty): \{T < t\} \in \mathscr{F}_t. \tag{1}$$

Proof. If T is optional relative to $\{\mathscr{F}_{t+}\}$, then by definition

$$\forall t \in (0, \infty): \{T \le t\} \in \mathscr{F}_{t+}.$$

Hence we have for $1/n < t$:

$$\left\{T \le t - \frac{1}{n}\right\} \in \mathscr{F}_{(t-1/n)+} \subset \mathscr{F}_t$$

and consequently

$$\{T < t\} = \bigcup_{n=1}^{\infty} \left\{T \le t - \frac{1}{n}\right\} \in \mathscr{F}_t.$$

Conversely if (1) is true, then for $t \in [0, \infty)$:

$$\left\{T < t + \frac{1}{n}\right\} \in \mathscr{F}_{t+1/n}$$

and consequently

$$\{T \le t\} = \bigcap_{n=1}^{\infty} \left\{T < t + \frac{1}{n}\right\} \in \bigwedge_{n=1}^{\infty} \mathscr{F}_{t+1/n} = \mathscr{F}_{t+}. \qquad \square$$

EXAMPLES.

1. The lifetime ζ defined in (14) of §1.2 is optional relative to $\{\mathscr{F}_{t+}^0\}$, because ∂ is absorbing which implies (Q is the set of rational numbers)

$$\{\zeta < t\} = \bigcup_{r \in Q \cap [0,t)} \{X_r = \partial\} \in \mathscr{F}_t^0.$$

2. Suppose $\mathscr{F}_0$ contains all P-null sets and $\forall t \colon P\{T = t\} = 0$. Then we have

$$\forall t \colon \{T \le t\} - \{T < t\} \in \mathscr{F}_t$$

so that optionality relative to $\{\mathscr{F}_t\}$ is the same as relative to $\{\mathscr{F}_{t+}\}$.

3. Consider a Poisson process with left continuous paths and $T = $ first jump time. Then T is optional relative to $\{\mathscr{F}_{t+}^0\}$ but not to $\{\mathscr{F}_t^0\}$, because

$$\{T = t\} \notin \mathscr{F}_t^0 [= \mathscr{F}_{t-}^0].$$

The reader is supposed to be familiar with elementary properties of the Poisson process which are "intuitively obvious", albeit sometimes tedious to establish rigorously.

From now on we fix the family $\{\mathscr{F}_t, t \in [0, \infty]\}$ and call T an "optional time" or "strictly optional time" according as it is optional relative to $\{\mathscr{F}_{t+}\}$ or $\{\mathscr{F}_t\}$. We shall prove most results for the former case but some of them extend to the latter.

Proposition 2. *If S and T are optional, then so are*

$$S \wedge T, \qquad S \vee T, \qquad S + T.$$

Proof. The first two are easy; to treat the third we consider for each t:

$$\{S + T < t\} = \bigcup_{r \in Q_t} \{S < r; T < t - r\}$$

where $Q_t = Q \cap (0, t)$. Clearly each member of the union is in $\mathscr{F}_t$; hence so is the union. $\qquad\square$

Proposition 3. T_n *is optional for each $n \geq 1$, then so are*

$$\sup_n T_n, \qquad \inf_n T_n, \qquad \limsup_n T_n, \qquad \liminf_n T_n. \qquad (2)$$

Proof. This follows from Proposition 1 and

$$\left\{\sup_n T_n \leq t\right\} = \bigcap_n \{T_n \leq t\} \in \mathscr{F}_{t+};$$

$$\left\{\inf_n T_n < t\right\} = \bigcup_n \{T_n < t\} \in \mathscr{F}_t;$$

$$\limsup_n T_n = \inf_{m \geq 1} \sup_{n \geq m} T_n;$$

$$\liminf_n T_n = \sup_{m \geq 1} \inf_{n \geq m} T_n. \qquad\square$$

For strictly optional T_n only the first item in (2) is strictly optional. In Example (3) above, if we set $T_n = T + 1/n$ then each T_n is strictly optional but $T = \inf_n T_n$ is not.

Associated with an optional time T, there is a σ-field defined as follows.

Definition. If T is strictly optional, $\mathscr{F}_T$ is the class of subsets of $\mathscr{F}$ such that

$$\forall t \in [0, \infty): \Lambda \cap \{T \leq t\} \in \mathscr{F}_t.$$

If T is optional, $\mathscr{F}_{T+}$ is the class of subsets of $\mathscr{F}$ such that

$$\forall t \in [0, \infty): \Lambda \cap \{T \leq t\} \in \mathscr{F}_{t+}. \qquad (3)$$

It follows from the proof of Proposition 1 that (3) is equivalent to

$$\forall t \in (0, \infty]: \Lambda \cap \{T < t\} \in \mathscr{F}_t.$$

Let us verify, for example, that $\mathscr{F}_{T+}$ is a σ-field. We see from (3) that $\Omega \in \mathscr{F}_{T+}$ because T is optional. It then follows that $\mathscr{F}_{T+}$ is closed under complementation. It is also closed under countable union; hence it is a

σ-field. For $T = a$ constant t, $\mathscr{F}_T$ reduces to $\mathscr{F}_t$, and $\mathscr{F}_{T+}$ reduces to $\mathscr{F}_{t+}$. Thus our definition and notation make sense and it is a leading idea below to extend properties concerning constant times to their analogues for optional times.

It is easy to see that if $\{\mathscr{F}_t\}$ is right continuous, then any optional T is strictly optional, and furthermore we have

$$\mathscr{F}_T = \mathscr{F}_{T+}.$$

Proposition 4. *If T is optional and $\Lambda \in \mathscr{F}_{T+}$, we define*

$$T_\Lambda = \begin{cases} T, & \text{on } \Lambda, \\ \infty, & \text{on } \Lambda^c. \end{cases} \tag{4}$$

Then T is also optional.

Proof. For $t < \infty$, we have

$$\{T_\Lambda < t\} = \Lambda \cap \{T < t\} \in \mathscr{F}_t.$$

This implies $\{T_\Lambda = \infty\} \in \mathscr{F}_\infty$ by the remark after Proposition 1. $\quad\square$

Theorem 5.

  (a)   If T is optional, then $T \in \mathscr{F}_{T+}$;
  (b)   If S and T are both optional and $S \leq T$, then $\mathscr{F}_{S+} \subset \mathscr{F}_{T+}$;
  (c)   If T_n is optional, $T_{n+1} \leq T_n$ for each $n \geq 1$, and $\lim_n T_n = T$, then

$$\mathscr{F}_{T+} = \bigwedge_{n=1}^{\infty} \mathscr{F}_{T_n+}. \tag{5}$$

In case each T_n is strictly optional, we may replace $\mathscr{F}_{T_n+}$ by $\mathscr{F}_{T_n}$ in (5).

Proof. We leave (a) and (b) as exercises. To prove (c), we remark that T is optional by Proposition 3. Hence we have by (b):

$$\mathscr{F}_{T+} \subset \bigwedge_{n=1}^{\infty} \mathscr{F}_{T_n+}. \tag{6}$$

Now let $\Lambda \in \bigwedge_{n=1}^{\infty} \mathscr{F}_{T_n+}$; then for each t:

$$\Lambda \cap \{T < t\} = \Lambda \cap \left[\bigcup_{n=1}^{\infty} \{T_n < t\} \right] = \bigcup_{n=1}^{\infty} [\Lambda \cap \{T_n < t\}] \in \mathscr{F}_t,$$

so that $\Lambda \in \mathscr{F}_{T+}$. Hence the inclusion in (6) may be reversed. $\quad\square$

Theorem 6. *If S and T are both optional, then the sets*

$$\{S < T\}, \qquad \{S \leq T\}, \qquad \{S = T\}$$

all belong to $\mathscr{F}_{S+} \wedge \mathscr{F}_{T+}$.

Proof. We have for each t:

$$\{S < T\} \cap \{T < t\} = \bigcup_{r \in Q_t} \{S < r < T < t\},$$

where $Q_t = Q \cap (0, t)$. For each r, $\{r < T < t\} \in \mathscr{F}_t$, $\{S < r\} \in \mathscr{F}_r \subset \mathscr{F}_t$. Hence the union above belongs to $\mathscr{F}_t$ and this shows $\{S < T\} \in \mathscr{F}_{T+}$. Next we have

$$\{S < T\} \cap \{S < t\} = \{S < T \wedge t\} = \bigcup_{r \in Q} \{S < r < T \wedge t\}$$

$$= \bigcup_{r \in Q_t} \{S < r < T\} \in \mathscr{F}_t,$$

since for each $r < t$, $\{r < T\} = \{T \leq r\}^c \in \mathscr{F}_{r+} \subset \mathscr{F}_t$. Hence, $\{S < T\} \in \mathscr{F}_{S+}$. Combining the two results we obtain $\{S < T\} \in \mathscr{F}_{S+} \wedge \mathscr{F}_{T+}$. Since

$$\{S \leq T\} = \{T < S\}^c,$$
$$\{S = T\} = \{T \leq S\} - \{T < S\},$$

the rest of the theorem follows.   □

We will now extend the notion of the σ-field $\mathscr{F}_{t-}$ to a random time T. It turns out that we can do so with an arbitrary positive function T, but $\mathscr{F}_{T-}$ will have nice properties when T is an optional time.

Definition. For any function T from Ω to $[0, \infty]$, $\mathscr{F}_{T-}$ is the σ-field generated by $\mathscr{F}_{0+}$ and the class of sets:

$$\{t < T\} \cap \Lambda, \quad \text{where } t \in [0, \infty) \text{ and } \Lambda \in \mathscr{F}_t. \tag{7}$$

It is easy to see that $\mathscr{F}_{T-}$ contains also sets of the form

$$\{T = \infty\} \cap \Lambda, \quad \text{where } \Lambda \in \mathscr{F}_\infty. \tag{8}$$

As defined above, $\mathscr{F}_{T-}$ need not be included in $\mathscr{F}$; but this is the case if $T \in \mathscr{F}$, as we shall assume in what follows. It is obvious that $T \in \mathscr{F}_{T-}$.

If T is optional, then $\mathscr{F}_{0+} \subset \mathscr{F}_{T+}$ by Theorem 5, (b). Next, for each $u \geq 0$, the set

$$[\{t < T\} \cap \Lambda] \cap \{T < u\}$$

is empty for $u \leq t$, and belongs to $\mathscr{F}_u$ if $t < u$ and $\Lambda \in \mathscr{F}_t$. Hence each generating set of $\mathscr{F}_{T-}$ in (7) belongs to $\mathscr{F}_{T+}$, by definition of the latter. In other words, we have for an optional T:

$$\mathscr{F}_{T-} \subset \mathscr{F}_{T+}. \tag{9}$$

Similarly if T is strictly optional, then $\mathscr{F}_{T-} \subset \mathscr{F}_T$. The next proposition is sharper; note that it reduces to $\mathscr{F}_{0+} \subset \mathscr{F}_{T-}$ for $S \equiv 0$, which is true by definition of $\mathscr{F}_{T-}$.

Proposition 7. *If S and T are both optional and $S \leq T$, then $\mathscr{F}_{S-} \subset \mathscr{F}_{T-}$. If moreover $S < T$ on $\{S < \infty\} \cap \{T > 0\}$, then*

$$\mathscr{F}_{S+} \subset \mathscr{F}_{T-}. \tag{10}$$

Proof. The first assertion is easy. Under the hypothesis of the second assertion, we have for any Λ:

$$\Lambda = \bigcup_{r \in Q} [\Lambda \cap \{S < r < T\}] \cup [\Lambda \cap \{S = \infty\}] \cup [\Lambda \cap \{T = 0\}].$$

If $\Lambda \in \mathscr{F}_{S+}$, then $\Lambda \cap \{S < r\} \in \mathscr{F}_r$ for each r, and so

$$[\Lambda \cap \{S < r\}] \cap \{r < T\} \in \mathscr{F}_{T-}.$$

Next, $\Lambda \cap \{S = \infty\} \in \mathscr{F}_\infty$ and so

$$\Lambda \cap \{S = \infty\} = [\Lambda \cap \{S = \infty\}] \cap \{T = \infty\} \in \mathscr{F}_{T-}.$$

Finally,

$$\Lambda \cap \{T = 0\} = [\Lambda \cap \{S = 0\}] \cap \{T = 0\} \in \mathscr{F}_{T-},$$

because $\Lambda \cap \{S = 0\} \in \mathscr{F}_{0+} \subset \mathscr{F}_{T-}$ and $\{T = 0\} \in \mathscr{F}_{T-}$. Combining these results we obtain $\Lambda \in \mathscr{F}_{T-}$. □

Next, we introduce a special class of optional times which plays an essential role in the general theory of stochastic processes. Although its use will be limited in this introductory text, the discussion below will serve the purpose of further strengthening the analogy between optional and constant times.

Definition. *T is a predictable time* iff there exists a sequence of optional times $\{T_n\}$ such that

$$\forall n: T_n < T \quad \text{on } \{T > 0\}; \quad \text{and} \quad T_n \uparrow T. \tag{11}$$

Here "$T_n \uparrow T$" is a standard notation which means: "$\forall n: T_n \le T_{n+1}$ and $\lim_n T_n = T$." The limit T is optional by Proposition 3. Observe also that $T_n < \infty$ for each n. The sequence $\{T_n\}$ is said to *announce* T.

Theorem 8. *If T_n is optional and $T_n \uparrow T$, then*

$$\mathscr{F}_{T-} = \bigvee_{n=1}^{\infty} \mathscr{F}_{T_n-}. \tag{12}$$

If T is predictable and $\{T_n\}$ announces T, then we have also

$$\mathscr{F}_{T-} = \bigvee_{n=1}^{\infty} \mathscr{F}_{T_n+}. \tag{13}$$

Proof. We have "$\supset$" in (12) by Proposition 7. To prove the reverse inclusion, we have if $\Lambda \in \mathscr{F}_t$:

$$\{t < T\} \cap \Lambda = \bigcup_{n=1}^{\infty} [\{t < T_n\} \cap \Lambda] \in \bigvee_{n=1}^{\infty} \mathscr{F}_{T_n-}.$$

Also $\mathscr{F}_{0+} \subset \mathscr{F}_{T_n-}$ for each n. Hence each generating set of $\mathscr{F}_{T-}$ belongs to the right member of (12), and (12) is proved. If $\{T_n\}$ announces the predictable T, then $\mathscr{F}_{T-} \supset \mathscr{F}_{T_n+}$ by Proposition 7, and so we have "$\supset$" in (13). Since $\mathscr{F}_{T_n+} \supset \mathscr{F}_{T_n-}$ for each n, the reverse inclusion follows from (12).   □

We now return to the Markov process $X = \{X_t, \mathscr{F}_t, t \in \mathbf{T}\}$.

Definition. X is said to be Borel measurable iff

$$(t, \omega) \to X(t, \omega)$$

is in $\mathscr{B} \times \mathscr{F}$, where $\mathscr{B}$ is the Euclidean Borel field on $\mathbf{T}$. X is said to have right limits iff for each ω, the sample function $t \to X(t, \omega)$ has right limits everywhere in $\mathbf{T}$; X is said to be right continuous iff for each ω, the sample function $t \to X(t, \omega)$ is right continuous in $\mathbf{T}$. Similarly, "left" versions of these properties are defined in $(0, \infty]$.

Proposition 9. *If X is right [or left] continuous, then it is Borel measurable.*

Proof. For each $n \ge 0$, define

$$X_n(t, \omega) = X\left(\frac{[2^n t] + 1}{2^n}, \omega\right). \tag{14}$$

It is clear that

$$(t, \omega) \to X_n(t, \omega) = \sum_{k=0}^{\infty} 1_{[k/2^n, \, k+1/2^n)}(t) X \left(\frac{k+1}{2^n}, \omega \right)$$

is Borel measurable. If X is right continuous, then

$$X(t, \omega) = \lim_n X_n(t, \omega)$$

for $(t, \omega) \in \mathbf{T} \times \Omega$. Hence X is Borel measurable; see the Lemma in §1.5 below for a proof. If X is left continuous we need only modify the definition in (14) by omitting the "$+1$" in the right member.   □

Definition. Let X be Borel measurable and T an $\mathscr{F}$-measurable function from Ω to $[0, \infty]$ (namely an extended-valued random variable). We define X_T, or $X(T)$, on the set $\{T < \infty\}$ by

$$X_T(\omega) = X(T(\omega), \omega). \tag{15}$$

If $X_\infty(\omega)$ is defined for all $\omega \in \Omega$, then X_T is defined on Ω by (15).

The next result extends the hypothesis "$X_t \in \mathscr{F}_t$".

Theorem 10. *Let X have right limits. If T is optional, then*

$$X_{T+}1_{\{T < \infty\}} \in \mathscr{F}_{T+}. \tag{16}$$

If X has left limits as well, and T is predictable, then

$$X_{T-}1_{\{T < \infty\}} \in \mathscr{F}_{T-} \tag{17}$$

where $X_{0-} = X_0$.

Note. If $T(\omega) = \infty$, the left members of (16) and (17) are defined to be zero by convention, even when $X_{\infty+}$ or $X_{\infty-}$ are not defined.

Proof. Define for $n \geq 0$:

$$T_n = \frac{[2^n T] + 1}{2^n}. \tag{18}$$

This is just the usual dyadic approximation of T from above. For each n, T_n is strictly optional (proof ?) and $T_n > T$, $\lim_n \downarrow T_n = T$. For convenience let us introduce the "dyadic set"

$$D = \left\{ \frac{k}{2^n} \,\middle|\, k \geq 0, n \geq 0 \right\}. \tag{19}$$

For each $d \in D$ and $B \in \mathscr{E}$ we have

$$\{T_n = d; X(T_n) \in B\} = \{T_n = d; X(d) \in B\} \in \mathscr{F}_d$$

because T_n is strictly optional. Hence for each t,

$$\{X(T_n) \in B\} \cap \{T_n < t\} = \bigcup_{d \in D_t} \{T_n = d; X(T_n) \in B\} \in \mathscr{F}_t$$

where $D_t = D \cap [0, t)$. This shows that $\{X(T_n) \in B; T_n < \infty\}$ belongs to $\mathscr{F}_{T_n+}$ (actually to $\mathscr{F}_{T_n}$); since B is arbitrary we have $X(T_n) 1_{\{T_n < \infty\}} \in \mathscr{F}_{T_n+}$. Since $X(T+) = \lim_n X(T_n)$ on $\{T < \infty\}$ which belongs to $\mathscr{F}_{T+}$, and $\mathscr{F}_{T_n+} \downarrow \mathscr{F}_{T+}$, it follows that

$$X(T+)1_{\{T < \infty\}} \in \bigwedge_{n=1}^{\infty} \mathscr{F}_{T_n+} = \mathscr{F}_{T+}$$

by Theorem 5.

If T is predictable, let $\{T_n\}$ announce T. Then

$$\lim_n X(T_n) = X(T-)$$

on $\{0 < T < \infty\}$ provided X has left limits in $(0, \infty)$, and trivially on $\{T = 0\}$ provided we define $X(0-) = X(0)$. Now for each n, we have just shown that $X(T_n) \in \mathscr{F}_{T_n+}$; hence

$$X(T-)1_{\{T < \infty\}} \in \bigvee_n \mathscr{F}_{T_n+} = \mathscr{F}_{T-}$$

by Theorem 8, if we note that $T \in \mathscr{F}_{T-}$. If perchance X has left limits in $(0, \infty]$ so that $X_{\infty-}$ is defined, then we have $X(T-) \in \mathscr{F}_{T-}$.   □

We can extend the shift θ_t to θ_T for any function T from Ω to $[0, \infty]$. Observe that the defining equations for the shift, (6) of §1.2, is an identity in s and t for each ω. It follows that on the set $\{T < \infty\}$ we have

$$X_t(\theta_T(\omega)) = X_{t+T(\omega)}(\omega) = X_{T+t}(\omega). \tag{20}$$

Note that $T + t$ is another function of the same type as T. Thus we have

$$X_t \circ \theta_T = X_{T+t}$$

as functions on $\{T < \infty\}$. If T is optional, then so is $T + t$; indeed $T + t$ is strictly optional for $t > 0$. Hence by Theorem 10, if X is right continuous:

$$X_t \circ \theta_T 1_{\{T < \infty\}} = X_{T+t} 1_{\{T < \infty\}} \in \mathscr{F}_{T+t+}. \tag{21}$$

So far so good, but the next step is to consider the inverse mapping θ_T^{-1}. This would be awkward if θ_T were to be defined on $\{T < \infty\}$ only. Here is the device to extend it to Ω. Put

$$\forall \omega \in \Omega : X_\infty(\omega) = \partial.$$

Postulate the existence of a point ω_∂ in Ω for which

$$\forall t \in \mathbf{T} : X_t(\omega_\partial) = \partial;$$

this then holds for $t \in [0, \infty]$. Finally, define

$$\forall \omega \in \Omega : \theta_\infty(\omega) = \omega_\partial.$$

After this fixing all three terms in (20) reduce to ∂ when $t = \infty$ or $T = \infty$ or both. Hence (20) now holds for each $t \in [0, \infty]$ and any T from Ω to $[0, \infty]$. The existence of ω_∂ is of course trivial in the case where $\Omega = \mathbf{E}_\partial^\mathbf{T}$, and can in general be arranged by enlarging the given Ω.

With this amendment, we can rewrite (16) simply as

$$X_T = X_{T+} \in \mathscr{F}_{T+}.$$

In equation (21) we can now omit the factor $1_{\{T < \infty\}}$, and the result is equivalent to

$$\theta_T^{-1}\{\sigma(X_t)\} \subset \mathscr{F}_{T+t+}$$

where $\sigma(X_t)$ is the σ-field generated by X_t. Since $\mathscr{F}_{T+t+}$ increases with t it follows by the usual extension that

$$\theta_T^{-1}\{\mathscr{F}_t^0\} \subset \mathscr{F}_{T+t+} \subset \mathscr{F}_\infty. \tag{22}$$

Let us give a useful application at once.

Theorem 11. *Let S be optional relative to $\{\mathscr{F}_t^0\}$ and T be optional relative to $\{\mathscr{F}_t\}$, then*

$$T + S \circ \theta_T \tag{23}$$

is optional relative to $\{\mathscr{F}_t\}$.

Proof. For each t, we have by (22)

$$\{S \circ \theta_T < t\} = \theta_T^{-1}\{S < t\} \in \mathscr{F}_{T+t+}. \tag{24}$$

Hence for each $r < t$:

$$\{S \circ \theta_T < t - r\} \cap \{T + t - r < t\} \in \mathscr{F}_t$$

by (24) with $t - r$ replacing t and the definition of $\mathscr{F}_{T+t-r}$. Now we have

$$\{T + S \circ \theta_T < t\} = \bigcup_{r \in Q_t} \{T < r; S \circ \theta_T < t - r\}$$

and each member of the union above belongs to $\mathscr{F}_t$. Hence the set above belongs to $\mathscr{F}_t$ and this proves the theorem.   □

Remark. Later we will extend the result to an S which is optional relative to $\{\mathscr{F}_t^\mu\}$ or $\{\tilde{\mathscr{F}}_t\}$; see §2.3.

What does the random variable in (23) mean? Let A be a set in $\mathscr{E}$ and define

$$D_A(\omega) = \inf\{t \geq 0 \,|\, X(t, \omega) \in A\}. \qquad (25)$$

This is called the "first entrance time into A", and is a prime example of an optional time. Indeed, it was also historically the first such time to be considered. The reader may well recall instances of its use in the classical theory of probability. However for a continuous time Markov process, it is not a trivial matter to verify that D_A is optional for a general A. This is easy only for an open set A, and we shall treat other sets in §3.3. Suppose this has been shown; we can identify the random variable in (23) when $S = D_A$ as follows:

$$T(\omega) + D_A \circ \theta_T(\omega) = \inf\{t \geq T(\omega) \,|\, X(t, \omega) \in A\}. \qquad (26)$$

In particular if $T(\omega) \equiv s$, $s + D_A \circ \theta_s$ is "the first entrance time into A on or after time s"; if $T(\omega) = D_B(\omega)$ where $B \in \mathscr{E}$, then $D_B + D_A \circ \theta_{D_B}$ is "the first entrance time into A on or after the first entrance time into B." Of course, a similar interpretation is meaningful for any optional time T. We shall see many applications to Markov processes involving such random times.

The definitions and results of this section have been given without mention of probability. It is usually obvious how to modify them when exceptional sets are allowed. For instance, if each T_n is defined only P-a.e., and $\lim_n T_n$ exists only P-a.e., then there is a set Ω_0 with $P(\Omega_0) = 1$ on which all T_n's are defined and $\lim_n T_n$ exists for all ω in Ω_0. Furthermore if $\{T_n \leq t\}$ differs from a set in $\mathscr{F}_t$ by a P-null sets, then $\Omega_0 \cap \{T_n \leq t\} \in \mathscr{F}_t$ provided that $(\mathscr{F}, P)$ is complete and $\mathscr{F}_0(\subset \mathscr{F}_t)$ is augmented (by all P-null sets). Thus T_n is optional considered on the trace of $(\Omega, \mathscr{F})$ on Ω_0. In this way we can apply the preceding discussions to a reduced probability space, and we shall often do so tacitly. On the other hand, there are deeper results concerning optionality in which probability considerations are expedient; see for example Problem 12 below.

Exercises

1. If T is optional and $S = \varphi(T) > T$ where φ is a Borel function, then S is strictly optional, indeed predictable. In general a strictly positive predictable time is strictly optional.

2. $\mathcal{F}_{T-}$ is also generated by $\mathcal{F}_{0+}$ and one of the two classes of sets below:
   (i) $\{t \leq T\} \cap \Lambda$, where $0 \leq t < \infty$ and $\Lambda \in \mathcal{F}_{t-}$;
   (ii) $\{t < T\} \cap \Lambda$, where $0 \leq t < \infty$ and $\Lambda \in \mathcal{F}_{t+}$.

3. If S and T are functions from Ω to $[0, \infty]$ such that $S \leq T$, then $\mathcal{F}_{S-} \subset \mathcal{F}_{T-}$ if and only if $S \in \mathcal{F}_{T-}$. This is the case if S is optional. On the other hand if S is arbitrary and T is optional, $\mathcal{F}_{S-} \subset \mathcal{F}_{T+}$ may be false.

4. Define for any function T from Ω to $[0, \infty]$ the σ-field
$$\mathcal{F}_{T+} = \bigwedge_{n=1}^{\infty} \mathcal{F}_{(T+1/n)-}.$$
   Prove that if T is optional, this coincides with the definition given in (3).

5. If T is optional then
$$\mathcal{F}_{T+} = \bigvee_{n=1}^{\infty} \mathcal{F}_{(T \wedge n)+}.$$

6. Let S be optional relative to $\{\mathcal{F}_t\}$ and define $\mathcal{G}_t = \mathcal{F}_{S+t}$, where $\mathcal{G}_0 = \mathcal{F}_{S+}$. Prove that if $T \geq S$, T is optional relative to $\{\mathcal{F}_t\}$ if and only if $T - S$ is optional relative to $\{\mathcal{G}_t\}$.

7. If T is predictable and $\Lambda \in \mathcal{F}_{T-}$, then the T_Λ defined in (4) is predictable.

8. Let $\{T^{(k)}\}$ be a sequence of predictable times. If $T^{(k)}$ increases to T, then T is predictable. If $T^{(k)}$ "settles down" to T, namely $T^{(k)}$ decreases and for each ω, there exists an integer $N(\omega)$ for which $T^{N(\omega)}(\omega) = T(\omega)$, then T is "almost surely predictable". [Hint: for the second assertion let $\{T_n^{(k)}\}$ announce $T^{(k)}$ such that $P(T_n^{(k)} + 2^{-n} < T) < 2^{-n-k}$ for all k and n. Let $S_n = \inf_k T_n^{(k)}$; then $\{S_n\}$ announces T almost surely.]

9. If S is optional and T is arbitrary positive, then
$$\mathcal{F}_{S+} \cap \{S < T\} \in \mathcal{F}_{T-};$$
   namely; for each $\Lambda \in \mathcal{F}_{S+}$, $\Lambda \cap \{S < T\} \in \mathcal{F}_{T-}$.

10. If S and T are both optional, then exactly one of the following four relations is in general false:
$$\mathcal{F}_{(S \wedge T)+} = \mathcal{F}_{S+} \wedge \mathcal{F}_{T+};$$
$$\mathcal{F}_{(S \wedge T)-} = \mathcal{F}_{S-} \wedge \mathcal{F}_{T-};$$
$$\mathcal{F}_{(S \vee T)+} = \mathcal{F}_{S+} \vee \mathcal{F}_{T+};$$
$$\mathcal{F}_{(S \vee T)-} = \mathcal{F}_{S-} \vee \mathcal{F}_{T-}.$$

11. Let $\{\mathscr{F}_t\}$ be an increasing family of σ-fields, and $\mathscr{G}_t$ be the σ-field generated by $\mathscr{F}_t$ and all P-null sets. Suppose S is optional relative to $\{\mathscr{G}_t\}$. Prove that there exists T which is optional relative to $\{\mathscr{F}_t\}$ and $P(S = T) = 1$. Furthermore if $\Lambda \in \mathscr{G}_S$ then there exists $M \in \mathscr{F}_{T+}$ such that $P(\Lambda \triangle M) = 0$. [Hint: approximate S as in (18); for the second assertion, consider $\{S_\Lambda = T < \infty\}$ where S_Λ is defined as in (4).]

12. If T is predictable, there exists an "almost surely" announcing sequence $\{T_n\}$ such that each T_n is countably-valued with values in the dyadic set. Namely, we have for each n, $T_n \leq T_{n+1} \leq T$, $T_n < T$ on $\{T > 0\}$, and $P\{\lim_n T_n = T\} = 1$. Furthermore each T_n is predictable. [Hint: let $T_n^{(k)} = ([2^k T_n] + 1)/2^k$. Show that there exists $\{k_n\}$ such that $P\{T_n^{(k_n)} \geq T$ i.o.$\} = 0$. Put $S_m = \inf_{n \geq m} T_n^{(k_n)}$ and show that S_m is dyadic-valued. For the last assertion use Problems 1 and 8.]

1.4. Martingale Theorems

Martingale theory is an essential tool in the analysis of Markov processes. In this section we give a brief account of some of the basic results in continuous time, with a view to applications in later sections. It is assumed that the reader is acquainted with the discrete time analogues of these results, such as discussed in Chapter 9 of *Course*. A theorem stated there for a submartingale will be cited for a supermartingale without comment, since it amounts only to a change of signs for the random variables involved.

Let $(\Omega, \mathscr{F}, P)$ be a probability space; $\{\mathscr{F}_t\}$ an increasing family of σ-fields (of sets in $\mathscr{F}$) to which a stochastic process $\{X_t\}$ is adapted, as in §1.1. The index set for t is $\mathbf{T} = [0, \infty)$ when this is not specified. We shall use the symbol $\mathbf{N}$ to denote the set of positive (≥ 0) integers, and $\mathbf{N}_m$ the subset of $\mathbf{N}$ not exceeding m, where $m \in \mathbf{N}$. As usual, indices such as m, n and k denote members of $\mathbf{N}$ without explicit mention.

Definition. $\{X_t, \mathscr{F}_t, t \in \mathbf{T}\}$ is a martingale iff

   (i) $\forall t : X_t$ is an integrable real-valued random variable;
  (ii) if $s < t$, then

$$X_s = E\{X_t | \mathscr{F}_s\}. \tag{1}$$

X is a supermartingale iff the "$=$" in (1) is replaced by "$\geq$"; and a submartingale iff it is replaced by "$\leq$".

We write $s \Downarrow t$ to mean "$s > t, s \to t$", and $s \Uparrow t$ to mean "$s < t, s \to t$". Let S be a countable dense subset of $\mathbf{T}$; for convenience we suppose $S \supset \mathbf{N}$.

Theorem 1. *Let* $\{X_t, \mathscr{F}_t\}$ *be a supermartingale. For* P *a.e.* ω, *the sample function* $X(\cdot, \omega)$ *restricted to the set* S *has a right limit at every* $t \in [0, \infty)$, *and* *a left limit at every $t \in (0, \infty)$. Thus we have:*

$$X(t+, \omega) = \lim_{\substack{s \in S \\ s \downarrow\downarrow t}} X(s, \omega) \quad \text{exists for } t \in [0, \infty);$$

$$X(t-, \omega) = \lim_{\substack{s \in S \\ s \uparrow\uparrow t}} X(s, \omega) \quad \text{exists for } t \in (0, \infty).$$

(2)

Furthermore, for each finite interval $I \subset \mathbf{T}$, the set of numbers

$$\{X(s, \omega), s \in S \cap I\}$$

(3)

is bounded. Hence the limits in (2) are finite numbers.

Proof. Let $\{X_n, \mathscr{F}_n, n \in \mathbf{N}_m\}$ be a discrete parameter supermartingale, and let

$$U(\omega; \mathbf{N}_m; [a, b]), \quad \text{where } a < b,$$

denote the number of upcrossings from strictly below a to strictly above b by the sample sequence $\{X_n(\omega), n \in \mathbf{N}_m\}$. We recall that the upcrossing inequality states that

$$E\{U(\mathbf{N}_m; [a, b])\} \leq \frac{E(X_m^-) + a}{b - a}.$$

(4)

See Theorem 9.4.2 of *Course*. Observe that this inequality really applies to a supermartingale indexed by any finite linearly ordered set with m as the last index. In other words, the bound given on the right side of (4) depends on the last random variable of the supermartingale sequence, not on the number of terms in it (as long as the number is finite). Hence if we consider the supermartingale $\{X_t\}$ with the index t restricted to $[0, m] \cap S'$, where S' is any finite subset of S containing m, and denote the corresponding upcrossing number by $U([0, m] \cap S'; [a, b])$, then exactly the same bound as in (4) applies to it. Now let S' increase to S, then the upcrossing number increases to $U([0, m] \cap S; [a, b])$. Hence by the monotone convergence theorem, we have

$$E\{U([0, m] \cap S; [a, b])\} \leq \frac{E(X_m^-) + a}{b - a} < \infty.$$

(5)

It follows that for P-a.e. ω.

$$U(\omega; [0, m] \cap S; [a, b]) < \infty.$$

(6)

This is true for each $m \in \mathbf{N}$ and each pair of rational numbers (a, b) such that $a < b$. Therefore (6) holds simultaneously for all $m \in \mathbf{N}$ and all such pairs in a set Ω_0 with $P(\Omega_0) = 1$. We claim that for each $\omega \in \Omega_0$, the limit in (2) must exist. It is sufficient to consider the right limit. Suppose that for some t, $\lim_{s \in S, \, s\downarrow\downarrow t} X(s, \omega)$ does not exist. Then we have firstly

$$\varliminf_{\substack{s \in S \\ s\downarrow\downarrow t}} X(s, \omega) < \varlimsup_{\substack{s \in S \\ s\downarrow\downarrow t}} X(s, \omega),$$

and hence secondly two rational numbers a and b such that

$$\varliminf_{\substack{s \in S \\ s\downarrow\downarrow t}} X(s, \omega) < a < b < \varlimsup_{\substack{s \in S \\ s\downarrow\downarrow t}} X(s, \omega).$$

But this means that the sample function $X(\cdot, \omega)$ crosses from a to b infinitely many times on the set S in any right neighborhood of t. Since $\omega \in \Omega_0$ this behavior is ruled out by definition of Ω_0.

To prove the second assertion of the theorem, we recall the following inequality for a discrete parameter supermartingale (Theorem 9.4.1 of *Course*). For any $\lambda > 0$:

$$\lambda P \left\{ \max_{n \in \mathbf{N}_m} |X_n| \geq \lambda \right\} \leq E(X_0) + 2E(X_m^-).$$

This leads by the same arguments as before to

$$\lambda P \left\{ \sup_{t \in [0,m] \cap S} |X_t| \geq \lambda \right\} \leq E(X_0) + 2E(X_m^-). \tag{7}$$

Dividing by λ and letting $\lambda \to \infty$, we conclude that

$$P \left\{ \sup_{t \in [0,m] \cap S} |X_t| = \infty \right\} = 0, \quad \text{or} \quad P \left\{ \sup_{t \in [0,m] \cap S} |X_t| < \infty \right\} = 1.$$

This means that the sample function $X(\cdot, \omega)$ restricted to $[0, m] \cap S$ is bounded P-a.e., say in a set Δ_m with $P(\Delta_m) = 1$. Then $P(\bigcap_{m=1}^{\infty} \Delta_m) = 1$ and if $\omega \in \bigcap_{m=1}^{\infty} \Delta_m$, $X(\cdot, \omega)$ is bounded in $I \cap S$ for each finite interval I. $\qquad \square$

We shall say "almost surely" as paraphrase for "for P-a.e. ω".

Corollary 1. *If X is right continuous, then it has left limits everywhere in $(0, \infty)$ and is bounded on each finite interval, almost surely.*

Proof. The boundedness assertion in Theorem 1 and the hypothesis of right continuity imply that X is bounded on each finite interval. Next, if a sample function $X(\cdot,\ )$ is right continuous but does not have a left limit at some t, then it must oscillate between two numbers a and b infinitely many times in a left neighborhood of t. Because the function is right continuous at the points where it takes the values a and b, these oscillations also occur on the countable dense set S. This is almost surely impossible by the theorem. $\square$

Corollary 2. *If X is right continuous and either* (a) $X_t \geq 0$ *for each t; or* (b) $\sup_t E(|X_t|) < \infty$, *then* $\lim_{t\to\infty} X_t$ *exists almost surely and is an integrable random variable.*

Proof. Under either condition we can let $m \to \infty$ in (5) to obtain

$$E\{U(S;[a,b])\} \leq \frac{A+a}{b-a}, \tag{8}$$

where $A = 0$ under (a), A denotes the sup under (b). Now let $U(\omega;[a,b])$ denote the total number of upcrossings of $[a,b]$ by the unrestricted sample function $X(\cdot,\omega)$. We will show that if $X(\cdot,\omega)$ is right continuous, then

$$U(\omega;S;[a,b]) = U(\omega;[a,b]). \tag{9}$$

For if the right member above be $\geq k$, then there exist $s_1 < t_1 < \cdots < s_k < t_k$ such that $X(s_j,\omega) < a$ and $X(t_j,\omega) > b$ for $1 \leq j \leq k$. Hence by right continuity at these $2k$ points, there exist $s_1' < t_1' < \cdots < s_k' < t_k'$, all members of S such that $s_1 < s_1' < t_1 < t_1' < \cdots < s_k < s_k' < t_k < t_k'$ such that $X(s_j',\omega) < a$ and $X(t_j',\omega) > b$ for $1 \leq j \leq k$. Thus the left member of (9) must also be $\geq k$. Of course it cannot exceed the right member. Hence (9) is proved. We have therefore

$$E\{U([a,b])\} \leq \frac{A+a}{b-a}. \tag{10}$$

It then follows as before that almost surely no sample function oscillates between any two distinct values infinitely many times in **T**, and so it must tend to a limit as $t \to \infty$, possibly $\pm\infty$. Let the limit be X_∞. Under condition (a), we have by Fatou's lemma and supermartingale property:

$$E(X_\infty) = E\left(\lim_{n\to\infty} X_n\right) \leq \varliminf_n E(X_n) \leq E(X_0) < \infty.$$

Under condition (b) we have similarly:

$$E(|X_\infty|) \leq \varliminf_n E(|X_n|) \leq \sup_t E(|X_t|).$$

Hence X_∞ is integrable in either case. $\square$

Theorem 2. *Let* $(X_t, \mathscr{F}_t)$ *be a supermartingale, and* X_{t+} *and* X_{t-} *be as in* (2). *We have*

$$\forall t \in [0, \infty): X_t \geq E\{X_{t+}|\mathscr{F}_t\}, \tag{11}$$

$$\forall t \in (0, \infty): X_{t-} \geq E\{X_t|\mathscr{F}_{t-}\}. \tag{12}$$

Furthermore $\{X_{t+}, \mathscr{F}_{t+}\}$ *is a supermartingale; it is a martingale if* $\{X_t, \mathscr{F}_t\}$ *is.*

Proof. Let $t_n \in S$, $t_n \downdownarrows t$, then

$$X_t \geq E\{X_{t_n}|\mathscr{F}_t\}. \tag{13}$$

Now the supermartingale $\{X_s\}$ with s in the index set $\{t, \ldots, t_n, \ldots, t_1\}$ is uniformly integrable; see Theorem 9.4.7(c) of *Course*. Hence we obtain (11) by letting $n \to \infty$ in (13). Next, let $s_n \in S$, $s_n \upuparrows t$; then

$$X_{s_n} \geq E\{X_t|\mathscr{F}_{s_n}\}.$$

Letting $n \to \infty$ we obtain (12) since $\mathscr{F}_{s_n} \uparrow \mathscr{F}_{t-}$; here we use Theorem 9.4.8, (15a) of *Course*. Finally, let $u_n > u > t_n > t$, $u_n \in S$, $t_n \in S$, $u_n \downdownarrows u$, $t_n \downdownarrows t$, and $\Lambda \in \mathscr{F}_{t+}$. Then we have

$$\int_\Lambda X_{u_n} dP \leq \int_\Lambda X_{t_n} dP.$$

Letting $n \to \infty$ and using uniform integrability as before, we obtain

$$\int_\Lambda X_{u+} dP \leq \int_\Lambda X_{t+} dP.$$

This proves that $\{X_{t+}, \mathscr{F}_{t+}\}$ is a supermartingale. The case of a martingale is similar.   □

Two stochastic processes $X = \{X_t\}$ and $Y = \{Y_t\}$ are said to be *versions* of each other iff we have

$$\forall t: P\{X_t = Y_t\} = 1. \tag{14}$$

It then follows that for any countable subset S of $\mathbf{T}$, we have also

$$P\{X_t = Y_t \text{ for all } t \in S\} = 1.$$

In particular, for any $(t_1, \ldots, t_n)$, the distributions of $(X_{t_1}, \ldots, X_{t_n})$ and $(Y_{t_1}, \ldots, Y_{t_n})$ are the same. Thus the two processes X and Y have identical finite-dimensional joint distributions. In spite of this they may have quite different sample functions, because the properties of a sample function $X(\cdot, \omega)$ or $Y(\cdot, \omega)$ are far from being determined by its values on a countable set. We shall not delve into this question but proceed to find a good version for a supermartingale, under certain conditions.

From now on we shall suppose that $(\Omega, \mathscr{F}, P)$ is a *complete* probability space. A sub-σ-field of $\mathscr{F}$ is called *augmented* iff it contains all P-null sets in $\mathscr{F}$. We shall assume for the increasing family $\{\mathscr{F}_t\}$ that $\mathscr{F}_0$ is augmented, then so is $\mathscr{F}_t$ for each t. Now if $\{Y_t\}$ is a version of $\{X_t\}$, and $X_t \in \mathscr{F}_t$, then also $Y_t \in \mathscr{F}_t$. Moreover if $\{X_t, \mathscr{F}_t\}$ is a (super) martingale, then so is $\{Y_t, \mathscr{F}_t\}$, as can be verified trivially.

Theorem 3. *Suppose that* $\{X_t, \mathscr{F}_t\}$ *is a supermartingale and* $\{\mathscr{F}_t\}$ *is right continuous, namely* $\mathscr{F}_t = \mathscr{F}_{t+}$ *for each* $t \in \mathbf{T}$. *Then the process* $\{X_t\}$ *has a right continuous version if and only if the function*

$$t \to E(X_t) \quad \text{is right continuous in } \mathbf{T}. \tag{15}$$

Proof. Suppose $\{Y_t\}$ is a right continuous version of $\{X_t\}$. Then if $t_n \downdownarrows t$ we have

$$\lim_n E(X_{t_n}) = \lim_n E(Y_{t_n}) = E(Y_t) = E(X_t),$$

where the second equation follows from the uniform integrability of $\{Y_{t_n}\}$ and the right continuity of Y. Hence (15) is true.

Conversely suppose (15) is true; let S be a countable dense subset of $\mathbf{T}$ and define X_{t+} as in (2). Since $X_{t+} \in \mathscr{F}_{t+} = \mathscr{F}_t$, we have by (11)

$$X_t \geq E(X_{t+} | \mathscr{F}_{t+}) = X_{t+} \qquad P\text{-a.e.} \tag{16}$$

On the other hand, let $t_n \in S$ and $t_n \downdownarrows t$; then

$$E(X_{t+}) = \lim_n E(X_{t_n}) = E(X_t) \tag{17}$$

where the first equation follows from uniform integrability and the second from (15). Combining (16) and (17) we obtain

$$\forall t: P(X_t = X_{t+}) = 1;$$

namely $\{X_{t+}\}$ is a version of $\{X_t\}$. Now for each ω, the function $t \to X(t+, \omega)$ as defined in (2) is right continuous in $\mathbf{T}$, by elementary analysis. Hence $\{X_{t+}\}$ is a right continuous version of $\{X_t\}$, and $\{X_{t+}, \mathscr{F}_t\}$ is a supermartingale by a previous remark. This is also given by Theorem 2, but the point is that there we did not know whether $\{X_{t+}\}$ is a version of $\{X_t\}$. $\square$

We shall say that the supermartingale $\{X_t, \mathscr{F}_t\}$ is right continuous iff both $\{X_t\}$ and $\{\mathscr{F}_t\}$ are right continuous. An immediate consequence of Theorem 3 is the following.

Corollary. *Suppose $\{\mathscr{F}_t\}$ is right continuous and $(X_t, \mathscr{F}_t)$ is a martingale. Then $\{X_t\}$ has a right continuous version.*

A particularly important class of martingales is given by

$$\{E(Y \mid \mathscr{F}_t), \mathscr{F}_t\} \tag{18}$$

where Y is an integrable random variable. According to the Corollary above, the process $\{E(Y \mid \mathscr{F}_t)\}$ has a right continuous version provided that $\{\mathscr{F}_t\}$ is right continuous. In this case we shall always use this version of the process.

The next theorem is a basic tool in martingale theory, called Doob's Stopping Theorem.

Theorem 4. *Let $\{X_t, \mathscr{F}_t\}$ be a right continuous supermartingale satisfying the following condition. There exists an integrable random variable Y such that*

$$\forall t: X_t \geq E(Y \mid \mathscr{F}_t). \tag{19}$$

Let S and T be both optional and $S \leq T$. Then we have

(a) $\lim_{t \to \infty} X_t = X_\infty$ *exists almost surely; X_S and X_T are integrable, where $X_S = X_\infty$ on $\{S = \infty\}$, $X_T = X_\infty$ on $\{T = \infty\}$;*
(b) $X_S \geq E(X_T \mid \mathscr{F}_{S+})$.

In case $\{X_t, \mathscr{F}_t\}$ is a martingale, there is equality in (b).

Proof. Let us first observe that condition (19) is satisfied with $Y \equiv 0$ if $X_t \geq 0$, $\forall t$. Next, it is also satisfied when there is equality in (19), namely for the class of martingales exhibited in (18). The general case of the theorem amounts to a combination of these two cases, as will be apparent below. We put

$$Z_t = X_t - E(Y \mid \mathscr{F}_t).$$

Then $\{Z_t, \mathscr{F}_t\}$ is a right continuous positive supermartingale by (19) and the fact that the martingale in (18) is right continuous by choice. Hence we have the decomposition

$$X_t = Z_t + E(Y \mid \mathscr{F}_t) \tag{20}$$

into a positive supermartingale and a martingale of the special kind, both right continuous. Corollary 2 to Theorem 1 applies to both terms on the right side of (20), since $Z_t \geq 0$ and $E\{|E(Y \mid \mathscr{F}_t)|\} \leq E(|Y|) < \infty$. Hence $\lim_{t \to \infty} X_t = X_\infty$ exists and is a finite random variable almost surely. Moreover it is easily verified that

$$\forall t: X_t \geq E(X_\infty \mid \mathscr{F}_t)$$

so that $\{X_t, \mathscr{F}_t, t \in [0, \infty]\}$ is a supermartingale. Hence for each n,

$$\left\{X\left(\frac{k}{2^n}\right), \mathscr{F}\left(\frac{k}{2^n}\right), k \in \mathbf{N}_\infty\right\}$$

is a discrete parameter supermartingale, where $\mathbf{N}_\infty = \{0, 1, 2, \ldots, \infty\}$. Let

$$S_n = \frac{[2^n S] + 1}{2^n}, \qquad T_n = \frac{[2^n T] + 1}{2^n}.$$

Then S_n and T_n are optional relative to the family $\{\mathscr{F}(k/2^n), k \in \mathbf{N}\}$, indeed strictly so, and $S_n \leq T_n$. Note that $X(S_n) = X(S) = X_\infty$ on $\{S = \infty\}$; $X(T_n) = X(T) = X_\infty$ on $\{T = \infty\}$. We now invoke Theorem 9.3.5 of *Course* to obtain

$$X(S_n) \geq E\{X(T_n) | \mathscr{F}(S_n)\}. \tag{21}$$

Since $\mathscr{F}_{S+} = \bigwedge_{n=1}^\infty \mathscr{F}_{S_n+}$ by Theorem 5 of §1.3, this implies: for any $\Lambda \in \mathscr{F}_{S+}$ we have

$$\int_\Lambda X(S_n)\,dP \geq \int_\Lambda X(T_n)\,dP. \tag{22}$$

Letting $n \to \infty$, then $X(S_n) \to X(S)$ and $X(T_n) \to X(T)$ by right continuity. Furthermore $\{X(S_n), \mathscr{F}(S_n)\}$ and $\{X(T_n), \mathscr{F}(T_n)\}$ are both supermartingales on the index set $\tilde{\mathbf{N}} = \{\ldots, n, \ldots, 1, 0\}$ which are uniformly integrable. To see this, e.g., for the second sequence, note that (21) holds when S_n is replaced by T_{n-1} since $T_{n-1} \geq T_n$; next if we put $S_n \equiv 0$ in (21) we deduce

$$\infty > E\{X(0)\} \geq E\{X(T_n)\},$$

and consequently

$$\infty > \lim_n \uparrow E\{X(T_n)\}.$$

This condition is equivalent to the asserted uniform integrability; see Theorem 9.4.7 of *Course*. The latter of course implies the integrability of the limits X_S and X_T, proving assertion (a). Furthermore we can now take the limits under the integrals in (22) to obtain

$$\int_\Lambda X(S)\,dP \geq \int_\Lambda X(T)\,dP. \tag{23}$$

The truth of this relation for each $\Lambda \in \mathscr{F}_{S+}$ is equivalent to the assertion (b). The final sentence of the theorem is proved by considering $\{-X_t\}$ as well as $\{X_t\}$. Theorem 4 is completely proved.

Let us consider the special case where

$$X_t = E(Y|\mathscr{F}_t). \tag{24}$$

We know $X_\infty = \lim_{t\to\infty} X_t$ exists and it can be easily identified as

$$X_\infty = E(Y|\mathscr{F}_\infty).$$

Now apply Theorem 4(b) with equality and with T for S, ∞ for T there. The result is, for any optional T:

$$X_T = E\{X_\infty|\mathscr{F}_{T+}\} = E\{E(Y|\mathscr{F}_\infty)|\mathscr{F}_{T+}\} = E\{Y|\mathscr{F}_{T+}\} \tag{25}$$

since $\mathscr{F}_{T+} \subset \mathscr{F}_\infty$. We may replace $\mathscr{F}_{T+}$ by $\mathscr{F}_T$ above since $\{\mathscr{F}_t\}$ is right continuous by hypothesis.

Next, let T be predictable, and $\{T_n\}$ announce T. Recalling that $\{X_t\}$ has left limits by Corollary 1 to Theorem 1, we have

$$X_{T-} = \lim_n X_{T_n} = \lim_n E\{Y|\mathscr{F}_{T_n+}\}, \tag{26}$$

where $X_{0-} = X_0$. As n increases, $\mathscr{F}_{T_n+}$ increases and by Theorem 9.4.8 of *Course*, the limits is equal to $E(Y|\bigvee_{n=1}^{\infty} \mathscr{F}_{T_n+})$. Using Theorem 8 of §1.3, we conclude that

$$X_{T-} = E\{Y|\mathscr{F}_{T-}\}. \tag{27}$$

Since $\mathscr{F}_{T-} \subset \mathscr{F}_{T+}$, we have furthermore by (25)

$$X_{T-} = E\{E(Y|\mathscr{F}_{T+})|\mathscr{F}_{T-}\} = E\{X_{T+}|\mathscr{F}_{T-}\}. \tag{28}$$

It is easy to show that these last relations do not hold for an arbitrary optional T (see Example 3 of §1.3). In fact, Dellacherie [2] has given an example in which X_{T-} is not integrable for an optional T.

The next theorem, due to P. A. Meyer [1], may be used in the study of excessive functions. Its proof shows the need for deeper notions of measurability and serves as an excellent introduction to the more advanced theory.

Theorem 5. *Let $\{X_t^{(n)}, \mathscr{F}_t\}$ be a right continuous positive supermartingale for each n, and suppose that almost surely we have*

$$\forall t: X_t^{(n)} \uparrow X_t. \tag{29}$$

If X_t is integrable for each t, then $\{X_t, \mathscr{F}_t\}$ is a right continuous supermartingale.

Proof. The fact that $\{X_t, \mathcal{F}_t\}$ is a positive supermartingale is trivial; only the right continuity of $t \to X_t$ is in question. Let D be the dyadic set, and put for each $t \geq 0$:

$$Y_t(\omega) = \lim_{\substack{s \downarrow \downarrow t \\ s \in D}} X_s(\omega).$$

[Note that Y_t is denoted by X_{t+} in (2) above when S is D.] By Theorem 1, the limit above exists for all $t \geq 0$ for $\omega \in \Omega_0$ where $P(\Omega_0) = 1$. From here on we shall confine ourselves to Ω_0.

Let $t_k = ([2^k t] + 1)/2^k$ for each $t \geq 0$ and $k \geq 0$. Then for each ω and t we have by (29) and the right continuity of $t \to X^{(n)}(t, \omega)$:

$$X(t, \omega) = \sup_n X^{(n)}(t, \omega) = \sup_n \lim_{k \to \infty} X^{(n)}(t_k, \omega)$$

$$\leq \lim_{k \to \infty} \sup_n X^{(n)}(t_k, \omega) = \lim_{k \to \infty} X(t_k, \omega) = Y(t, \omega). \tag{30}$$

Next let T be any optional time and $T_k = ([2^k T] + 1)/2^k$. Then we have by Theorem 4,

$$E\{X^{(n)}(T)\} \geq E\{X^{(n)}(T_k)\}$$

for each $k \geq 1$. Letting $n \to \infty$ we obtain by monotone convergence:

$$E\{X(T)\} \geq E\{X(T_k)\}. \tag{31}$$

Since $X(T_k)$ converges to $Y(T)$ by the definition of the latter, it follows from (31) and Fatou's lemma that

$$E\{X(T)\} \geq \lim_k E\{X(T_k)\} \geq E\{Y(T)\}. \tag{32}$$

Now for any $\varepsilon > 0$ define

$$T(\omega) = \inf\{t \geq 0 \,|\, Y_t(\omega) - X_t(\omega) \geq \varepsilon\} \tag{33}$$

where $\inf \varnothing = +\infty$ as usual. The verification that T is an optional time is a delicate matter and will be postponed until the next section. Suppose this is so; then if $T(\omega) < \infty$ we have $\delta_j \geq 0$, $\delta_j \to 0$ such that

$$Y(T(\omega) + \delta_j, \omega) - X(T(\omega) + \delta_j, \omega) \geq \varepsilon.$$

Hence we have for all $n \geq 1$:

$$Y(T(\omega) + \delta_j, \omega) - X^{(n)}(T(\omega) + \delta_j, \omega) \geq \varepsilon$$

because $X \geq X^{(n)}$. Since both $t \to Y(t, \omega)$ and $t \to X^{(n)}(t, \omega)$ are right continuous the inequality above implies that

$$Y(T(\omega), \omega) - X^{(n)}(T(\omega), \omega) \geq \varepsilon$$

as $\delta_j \to 0$. Note that the case where all $\delta_j = 0$ is included in this argument. Letting $n \to \infty$ we obtain

$$Y(T(\omega), \omega) - X(T(\omega), \omega) \geq \varepsilon. \tag{34}$$

Therefore (34) holds on the set $\{T < \infty\}$, and consequently (32) is possible only if $P\{T < \infty\} = 0$. Since ε is arbitrary, this means we have almost surely:

$$\forall t \geq 0: \ Y(t, \omega) \leq X(t, \omega). \tag{35}$$

Together with (30) we conclude that equality must hold in (35), namely that $P\{\forall t \geq 0: Y_t = X_t\} = 1$. The theorem is proved since $t \to Y(t, \omega)$ is right continuous for $\omega \in \Omega_0$. $\qquad\square$

Remark. If we do not assume in the theorem that X_t is integrable for each t, of course $\{X_t\}$ cannot be a supermartingale. But it is still true that $t \to X(t, \omega)$ is right continuous as an extended-valued function, for P-a.e. ω. The assumption that each $X^{(n)}$ is positive may also be dropped. Both these extensions are achieved by the following reduction. Fix two positive integers u and k and put

$$Y_t^{(n)} = (X_t^{(n)} - E\{X_u^{(1)} | \mathscr{F}_t\}) \wedge k, \qquad 0 \leq t \leq u$$

where the conditional expectation is taken in its right continuous version. Then $\{Y_t^{(n)}\}$ satisfies the conditions of the theorem, hence

$$Y_t = \sup_n Y_t^{(n)}$$

is right continuous almost surely in $[0, u]$. Now let first $k \uparrow \infty$, then $u \uparrow \infty$.

An important complement to Theorem 5 under a stronger hypothesis will next be given.

Theorem 6. *Under the hypotheses of Theorem 5, suppose further that the index set is $[0, \infty]$ and (29) holds for $t \in [0, \infty]$. Suppose also that for any sequence of optional times $\{T_n\}$ increasing to T almost surely, we have*

$$\lim \ E\{X(T_n)\} = E\{X(T)\}. \tag{36}$$

Then almost surely the convergence of $X^{(n)}$ to X_t is uniform in $t \in [0, \infty]$.

Proof. For any $\varepsilon > 0$, define

$$T_n^\varepsilon(\omega) = \inf\{t \geq 0 \,|\, X(t, \omega) - X^{(n)}(t, \omega) \geq \varepsilon\}.$$

Since X as well as X_n is right continuous by Theorem 5, this time it is easy to verify that T_n^ε is optional directly without recourse to the next section. Since $X_t^{(n)}$ increases with n for each t, T_n^ε increases with n, say to the limit $T_\infty^\varepsilon(\leq \infty)$. Now observe that:

$X^{(n)} \uparrow X$ uniformly in $[0, \infty]$, P-a.e.

$\quad \Leftrightarrow \forall \varepsilon > 0$: for P-a.e. ω, $\exists n(\omega)$ such that $T_{n(\omega)}^\varepsilon(\omega) = \infty$

$\quad \Leftrightarrow \forall \varepsilon > 0$: $P\{\exists n : T_n^\varepsilon = \infty\} = 1$

$\quad \Leftrightarrow \forall \varepsilon > 0$: $P\{\forall n : T_n^\varepsilon < \infty\} = 0$

$\quad \Leftrightarrow \forall \varepsilon > 0$: $\lim_n P\{T_n^\varepsilon < \infty\} = 0$

By right continuity, we have:

$$X(T_n^\varepsilon) - X_n(T_n^\varepsilon) \geq \varepsilon \quad \text{on } \{T_n^\varepsilon < \infty\}.$$

Hence for $m \leq n$:

$$\varepsilon P\{T_n^\varepsilon < \infty\} \leq E\{X(T_n^\varepsilon) - X_n(T_n^\varepsilon); \, T_n^\varepsilon < \infty\}$$
$$\leq E\{X(T_n^\varepsilon) - X_m(T_n^\varepsilon)\}.$$

Note that for any optional T, $E\{X(T)\} \leq E\{X(0)\} < \infty$ by Theorem 4. By the same theorem,

$$E\{X_m(T_n^\varepsilon)\} \geq E\{X_m(T_\infty^\varepsilon)\}.$$

Thus

$$\varepsilon P\{T_n^\varepsilon < \infty\} \leq E\{X(T_n^\varepsilon) - X_m(T_\infty^\varepsilon)\}.$$

Letting $n \to \infty$ and using (36), we obtain:

$$\varepsilon \lim_n P\{T_n^\varepsilon < \infty\} \leq E\{X(T_\infty^\varepsilon) - X_m(T_\infty^\varepsilon)\}. \tag{37}$$

Under the hypothesis (29) with $t = \infty$ included, $X_m(T_\infty^\varepsilon) \uparrow X(T_\infty^\varepsilon)$, P-a.e. Since $X(T_\infty^\varepsilon)$ is integrable, the right member of (37) converges to zero by dominated convergence. Hence so does the left member, and this is equivalent to the assertion of the theorem, as analyzed above. $\square$

The condition (36) is satisfied for instance when all sample functions of X are left continuous, therefore continuous since they are assumed to be right continuous in the theorem. In general, the condition supplies a kind of left continuity, not for general approach but for "optional approaches". A similar condition will be an essential feature of the Markov processes we are going to study, to be called "quasi left continuity" of Hunt processes. We will now study the case where $T_n \to \infty$ in (36), and for this discussion we introduce a new term. The reader should be alerted to the abuse of the word "potential", used by different authors to mean different things. For this reason we give it a prefix to indicate that it is a special case of supermartingales. [We do not define a subpotential!]

Definition. A *superpotential* is a right continuous positive supermartingale $\{X_t\}$ satisfying the condition

$$\lim_{t \to \infty} E(X_t) = 0. \tag{38}$$

Since $X_\infty = \lim_{t \to \infty} X_t$ exists almost surely by Corollary 2 to Theorem 1, the condition (38) implies that $X_\infty = 0$ almost surely. Furthermore, it follows that $\{X_t\}$ is uniformly integrable (Theorem 4.5.4 of *Course*). Nonetheless this does not imply that for any sequence of optional times $T_n \to \infty$ we have

$$\lim_n E(X_{T_n}) = 0. \tag{39}$$

When this is the case, the superpotential is said to be "of class D"—a nomenclature whose origin is obscure. This class plays a fundamental role in the advanced theory, so let us prove one little result about it just to make the acquaintance.

Theorem 7. *A superpotential is of class D if and only if the class of random variables $\{X_T : T \in \mathcal{O}\}$ is uniformly integrable, where $\mathcal{O}$ is the class of all optional times.*

Proof. If $T_n \to \infty$, then $X_{T_n} \to 0$ almost surely; hence uniform integrability of $\{X_{T_n}\}$ implies (39). To prove the converse, let $n \in \mathbf{N}$ and

$$T_n = \inf\{t \geq 0 \mid X_t \geq n\}.$$

Since almost surely $X(t)$ is bounded in each finite t-interval by Corollary 1 to Theorem 1, and converges to zero as $t \to \infty$, we must have $T_n \uparrow \infty$. Hence (39) holds for this particular sequence by hypothesis. Now for any optional T, define

$$S = \begin{cases} T & \text{on } \{T \geq T_n\}, \\ \infty & \text{on } \{T < T_n\}. \end{cases}$$

Then $S \geq T_n$ and we have

$$\int_{\{X_T \geq n\}} X_T \, dP \leq \int_{\{T \geq T_n\}} X_T \, dP = \int_\Omega X_S \, dP \leq \int_\Omega X_{T_n} \, dP$$

where the middle equation is due to $X_\infty \equiv 0$ and the last inequality due to Doob's stopping theorem. Since the last term above converges to zero as $n \to \infty$, $\{X_T : T \in \mathcal{O}\}$ is uniformly integrable by definition. $\qquad\square$

1.5. Progressive Measurability and the Projection Theorem

For $t \geq 0$ let $\mathbf{T}_t = [0, t]$ and $\mathcal{B}_t$ be the Euclidean (classical) Borel field on $\mathbf{T}_t$. Let $\{\mathcal{F}_t\}$ be an increasing family of σ-fields.

Definition. $X = \{X_t, t \in \mathbf{T}\}$ is said to be *progressively measurable* relative to $\{\mathcal{F}_t, t \in \mathbf{T}\}$ iff the mapping

$$(s, \omega) \to X(s, \omega)$$

restricted to $\mathbf{T}_t \times \Omega$ is measurable $\mathcal{B}_t \times \mathcal{F}_t$ for each $t \geq 0$. This implies that $X_t \in \mathcal{F}_t$ for each t, namely $\{X_t\}$ is adapted to $\{\mathcal{F}_t\}$ (exercise).

This new concept of measurability brings to the fore the fundamental nature of $X(\cdot, \cdot)$ as a function of the pair (s, ω). For each s, $X(s, \cdot)$ is a random variable; for each ω, $X(\cdot, \omega)$ is a sample function. But the deeper properties of the process X really concern it as function of two variables on the domain $\mathbf{T} \times \Omega$. The structure of this function is complicated by the fact that Ω is not generally supposed to be a topological space. But particular cases of Ω such that $(-\infty, +\infty)$ or $[0, 1]$ should make the comparison with a topological product space meaningful. For instance, from this point of view a random time such as an optional time represents a "curve" as illustrated below:

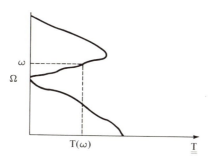

Thus the introduction of such times is analogous to the study of plane geometry by means of curves—a venerable tradition. This apparent analogy has been made into a pervasive concept in the "general theory of stochastic processes;" see Dellacherie [2] and Dellacherie-Meyer [1].

According to the new point of view, a function $X(\cdot, \cdot)$ on $\mathbf{T} \times \Omega$ which is in $\mathcal{B} \times \mathcal{F}$ is precisely a Borel measurable process as defined in §1.3. In particular, for any subset H of $\mathbf{T} \times \Omega$, its indicator function 1_H may be identified as a Borel measurable process $\{1_H(t), t \in \mathbf{T}\}$ in the usual notation, where $1_H(t)$ is just the abbreviation of $\omega \to 1_H(t, \omega)$. Turning this idea around, we call the set H progressively measurable iff the associated process $\{1_H(t)\}$ is so. This amounts to the condition that $H \cap ([0, t] \times \Omega) \in \mathscr{B}_t \times \mathscr{F}_t$ for each $t \geq 0$. Now it is easy to verify that the class of progressively measurable sets forms a σ-field $\mathscr{G}$, and that X is progressively measurable if and only if $X \in \mathscr{G}/\mathscr{E}$ in the notation of §1.1. It follows that progressive measurability is preserved by certain operations such as sequential pointwise convergence. The same is not true for, e.g., right continuous process.

Recall that $\mathscr{F}_t^0 = \sigma(X_s, 0 \leq s \leq t)$. If X is progressively measurable relative to $\{\mathscr{F}_t^0\}$, then it is so relative to any $\{\mathscr{F}_t\}$ to which it is adapted.

Theorem 1. *If $\{X_t\}$ is right [or left] continuous, then it is progressively measurable relative to $\{\mathscr{F}_t^0\}$.*

Proof. We will prove the result for right continuity, the left case being similar. Fix t and define for $n \geq 1$:

$$X^{(n)}(s, \omega) = X\left(\frac{k+1}{2^n} t, \omega\right) \quad \text{if } s \in \left[\frac{k}{2^n} t, \frac{k+1}{2^n} t\right), 0 \leq k \leq 2^n - 1;$$

$$X^{(n)}(t, \omega) = X(t, \omega).$$

Then $X^{(n)}$ is defined on $\mathbf{T}_t \times \Omega$ and converges pointwise by right continuity. For each $B \in \mathscr{E}$, we have

$$\{(s, \omega) \mid X^{(n)}(s, \omega) \in B\} = \bigcup_{k=0}^{2^n - 1} \left(\left[\frac{k}{2^n} t, \frac{k+1}{2^n} t\right) \times \left\{\omega \mid X\left(\frac{k+1}{2^n} t, \omega\right) \in B\right\}\right)$$

$$\cup (\{t\} \times \{\omega \mid X(t, \omega) \in B\}) \in \mathscr{B}_t \times \mathscr{F}_t^0.$$

Hence the limit, which is X restricted to $\mathbf{T}_t \times \Omega$, is likewise measurable by the following lemma which is spelled out here since it requires a little care in a general state space. The latter is assumed to be separable metric and locally compact, so that each open set is the union of a countable family of closed sets.

Lemma. *Let $\mathscr{A}$ be any σ-field of subsets of $\mathbf{T} \times \Omega$. Suppose that $X^{(n)} \in \mathscr{A}$ for each n and $X^{(n)} \to X$ everywhere. Then $X \in \mathscr{A}$.*

Proof. Let $\mathbb{C}$ be the class of sets B in $\mathscr{E}$ such that $X^{-1}(B) \in \mathscr{A}$. Let G be open and $G = \bigcup_{k=1}^{\infty} F_k$ where $F_k = \{x \mid d(x, G^c) \geq 1/k\}$ an d is a metric. Then we have by pointwise convergence of $X^{(n)}$ to X:

$$X^{-1}(G) = \bigcup_{k=1}^{\infty} \bigcup_{n=1}^{\infty} \bigcap_{m=n}^{\infty} (X^{(m)})^{-1}(F_k).$$

This belongs to $\mathscr{A}$, and so the class $\mathbb{C}$ contains all open sets. Since it is a σ-field by properties of X^{-1}, it contains the minimal σ-field containing all open sets, namely $\mathscr{E}$. Thus $X \in \mathscr{A}/\mathscr{E}$. □

The next result is an extension of Theorem 10 of §1.3.

Theorem 2. *If X is progressively measurable and T is optional, then $X_T 1_{\{T < \infty\}} \in \mathscr{F}_{T+}$.*

Proof. Consider the two mappings:

$$\varphi : (s, \omega) \to X(s, \omega)$$
$$\psi : \omega \to (T(\omega), \omega).$$

Their composition is

$$\varphi \circ \psi : \omega \to X(T(\omega), \omega) = X_T(\omega).$$

It follows from the definition of optionality that for each $s < t$:

$$\psi^{-1}(\mathscr{B}_s \times \mathscr{F}_s) \subset \mathscr{F}_t. \tag{1}$$

For if $A \in \mathscr{B}_s$ and $\Lambda \in \mathscr{F}_s$, then

$$\{\omega \,|\, T(\omega) \in A; \omega \in \Lambda\} = \{T \in A\} \cap \Lambda \in \mathscr{F}_{s+} \subset \mathscr{F}_t.$$

Next if we write for a fixed s the restriction of φ to $\mathbf{T}_s \times \Omega$ as $\tilde{\varphi}$, we have by progressive measurability

$$\tilde{\varphi}^{-1}(\mathscr{E}) \subset \mathscr{B}_s \times \mathscr{F}_s. \tag{2}$$

Combining (1) and (2), we obtain

$$(\tilde{\varphi} \circ \psi)^{-1}(\mathscr{E}) = \psi^{-1} \circ \tilde{\varphi}^{-1}(\mathscr{E}) \subset \mathscr{F}_t.$$

In particular if $B \in \mathscr{E}$, then for each $s < t$:

$$\{\omega \,|\, T(\omega) \le s; X_T(\omega) \in B\} \in \mathscr{F}_t.$$

Taking a sequence of s increasing strictly to t, we deduce that for each t:

$$\{X_T \in B\} \cap \{T < t\} \in \mathscr{F}_t.$$

This is equivalent to the assertion of the theorem. □

Remark. A similar argument shows that if T is strictly optional, then $X_T 1_{\{T < \infty\}} \in \mathscr{F}_T$.

Let H be a subset of $\mathbf{T} \times \Omega$. We define the *debut* of H as follows:

$$D_H(\omega) = \inf\{t \geq 0 \,|\, (t, \omega) \in H\} \tag{3}$$

where $\inf \emptyset = \infty$ as usual.

Theorem 3 (The Projection Theorem). *If for each t, $(\Omega, \mathscr{F}_t, P)$ is a complete measure space, then D_H as defined in* (3) *is an optional time for each progressively measurable set H.*

Proof. Let $\mathbf{T}_{t-} = [0, t)$ and consider the set

$$H_t = H \cap (\mathbf{T}_{t-} \times \Omega).$$

Then $H_t \in \mathscr{B}_t \times \mathscr{F}_t$ since H is progressively measurable. Let π_Ω denote the projection mapping of $\mathbf{T} \times \Omega$ onto Ω, namely for any $H \subset \mathbf{T} \times \Omega$:

$$\pi_\Omega(H) = \{\omega \in \Omega \,|\, \exists t \in \mathbf{T} \text{ such that } (t, \omega) \in H\}.$$

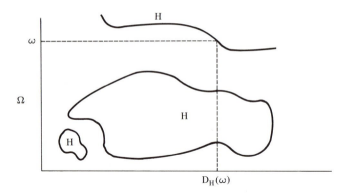

It is clear that

$$\begin{aligned}
\pi_\Omega(H_t) &= \{\omega \,|\, \exists s \in [0, t) \text{ such that } (s, \omega) \in H\} \\
&= \{\omega \,|\, D_H(\omega) < t\}.
\end{aligned} \tag{4}$$

[Observe that if we replace $[0, t)$ by $[0, t]$ in the second term above we cannot conclude that it is equal to the third term with "$<$" replaced by "$\leq$".] By the theory of analytical sets (see Dellacherie-Meyer [1], Chapter 3) the projection of each set in $\mathscr{B}_t \times \mathscr{F}_t$ on Ω is an "$\mathscr{F}_t$-analytic" set, hence $\mathscr{F}_t$-measurable when $(\Omega, \mathscr{F}_t, P)$ is complete. Thus for each t we have $\{D_H < t\} \in \mathscr{F}_t$, namely D_H is optional. $\square$

The following "converse" to Theorem 3 is instructive and is a simple illustration of the general methodology. Let T be optional and consider the set

$$H = \{(t, \omega) \mid T(\omega) < t\}.$$

Then

$$1_H(t, \omega) = \begin{cases} 1 & \text{on } \{(t, \omega) \mid T(\omega) < t\}, \\ 0 & \text{on } \{(t, \omega) \mid T(\omega) \geq t\}. \end{cases}$$

Since T is optional, we have $1_H(t) \in \mathscr{F}_t$ for each t. Furthermore, it is *trivial* that for each ω, the sample function $1_H(\cdot, \omega)$ is left continuous. Hence 1_H is progressively measurable by Theorem 1 above. Now a picture shows that

$$D_H(\omega) = T(\omega).$$

Thus each optional time is the debut of a progressively measurable set.

We now return to the T in (33) of §1.4 and show that it is optional. Since each $X^{(n)}$ is right continuous, it is progressively measurable relative to $\{\mathscr{F}_t\}$ by Theorem 1. Therefore X is progressively measurable as the limit (or supremum) of $X^{(n)}$ by the Lemma above in which we take $\mathscr{A}$ to be the progressively measurable σ-field. On the other hand, the process $(t, \omega) \to Y(t, \omega)$, being right continuous, is progressively measurable by Theorem 1. Hence the set

$$\{(t, \omega) \mid Y(t, \omega) - X(t, \omega) \geq \varepsilon\}$$

is progressively measurable, and T is optional by Theorem 3.

As an illustration of the fine analysis of sample function behavior, we prove the following theorem which is needed in Chapter 2. Recall that for a discrete time positive supermartingale, almost surely every sample sequence remains at the value zero if it is ever taken. In continuous time the result has a somewhat delicate ramification.

Theorem 4. *Let $\{X_t, \mathscr{F}_t\}$ be a positive supermartingale having right continuous paths. Let*

$$T_1(\omega) = \inf\{t \geq 0 \mid X_t(\omega) = 0\},$$
$$T_2(\omega) = \inf\{t \geq 0 \mid X_{t-}(\omega) = 0\},$$
$$T = T_1 \wedge T_2.$$

Then we have almost surely $X(T + t) = 0$ for all $t \geq 0$ on the set $\{T < \infty\}$.

Proof. Since $\{X_t\}$ is right continuous, it has left limits by Corollary 1 to Theorem 1 of §1.4. By Theorem 1, both $\{X_t\}$ and $\{X_{t-}\}$ are progressively measurable relative to $\{\mathscr{F}_t\}$. If we put

$$H_1 = \{(t,\omega)\,|\,X(t,\omega) = 0\}, \qquad H_2 = \{(t,\omega)\,|\,X(t-,\omega) = 0\},$$

then $T_1 = D_{H_1}$, $T_2 = D_{H_2}$; hence both are optional. It follows from Doob's stopping theorem that for each $t \geq 0$:

$$E\{X(T_1); T_1 < \infty\} \geq E\{X(T_1 + t); T_1 < \infty\}. \tag{5}$$

But $X(T_1) = 0$ on $\{T_1 < \infty\}$ by the definition of T_1 and right continuity of paths. Hence the right member of (5) is also equal to zero. Since $X \geq 0$ this implies that

$$P\{X(T_1 + t) = 0 \text{ for all } t \in Q \cap [0,\infty); T_1 < \infty\} = P\{T_1 < \infty\}.$$

We may omit Q in the above by right continuity, and the result is equivalent to the assertion of the theorem when T is replaced by T_1.

The situation is different for T_2 because we can no longer say at once that $X(T_2) = 0$, although this is part of the desired conclusion. It is conceivable that $X(t) \to 0$ as $t \uparrow\uparrow T_2$ but jumps to a value different from 0 at T_2. To see that this does not happen, we must make a more detailed analysis of sample functions of the kind frequently needed in the study of Markov processes later. We introduce the approximating optional times as follows:

$$S_n(\omega) = \inf\left\{t \geq 0\,\big|\,X_t(\omega) < \frac{1}{n}\right\}.$$

It is easy to show S_n is optional because $(-\infty, 1/n)$ is an open set without using progressive measurability. Clearly S_n increases with n, and $S_n \uparrow S \leq T$ where S is optional. Now we must consider two cases.

Case 1. $\forall n : S_n < S$. In this case $X(S-) = \lim_n X(S_n) = 0$. Since $X(t) \geq 1/n$ for $t \in [0, S_n)$, so also $X(t-) \geq 1/n$ for $t \in [0, S_n)$, it is clear that $S = T_2 = T$.

Case 2. $\exists n_0 : S_{n_0} = S_{n_0+1} = \cdots = S$. In this case $X(S) = 0$ because $X(S_n) \leq 1/n$ for all n, and we have $S = T_1 = T$. Unless $S_{n_0} = 0$, X jumps to 0 at S_{n_0}.

We have therefore proved that $T = S$ is optional (without proving that T_1 and T_2 are). Now applying Doob's stopping theorem as follows:

$$E\{X(S_n); S_n < \infty\} \geq E\{X(S); S_n < \infty\}. \tag{6}$$

Since the left member does not exceed $1/n$, letting $n \to \infty$ we obtain

$$E\{X(S), S < \infty\} = 0.$$

The rest is as before. $\square$

We used the fact that $\{S < \infty\} \subset \bigcap_n \{S_n < \infty\}$. The reverse inclusion is not true!

Exercises

1. Let $\{X_t, \mathscr{F}_t\}$ be a martingale with right continuous paths. Show that there exists an integrable Y such that $X_t = E(Y|\mathscr{F}_t)$ for each t, if and only if $\{X_t\}$ is uniformly integrable.

2. Show that if T is the first jump in a Poisson process (see Example 3 of §1.3), then $E\{X_T|\mathscr{F}_{T-}\} \neq X_{T-}$.

3. In the notation of the proof of Theorem 6 of §1.4, show that $X_t^{(n)}$ converges to X_t uniformly in each finite t-interval, P-a.e., if and only if for each $\varepsilon > 0$ we have $P\{\lim_n T_n^\varepsilon = \infty\} = 1$.

4. If $(t, \omega) \to X(t, \omega)$ is in $\mathscr{B} \times \mathscr{F}^0$ then for each $A \in \mathscr{E}$, the function $(t, x) \to P_t(x, A)$ is in $\mathscr{B} \times \mathscr{E}$. Hence this is the case if $\{X_t\}$ is right [or left] continuous. [Hint: consider the class of functions φ on $\mathbf{T} \times \Omega$ such that $(t, x) \to E^x\{\varphi\}$ belongs to $\mathscr{B} \times \mathscr{E}$. It contains functions of the form $1_B(t)1_A(\omega)$ where $B \in \mathscr{B}$ and $\Lambda \in \mathscr{F}^0$.]

5. If $\{X_t\}$ is adapted to $\{\mathscr{F}_t\}$, and progressively measurable relative to $\{\mathscr{F}_{t+\varepsilon}\}$ for each $\varepsilon > 0$, then $\{X_t\}$ is progressively measurable relative to $\{\mathscr{F}_t\}$.

6. Give an example of a process $\{X_t\}$ which is adapted to $\{\mathscr{F}_t\}$ but not progressively measurable relative to $\{\mathscr{F}_t\}$.

7. Suppose that $\{\mathscr{F}_t\}$ is right continuous and S and T are optional relative to $\{\mathscr{F}_t\}$, with $S \leq T$. Put

$$[[S, T)) = \{(t, \omega) \in \mathbf{T} \times \Omega \,|\, S(\omega) \leq t < T(\omega)\};$$

similarly for $[[S, T]]$, $((S, T))$, $((S, T]]$. Show that all these four sets are progressively measurable relative to $\{\mathscr{F}_t\}$.

The σ-field generated by $[[S, T))$ when S and T range over all optional times such that $S \leq T$ is called the *optional field*; the σ-field generated by $[[S, T))$ when S and T range over all predictable times such that $S \leq T$ is called the *predictable field*. Thus we have

$$\text{predictable field} \subset \text{optional field}$$
$$\subset \text{progressively measurable field.}$$

These are the fundamental σ-fields for the general theory of stochastic processes.

NOTES ON CHAPTER 1

§1.1. The basic notions of the Markov property, as well as optionality, in the discrete parameter case are treated in Chapters 8 and 9 of the *Course*. A number of proofs carry over to the continuous parameter case, without change.

§1.2. Among the examples of Markov processes given here, only the case of Brownian motion will be developed in Chapter 4. But for dimension $d = 1$ the theory is somewhat special and will not be treated on its own merits. The case of Markov chains is historically the oldest, but its modern development is not covered by the general theory. It will only be mentioned here occasionally for peripheral illustrations. The class of spatially homogeneous Markov processes, sometimes referred to as Lévy or additive processes, will be briefly described in §4.1.

§1.3. Most of the material on optionality may be found in Chung and Doob [1] in a more general form. For deeper properties, which are sparingly used in this book, see Meyer [1]. The latter is somewhat dated but in certain respects more readable than the comprehensive new edition which is Dellacherie and Meyer [1].

§1.4. It is no longer necessary to attribute the foundation of martingale theory to Doob, but his book [1] is obsolete especially for the treatment of the continuous parameter case. The review here borrows much from Meyer [1] and is confined to later needs, except for Theorems 6 and 7 which are given for the sake of illustration and general knowledge. Meyer's proof of Theorem 5 initiated the method of projection in order to establish the optionality of the random time T defined in (33). This idea was later developed into a powerful methodology based on the two σ-fields mentioned at the end of the section. A very clear exposition of this "general theory" is given in Dellacherie [2].

A curious incident happened in the airplane from Zürich to Beijing in May of 1979. At the prodding of the author, Doob produced a new proof of Theorem 5 without using projection. Unfortunately it is not quite simple enough to be included here, so the interested reader must await its appearance in Doob's forthcoming book.

Chapter 2

Basic Properties

2.1. Martingale Connection

Let a homogeneous Markov process $\{X_t, \mathscr{F}_t, t \in \mathbf{T}\}$ with transition semi-group (P_t) be given. We seek a class of functions f on $\mathbf{E}$ such that $\{f(X_t), \mathscr{F}_t\}$ is a supermartingale.

Definition. Let $f \in \mathscr{E}, 0 \leq f \leq \infty$; f is *α-superaveraging* relative to (P_t) iff

$$\forall t \geq 0 : f \geq e^{-\alpha t} P_t f; \tag{1}$$

f is *α-excessive* iff in addition we have

$$f = \lim_{t \downarrow 0} e^{-\alpha t} P_t f. \tag{2}$$

It follows from (1) that its right member is a decreasing function of t, hence limit in (2) always exists.

Excessive functions will play a major role in Chapter 3. The connection with martingale theory is given by the proposition below.

Proposition 1. *If f is α-superaveraging and $f(X_t)$ is integrable for each t, then $\{e^{-\alpha t} f(X_t), \mathscr{F}_t\}$ is a supermartingale.*

Proof. Let $s > 0, t \geq 0$. We have by the Markov property:

$$f(X_s) \geq e^{-\alpha t} P_t f(X_s) = e^{-\alpha t} E^{X_s} \{f(X_t)\} = e^{-\alpha t} E\{f(X_{s+t}) | \mathscr{F}_s\}.$$

Hence

$$e^{-\alpha s} f(X_s) \geq E\{e^{-\alpha(s+t)} f(X_{s+t}) | \mathscr{F}_s\}$$

and this establishes the proposition. □

Next we present a subclass of α-superaveraging functions which plays a dominant role in the theory. We begin by stating a further condition on the transition function.

Definition. (P_t) is said to be *Borelian* iff for each $A \in \mathscr{E}$:

$$(t, x) \to P_t(x, A)$$

is measurable $\mathscr{B} \times \mathscr{E}$. This is equivalent to: for each $f \in b\mathscr{E}$:

$$(t, x) \to P_t f(x)$$

is measurable $\mathscr{B} \times \mathscr{E}$. According to Exercise 4 of §1.5, if e.g. $\{X_t\}$ is right continuous, then (P_t) is Borelian. The next definition applies to a Borelian (P_t) in general.

Definition. Let $f \in b\mathscr{E}$, $\alpha > 0$, then the α-potential of f is the function $U^\alpha f$ given by

$$U^\alpha f(x) = \int_0^\infty e^{-\alpha t} P_t f(x)\, dt$$

$$= E^x \left\{ \int_0^\infty e^{-\alpha t} f(X_t)\, dt \right\}. \tag{3}$$

The integration with respect to t is possible because of the Borelian assumption above and the equality of the two expressions in (3) is due to Fubini's theorem. In particular $U^\alpha f \in \mathscr{E}$. If we denote the sup-norm of f by $\|f\|$, then we have

$$\|U^\alpha f\| \le \frac{1}{\alpha} \|f\|. \tag{4}$$

Taking $f \equiv 1$ we see that the "operator norm" satisfies

$$\|U^\alpha\| = \frac{1}{\alpha}. \tag{5}$$

U^α is also defined for $f \in \mathscr{E}_+$ but may take the value $+\infty$. The family of operators $\{U^\alpha, \alpha > 0\}$ is also known as the "resolvent" of the semigroup (P_t), or by abuse of language, of the process (X_t). We postpone a discussion of the relevant facts. For the present the next two propositions are important.

Proposition 2. *If $f \in b\mathscr{E}_+$, then for each $\alpha > 0$, $U^\alpha f$ is α-excessive; and $\{e^{-\alpha t} U^\alpha f(X_t), \mathscr{F}_t\}$ is a supermartingale.*

Proof. We have

$$e^{-\alpha t} P_t(U^\alpha f) = e^{-\alpha t} \int_0^\infty e^{-\alpha s} P_{t+s} f\, ds = \int_t^\infty e^{-\alpha s} P_s f\, ds$$

$$\le \int_0^\infty e^{-\alpha s} P_s f\, ds = U^\alpha f.$$

This shows that $U^\alpha f$ is α-superaveraging. As $t \downarrow 0$, the third term above converges to the fourth term, hence $U^\alpha f$ is α-excessive—a fact we store for later use. Since $U^\alpha f$ is bounded the last assertion of the proposition follows from Proposition 1. ☐

Proposition 3. *Suppose $\{X_t\}$ is progressively measurable relative to $\{\mathscr{F}_t\}$. For $f \in b\mathscr{E}_+$ and $\alpha > 0$, define*

$$Y_t = \int_0^t e^{-\alpha s} f(X_s)\, ds + e^{-\alpha t} U^\alpha f(X_t). \tag{6}$$

Then $\{Y_t, \mathscr{F}_t\}$ is a uniformly integrable martingale which is progressively measurable relative to $\{\mathscr{F}_t\}$.

Proof. Let

$$Y_\infty = \int_0^\infty e^{-\alpha s} f(X_s)\, ds$$

which is a bounded random variable; then we have for each x:

$$E^x\{Y_\infty\} = U^\alpha f(x).$$

We have

$$E\{Y_\infty | \mathscr{F}_t\} = \int_0^t e^{-\alpha s} f(X_s)\, ds + E\left\{\int_t^\infty e^{-\alpha s} f(X_s)\, ds \Big| \mathscr{F}_t\right\}. \tag{7}$$

The first term on the right belongs to $\mathscr{F}_t$, because $\{X_t\}$ is progressively measurative relative to $\{\mathscr{F}_t\}$ (exercise). The second term on the right side of (7) is equal to

$$E\left\{\int_0^\infty e^{-\alpha(t+u)} f(X_{t+u})\, du \Big| \mathscr{F}_t\right\}$$

$$= E\left\{e^{-\alpha t} \int_0^\infty e^{-\alpha u} f(X_u \circ \theta_t)\, du \Big| \mathscr{F}_t\right\}$$

$$= E\{e^{-\alpha t} Y_\infty \circ \theta_t | \mathscr{F}_t\} = e^{-\alpha t} E^{X_t}\{Y_\infty\} = e^{-\alpha t} U^\alpha f(X_t).$$

Hence we have

$$E\{Y_\infty | \mathscr{F}_t\} = Y_t. \tag{8}$$

Since Y_∞ is bounded, $\{Y_t, \mathscr{F}_t\}$ is a martingale. It is an easy proposition that for any integrable Y, the family $E\{Y | \mathscr{G}\}$ where $\mathscr{G}$ ranges over all σ-fields of

$(\Omega, \mathscr{F}, P)$, is uniformly integrable. Here is the proof:

$$\int_{\{|E(Y|\mathscr{G})| \geq n\}} |E(Y|\mathscr{G})|\, dP \leq \int_{\{|E(Y|\mathscr{G})| \geq n\}} |Y|\, dP;$$

$$P\{|E(Y|\mathscr{G})| \geq n\} \leq \frac{1}{n}\, E(|Y|) \to 0.$$

Hence $\{Y_t\}$ is uniformly integrable by (8). Finally, the first term on the right side of (7) is continuous in t, hence it is progressively measurable as a process. On the other hand since $U^\alpha f \in \mathscr{E}$, it is clear that $\{U^\alpha f(X_t)\}$ as a process is progressively measurable if $\{X_t\}$ is, relative to $\{\mathscr{F}_t\}$.   □

Remark. Let us change notation and put

$$A_t = \int_0^t e^{-\alpha s} f(X_s)\, ds.$$

Then $\{A_t\}$ is an increasing process, namely for each ω, $A_t(\omega)$ increases with t; and $A_\infty = Y_\infty$ by definition of the latter. The decomposition given in (6), together with (8), may be rewritten as:

$$e^{-\alpha t} U^\alpha f(X_t) = E\{A_\infty | \mathscr{F}_t\} - A_t. \tag{9}$$

This is a simple case of the Doob-Meyer decomposition of a supermartingale into a uniformly integrable martingale minus an increasing process. If we assume that the family $\{\mathscr{F}_t\}$ is right continuous, then by the corollary to Theorem 3 of §1.4, the supermartingale in (9) has a right continuous version. It is in fact a superpotential of class D (exercise).

2.2. Feller Process

Before further development of the machinery in §2.1, we turn our attention to a class of transition semigroups having nice analytic properties. We show that the sample functions of the associated Markov process have desirable regularity properties, if a proper version is taken. This will serve as a model for a more general setting in Chapter 3.

Let $\mathbf{E}$ be the state space as in §1.1, and let $\mathbf{E}_\partial = \mathbf{E} \cup \{\partial\}$ (where $\partial \notin \mathbf{E}$) be the Alexandroff one-point compactification of $\mathbf{E}$. Namely, if $\mathbf{E}$ is not compact, then a neighborhood system for the "point at infinity ∂" is the class of the complements of all compact subsets of $\mathbf{E}$; if $\mathbf{E}$ is compact, then ∂ is an isolated point in $\mathbf{E}_\partial$. $\mathbf{E}_\partial$ is a compact separable metric space. Let $\mathbb{C}$ denote the class of all continuous functions on $\mathbf{E}_\partial$. Since $\mathbf{E}_\partial$ is compact, each f in $\mathbb{C}$ is bounded. We define the usual sup-norm of f as follows:

$$\|f\| = \sup_{x \in \mathbf{E}_\partial} |f(x)|.$$

The constant function $x \to f(\partial)$ on $\mathbf{E}_\partial$ is of course in $\mathbb{C}$, and if we put

$$\forall x \in \mathbf{E}_\partial\colon f_0(x) = f(x) - f(\partial),$$

then $\lim_{x \to \partial} f_0(x) = 0$. Let $\mathbb{C}_0$ denote the subclass of $\mathbb{C}$ vanishing at ∂ ("vanishing at infinity"); $\mathbb{C}_c$ denote the subclass of $\mathbb{C}_0$ having compact supports. Recall that $\mathbb{C}$ and $\mathbb{C}_0$ are both Banach spaces with the norm $\| \ \|$; and that $\mathbb{C}_0$ is the uniform closure (or completion) of $\mathbb{C}_c$.

Let (P_t) be a submarkovian transition semigroup on $(\mathbf{E}, \mathscr{E})$. It is extended to be Markovian on $(\mathbf{E}_\partial, \mathscr{E}_\partial)$, as shown in §1.2. The following definition applies to this extension.

Definition. (P_t) is said to have the "Feller property" or "Fellerian" iff $P_0 =$ identity mapping, and

(i) $\forall f \in \mathbb{C}, t \geq 0,$ $\qquad P_t f \in \mathbb{C};$
(ii) $\forall f \in \mathbb{C},$

$$\lim_{t \to 0} \| P_t f - f \| = 0. \tag{1}$$

It turns out that the condition in (1), which requires convergence in norm, is equivalent to the apparently weaker condition below, which requires only pointwise convergence:

(ii') $\forall f \in \mathbb{C}, x \in E_\partial:$

$$\lim_{t \to 0} P_t f(x) = f(x). \tag{2}$$

The proof is sketched in Exercise 4 below.

Since each member of $\mathbb{C}$ is the sum of a member of $\mathbb{C}_0$ plus a constant, it is easy to see that in the conditions (i), (ii) and (ii') above we may replace $\mathbb{C}$ by $\mathbb{C}_0$ without affecting their strength.

Theorem 1. *The function*

$$(t, x, f) \to P_t f(x)$$

on $\mathbf{T} \times \mathbf{E}_\partial \times \mathbb{C}$, *is continuous.*

Proof. Consider (t, x, f) fixed and (s, y, g) variable in the inequality below:

$$|P_t f(x) - P_s g(y)| \leq |P_t f(x) - P_t f(y)| + |P_t f(y) - P_s f(y)| + |P_s f(y) - P_s g(y)|.$$

Since $P_t f \in \mathbb{C}$, the first term converges to zero as $y \to x$; since (P_u) is Markovian, $\|P_u\| = 1$ for each u, the second term is bounded by

$$\|P_{|t-s|} f - f\|$$

which converges to zero as $|t - s| \to 0$ by (1); the third term is bounded by

$$\|P_s f - P_s g\| \le \|P_s\| \, \|f - g\| = \|f - g\|$$

which converges to zero as $g \to f$ in $\mathbb{C}$. Theorem 1 follows. $\qquad\square$

A (homogeneous) Markov process $(X_t, \mathscr{F}_t, t \in \mathbf{T})$ on $(\mathbf{E}_\partial, \mathscr{E}_\partial)$ whose semigroup (P_t) has the Feller property is called a *Feller process*. We now begin the study of its sample function properties.

Proposition 2. $\{X_t, t \in \mathbf{T}\}$ *is stochastically continuous, namely for each $t \in \mathbf{T}$, $X_s \to X_t$ in probability as $s \to t$, $s \in \mathbf{T}$.*

Proof. Let $f \in \mathbb{C}$, $g \in \mathbb{C}$, then if $t > 0$ and $\delta > 0$ we have

$$E\{f(X_t)g(X_{t+\delta})\} = E\{f(X_t)E^{X_t}[g(X_\delta)]\} = E\{f(X_t)P_\delta g(X_t)\}$$

by the Markov property. Since $P_\delta g \in \mathbb{C}$ and $P_\delta g \to g$, we have by bounded convergence, as $\delta \downarrow 0$:

$$E\{f(X_t)g(X_{t+\delta})\} \to E\{f(X_t)g(X_t)\}. \tag{3}$$

Now if h is a continuous function on $\mathbf{E}_\partial \times \mathbf{E}_\partial$, then there exists a sequence of functions $\{h_n\}$ each of the form $\sum_{j=1}^{l_n} f_{n_j}(x)g_{n_j}(y)$ where $f_{n_j} \in \mathbb{C}$, $g_{n_j} \in \mathbb{C}$ such that $h_n \to h$ uniformly on $\mathbf{E}_\partial \times \mathbf{E}_\partial$. This is a consequence of the Stone-Weierstrass theorem, see e.g., Royden [1]. It follows from this and (3) that

$$E\{h(X_t, X_{t+\delta})\} \to E\{h(X_t, X_t)\}.$$

Take h to be a metric of the space $\mathbf{E}_\partial$. Then the limit above is equal to zero, and the result asserts that $X_{t+\delta}$ converges to X_t in probability. Next if $0 < \delta < t$, then we have for each x:

$$\begin{aligned} E^x\{f(X_{t-\delta})g(X_t)\} &= E^x\{f(X_{t-\delta})E^{X_{t-\delta}}[g(X_\delta)]\} \\ &= E^x\{f(X_{t-\delta})P_\delta g(X_{t-\delta})\} = P_{t-\delta}(fP_\delta g)(x). \end{aligned} \tag{4}$$

By Theorem 1, the last term converges as $\delta \downarrow 0$ to

$$P_t(f \cdot g)(x) = E^x\{f(X_t)g(X_t)\}.$$

It follows as before that $\forall x$:

$$E^x\{h(X_{t-\delta}, X_t)\} \to 0,$$

hence also

$$E\{h(X_{t-\delta}, X_t)\} = E\{E[h(X_{t-\delta}, X_t)|\mathscr{F}_0]\} = E\{E^{X_0}[h(X_{t-\delta}, X_t)]\} \to 0. \quad \square$$

The preceding proof affords an occasion to clarify certain obscurities. When the symbol P or E is used without a superscript, the distribution of of X_0 is unspecified. Let it be denoted by μ; thus

$$\mu(A) = P(X_0 \in A), \qquad A \in \mathscr{E}.$$

For any $\varLambda \in \mathscr{F}^0$ (not $\mathscr{F}$!), we have

$$P(\varLambda) = E\{P(\varLambda|\mathscr{F}_0)\} = E\{P^{X_0}(\varLambda)\} = \int_E \mu(dx)P^x(\varLambda) = P^\mu(\varLambda);$$

see (10) of §1.2. More generally if $Y \in L^1(\mathscr{F}^0, P)$, then

$$E(Y) = E^\mu(Y).$$

Given any probability measure μ on $\mathscr{E}$, and the transition function $\{P_t(\cdot, \cdot),$ $t \in \mathbf{T}\}$, we can construct a Markov process with μ as its initial distribution, as reviewed in §1.2. It is assumed that such a process exists in the probability space $(\Omega, \mathscr{F}, P)$. Any statement concerning the process given in terms of P and E is therefore *ipso facto* true for any initial μ, without an explicit claimer to this effect. One may regard this usage as an editorial license to save print! For instance, with this understanding, formula (3) above contains its duplicate when E there is replaced by E^x, for each x. The resulting relation may then be written as

$$P_t(fP_\delta g)(x) \to P_t(fg)(x),$$

which may be easier to recognize than (3) itself. This is what is done in (4), since the corresponding argument for convergence as $\delta \downarrow 0$, under E instead of E^x, may be less obvious.

On the other hand, suppose that we have proved a result under P^x and E^x, for each x. Then integrating with respect to μ we obtain the result under P^μ and E^μ, for any μ. Now the identification of $P(\varLambda)$ with $P^\mu(\varLambda)$ above shows that the result is true under P and E. This is exactly what we did at the end of the proof of Proposition 2, but we have artfully concealed μ from the view there.

To sum up, for any $\varLambda \in \mathscr{F}^0$, $P(\varLambda) = 1$ is just a cryptic way of writing

$$\forall x \in \mathbf{E}: P^x(\varLambda) = 1.$$

We shall say in this case that $\varLambda$ is *almost sure*. The apparently stronger statement also follows: for any probability measure μ on $\mathscr{E}$, we have $P^\mu(\varLambda) = 1$. *After Theorem 5 of §2.3, we can extend this to any $\varLambda \in \mathscr{F}\,\widetilde{\ }$.*

The next proposition concerns the α-potential $U^\alpha f$. Let us record the following basic facts. If (P_t) is Fellerian, then

$$\forall f \in \mathbb{C}: U^\alpha f \in \mathbb{C}; \tag{5}$$

$$\forall f \in \mathbb{C}: \lim_{\alpha \to \infty} \|\alpha U^\alpha f - f\| = 0. \tag{6}$$

Since $P_t f \in \mathbb{C}$ for each t, (5) follows by dominated convergence as x tends to a limit in the first expression for $U^\alpha f$ in (3) of §2.1. To show (6), we have by a change of variables:

$$\sup_x \|\alpha U^\alpha f(x) - f(x)\| \le \sup_x \int_0^\infty \alpha e^{-\alpha t} |P_t f(x) - f(x)| \, dt$$

$$\le \int_0^\infty e^{-u} \|P_{u/\alpha} f - f\| \, du,$$

which converges to zero as $\alpha \to \infty$ by (1) and dominated convergence. In particular, we have for each x

$$\lim_{\alpha \to \infty} \alpha U^\alpha f(x) = f(x); \tag{7}$$

this should be compared with (2). As a general guide, the properties of αU^α as $\alpha \to \infty$ and as $\alpha \to 0$ are respectively reflected in those of P_t as $t \to 0$ and $t \to \infty$. This is a known folklore in the theory of Laplace transforms, and U^α is nothing but the Laplace transform of P_t in the operator sense.

A class of functions defined in $\mathscr{E}_\partial$ is said to "separate points" iff for any two distinct points x and y in $\mathscr{E}_\partial$, there exists a member f of the class such that $f(x) \ne f(y)$. This concept is used in the Stone-Weierstrass theorem. Let $\{O_n, n \in \mathbf{N}\}$ be a countable base of the open sets of $\mathscr{E}_\partial$, and put

$$\forall x \in \mathscr{E}_\partial: \varphi_n(x) = d(x, \bar{O}_n).$$

We know $\varphi_n \in \mathbb{C}$.

Proposition 3. *The countable subset D of $\mathbb{C}$ below separates points*:

$$D = \{U^\alpha \varphi_n: \alpha \in \mathbf{N}, n \in \mathbf{N}\}. \tag{8}$$

Proof. For any $x \ne y$, there exists O_n such that $x \in \bar{O}_n$ and $y \notin \bar{O}_n$. Hence $\varphi_n(x) = 0 < \varphi_n(y)$, namely φ_n separates x and y. We have by (7), for sufficiently large $\alpha \in \mathbf{N}$:

$$|\alpha U^\alpha \varphi_n(x) - \varphi_n(x)| < \tfrac{1}{2} \varphi_n(y),$$
$$|\alpha U^\alpha \varphi_n(y) - \varphi_n(y)| < \tfrac{1}{2} \varphi_n(y).$$

Hence $\alpha U^\alpha \varphi_n(x) \ne \alpha U^\alpha \varphi_n(y)$, namely $U^\alpha \varphi_n$ separates x and y. [Of course we can use (6) to get $\|\alpha U^\alpha \varphi_n - \varphi_n\| < \tfrac{1}{2} \varphi_n(y)$.]   □

The next is an analytical lemma separated out for the sake of clarity. We use $g \circ h$ to denote the composition of g and h; and $(g \circ h)|_S$ to denote its restriction to S.

Proposition 4. *Let $\mathbb{D}$ be a class of continuous functions from $\mathscr{E}_\partial$ to $R = (-\infty, +\infty)$ which separates points. Let h be any function on R to $\mathscr{E}_\partial$. Suppose that S is a dense subset of R such that for each $g \in \mathbb{D}$,*

$$(g \circ h)|_S \quad \text{has right and left limits in } R. \qquad (9)$$

Then $h|_S$ has right and left limits in R.

Proof. Suppose for some t, $h|_S$ does not have a right limit at t. This means that there exist $t_n \in S$, $t_n \downdownarrows t$, $t'_n \in S$, $t''_n \downdownarrows t$ such that

$$h(t_n) \to x \neq y \leftarrow h(t'_n).$$

There exists $g \in \mathbb{D}$ such that $g(x) \neq g(y)$. Since g is continuous, we have

$$(g \circ h)(t_n) \to g(x) \neq g(y) \leftarrow (g \circ h)(t'_n).$$

This contradicts (9) and proves the proposition since the case of the left limit is similar. ☐

We are ready to regulate the sample functions of a Feller process. The trick, going back to Doob, is to relate the latter with those of supermartingales obtained by composition with potentials of functions.

Proposition 5. *Let $\{X_t, \mathscr{F}_t, t \in \mathbf{T}\}$ be a Feller process, and S be any countable dense subset of $\mathbf{T}$. Then for almost every ω, the sample function $X(\cdot, \omega)$ restricted to S has right limits in $[0, \infty)$ and left limits in $(0, \infty)$.*

Proof. Let g be a member of the class $\mathbb{D}$ in (8), thus $g = U^k \varphi$ where $k \in \mathbf{N}$, $\varphi \in \mathbb{C}$. By Proposition 2 of §2.1,

$$\{e^{-kt} g(X_t), \mathscr{F}_t\}$$

is a supermartingale. Hence by Theorem 1 of §1.4, there exists Ω_g with $P(\Omega_g) = 1$, such that if $\omega \in \Omega_g$, then the sample function

$$t \to e^{-kt} g(X_t(\omega))$$

restricted to S has right and left limits as asserted. Clearly the factor e^{-kt} may be omitted in the above. Let $\Omega_* = \bigcap_{g \in \mathbb{D}} \Omega_g$, then $P(\Omega_*) = 1$. If $\omega \in \Omega_*$, then the preceding statement is true for each $g \in \mathbb{D}$. Since $\mathbb{D}$ separates points by Proposition 3, it now follows from Proposition 4 that $t \to X_t(\omega)$ restricted to S has the same property.

Now we define, for each $\omega \in \Omega_*$:

$$\forall t \geq 0: \tilde{X}_t(\omega) = \lim_{\substack{s \in S \\ s \downarrow\downarrow t}} X_s(\omega); \qquad \forall t > 0: \hat{X}_t(\omega) = \lim_{\substack{s \in S \\ s \uparrow\uparrow t}} X_s(\omega). \qquad (10)$$

Elementary analysis shows that for each ω, $\tilde{X}_t(\omega)$ is right continuous in t and has left limit $\hat{X}_t$ at each $t > 0$. Similarly $\hat{X}_t$ is left continuous in t and has right limit $\tilde{X}_t$ at each $t \geq 0$. $\square$

Theorem 6. *Suppose that each $\mathscr{F}_t$ is augmented. Then each of the processes $\{\tilde{X}_t\}$ and $\{\hat{X}_t\}$ is a version of $\{X_t\}$; hence it is a Feller process with the same transition semigroup (P_t) as $\{X_t\}$.*

Proof. By Proposition 2, for each fixed t, there exists $s_n \in S$, $s_n \downarrow\downarrow t$ such that

$$P\left\{\lim_n X_{s_n} = X_t\right\} = 1. \qquad (11)$$

This is because convergence in probability implies almost sure convergence along a subsequence. But by Proposition 5, the limit in (11) is equal to $\tilde{X}_t$. Hence $P\{\tilde{X}_t = X_t\} = 1$ for each t, namely $\{\tilde{X}_t\}$ is a version of $\{X_t\}$. It follows that for each $f \in \mathbb{C}$:

$$E\{f(\tilde{X}_{s+t})|\mathscr{F}_t\} = E\{f(X_{s+t})|\mathscr{F}_t\} = P_s f(X_t) = P_s f(\tilde{X}_t)$$

almost surely. Augmentation of $\mathscr{F}_t$ is needed to ensure that $\tilde{X}_t \in \mathscr{F}_t$, for each t. This yields the assertion concerning $\{\tilde{X}_t\}$; the case of $\{\hat{X}_t\}$ is similar. $\square$

Let us concentrate on the right continuous version $\tilde{X}$, and write it simply as X. Its sample functions may take ∂, namely infinity, either as a value or as a left limiting value. Even if (P_t) is strictly Markovian so that $P^x\{X_t \in E\} = P_t(x, E) = 1$ for each x and t, it does not follow that we have $X(t, \omega) \in E$ *for all* $t \in \mathbf{T}$ for any ω. In other words, it may not be bounded in each finite interval of t. The next proposition settles this question for a general sub-markovian (P_t).

Theorem 7. *Let $\{X_t, \mathscr{F}_t\}$ be a Feller process with right continuous paths having left limits; and let*

$$\zeta(\omega) = \inf\{t \geq 0 | X_{t-}(\omega) = \partial \text{ or } X_t(\omega) = \partial\}. \qquad (12)$$

Then we have almost surely $X(\zeta + t) = \partial$ for all $t \geq 0$, on the set $\{\zeta < \infty\}$.

Proof. As before let d be a metric for $\mathbf{E}_\partial$ and put

$$\varphi(x) = d(x, \partial), \qquad x \in \mathbf{E}_\partial.$$

Then φ vanishes only at ∂. Since $\varphi \in \mathbb{C}$, we have $P_t\varphi \to \varphi$ as $t \downarrow 0$, and consequently

$$U^1\varphi(x) = \int_0^\infty e^{-t} P_t\varphi(x)\,dt$$

vanishes only at ∂, like φ. Write

$$Z_t = e^{-1}U^1\varphi(X_t);$$

then $\{Z_t, \mathscr{F}_t\}$ is a positive supermartingale by Proposition 2 of §2.1, and $\{Z_t\}$ has right continuous paths with left limits, because $\{X_t\}$ does and $U^1\varphi$ is continuous. Furthermore $Z_{t-} = 0$ if and only if $X_{t-} = \partial$; $Z_t = 0$ if and only if $X_t = \partial$. Therefore we have

$$\zeta(\omega) = \inf\{t \geq 0 \,|\, Z_{t-}(\omega) = 0 \text{ or } Z_t(\omega) = 0\},$$

and the theorem follows from Theorem 4 of §1.5.   □

Corollary. *If* (P_t) *is strictly Markovian, then almost surely the sample function is bounded in each finite t-interval. Namely not only* $X(t, \omega) \in \mathbf{E}$ *for all* $t \geq 0$, *but* $X(t-, \omega) \in \mathbf{E}$ *for all* $t > 0$ *as well.*

Proof. Under the hypothesis we have

$$P\{X_t \in \mathbf{E} \text{ for all } t \in \mathbf{T} \cap Q\} = 1,$$

It follows from the theorem that for each $t \in Q$:

$$P\{\zeta < t\} = P\{\zeta < t; X_t \in \mathbf{E}\} = 0.$$

Hence $P\{\zeta < \infty\} = 0$, which is equivalent to the assertion of the Corollary.
                                                                                 □

Theorem 7 is a case of our discussion is §1.2 regarding ∂ as an absorbing state: cf. (13) there.

Exercises

1. Consider the Markov chain in Example 1 of §1.2, with $\mathbf{E}$ the set of positive integers, no ∂, and $\mathscr{E}$ the discrete topology on $\mathbf{E}$. Show that if $\lim_{t \downarrow 0} p_{ii}(t) = 1$ uniformly in all i, then (P_t) is a Feller semigroup. Take a version which is right continuous and has left limits. Show that for each sample function

$X(\cdot, \omega)$, there is a discrete set of t at which $X(t-, \omega) \neq X(t, \omega)$, and $X(\cdot, \omega)$ is constant between two consecutive values of t in this set.

2. Show that the semigroups of Examples 2, 3 and 4 in §1.2 are all Fellerian.

3. For a Markov chain suppose that there exists an i such that $\lim_{t \downarrow 0} [1 - p_{ii}(t)]/t = +\infty$. (Such a state is called *instantaneous*. It is not particularly easy to construct such an example; see Chung [2], p. 285.) Show that for any $\delta > 0$ we have $P^i\{X(t) = i \text{ for all } t \in [0, \delta]\} = 0$. Thus under P^i, no version of the process can be right continuous at $t = 0$. Hence (P_t) cannot be Fellerian. Prove the last assertion analytically.

4. Prove that in the definition of Feller property, condition (2) implies condition (1). [Hint: by the Riesz representation theorem the dual space of $\mathbb{C}$ is the space of finite measures on $\mathbf{E}_\partial$. Use this and the Hahn-Banach thorem to show that the set of functions of the form $\{U^\alpha \varphi\}$, where $\alpha > 0$ and $\varphi \in \mathbb{C}$, is dense in $\mathbb{C}$. If f belongs to this set, (1) is true.]

2.3. Strong Markov Property and Right Continuity of Fields

We proceed to derive several fundamental properties of a Feller process. It is assumed in the following theorems that the sample functions are right continuous. The existence of left limits will not be needed until the next section. We continue to adopt the convention in §1.3 that $X_\infty = \partial$, which makes it possible to write certain formulas without restriction. But such conventions should be watched carefully.

Theorem 1. *For each optional T, we have for each $f \in \mathbb{C}$, and $u \geq 0$:*

$$E\{f(x_{T+u}) | \mathscr{F}_{T+}\} = P_u f(X_T). \tag{1}$$

Proof. Observe first that on $\{T = \infty\}$, both sides of (1) reduce to $f(\partial)$. Let

$$T_n = \frac{[2^n T] + 1}{2^n},$$

so that $T_n \downarrow\downarrow T$, each T_n is strictly optional, and takes values in the dyadic set D (see (19) of §1.3). Recall from Theorem 5 of §1.3 that

$$\mathscr{F}_{T+} = \bigwedge_{n=1}^{\infty} \mathscr{F}_{T_n}.$$

Hence for any $\Lambda \in \mathscr{F}_{T+}$, if we write $\Lambda_d = \Lambda \cap \{T_n = d\}$ for $d \in D$, then $\Lambda_d \in \mathscr{F}_d$. Now the Markov property applied at $t = d$ yields

$$\int_{\Lambda_d} f(X_{d+u}) \, dP = \int_{\Lambda_d} P_u f(X_d) \, dP,$$

see (2) of §1.2. Since $T_n = d$ on Λ_d, we have by enumerating the possible values of T_n:

$$\int_\Lambda f(X_{T_n + u}) \, dP = \sum_{d \in D} \int_{\Lambda_d} f(X_{d+u}) \, dP$$

$$= \sum_{d \in D} \int_{\Lambda_d} P_u f(X_d) \, dP = \int_\Lambda P_u f(X_{T_n}) \, dP. \tag{2}$$

Since both f and $P_t f$ are bounded continuous, and $X(\cdot)$ is right continuous, we obtain by letting $n \to \infty$ in the first and last members of (2):

$$\int_\Lambda f(X_{T+u}) \, dP = \int_\Lambda P_u f(X_T) \, dP.$$

The truth of this for each $\Lambda \in \mathscr{F}_{T+}$ is equivalent to the assertion in (1). $\square$

When T is a constant t, the equation (1) reduces to the form of Markov property given as (iic) in §1.1, with the small but important improvement that the $\mathscr{F}_t$ there has now become $\mathscr{F}_{t+}$. The same arguments used in §1.1 to establish the equivalence of (iic) with (iib) now show that (1) remains true for $f \in b\mathscr{E}$. More generally, let us define the "post-T field" $\mathscr{F}'_T$ as follows:

$$\mathscr{F}'_T = \sigma(X_{T+u}, u \geq 0);$$

which reduces to $\mathscr{F}'_t$ when $T = t$. Then we have for each integrable Y in $\mathscr{F}'_T$:

$$E\{Y \mid \mathscr{F}_{T+}\} = E\{Y \mid X_T\}.$$

This is the extension of (iia) in §1.1. Alternatively, we have for each integrable Y in $\mathscr{F}^0$:

$$E\{Y \circ \theta_T \mid \mathscr{F}_{T+}\} = E^{X_T}\{Y\}. \tag{3}$$

Definition. The Markov process $\{X_t, \mathscr{F}_t\}$ is said to have the strong Markov property iff (3) is true for each optional time T.

Thus Theorem 1 is equivalent to the assertion that:

A Feller process with right continuous paths has the strong Markov property.

An immediate consequence will be stated as a separate theorem, which contains the preceding statement when $T \equiv 0$.

Theorem 2. *For each optional T, the process $\{X_{T+t}, \mathscr{F}_{T+t+}, t \in \mathbf{T}\}$ is a Markov process with (P_t) as transition semigroup. Furthermore, it has the strong Markov property.*

Proof. Recall that $T + t$ is optional, in fact strictly so for $t > 0$; and $X_{T+t} \in \mathscr{F}_{T+t+}$ by Theorem 10 of §1.3. Now apply (1) with T replaced by $T + t$:

$$E\{f(X_{T+t+u})|\mathscr{F}_{T+t+}\} = P_u f(X_{T+t}). \tag{4}$$

This proves the first assertion of the theorem. As for the second, it is a consequence of the re-statement of Theorem 1 because $\{X_{T+t}, \mathscr{F}_{T+t+}\}$ is a Feller process with right continuous paths.   □

There is a further extension of Theorem 1 which is useful in many applications. Indeed, the oldest case of the strong Markov property, known as Désiré André's reflection principle, requires such an extension (see Exercise 12 of §4.2).

Theorem 3. *Let $S \geq T$ and $S \in \mathscr{F}_{T+}$. Then we have for each $f \in \mathbb{C}$:*

$$E\{f(X_S)|\mathscr{F}_{T+}\} = E^{X_T}\{f(X_{S-T})\} \tag{5}$$

where $S - T$ is defined to be ∞ if $S = \infty$ (even if $T = \infty$); and $X_\infty = \partial, f(\partial) = 0$.

The quantity in the right member of (5) is the value of the function $(t, x) \to E^x\{f(X_t)\} = P_t f(x)$ at $(S(\omega) - T(\omega), X_T(\omega))$, with ω omitted. It may be written out more explicitly as

$$\int_\Omega f(X_{S(\omega)-T(\omega)}(\omega'))P^{X_T(\omega)}(d\omega'). \tag{6}$$

It is tempting to denote the right member of (5) by $P_{S-T}f(X_T)$ in analog with (1), but such a notation would conflict with our later, and established, usage of the balayage operator P_T in §3.4.

Proof of Theorem 3. Observe first that (5) reduces to $f(\partial) = f(\partial)$ on $\{S = \infty\}$ by the various conventions, so we may suppose $S < \infty$ in what follows. Let

$$U_n = \frac{[2^n(S - T)] + 1}{2^n};$$

then $U_n \downarrow (S - T)$. Since S and T belong to $\mathscr{F}_{T+}$, so does U_n for each n. Consequently, for each $d \in D$ (the dyadic set) we have $\{U_n = d\} \in \mathscr{F}_{T+}$. Now let Λ be any set in $\mathscr{F}_{T+}$, and put $\Lambda_d = \Lambda \cap \{U_n = d\} \in \mathscr{F}_{T+}$. We have for each $f \in \mathbb{C}$:

$$\int_\Lambda f(X_{T+U_n}) dP = \sum_{d \in D} \int_{\Lambda_d} f(X_{T+d}) dP$$

$$= \sum_{d \in D} \int_{\Lambda_d} P_d f(X_T) dP = \int_\Lambda P_{U_n} f(X_T) dP \tag{7}$$

where the second equation is by (1). Now $t \to P_t f(x)$ is continuous for each x and $f \in \mathbb{C}$ by Theorem 1 of §2.2. Letting $n \to \infty$ in the first and last members of (7), we obtain by right continuity:

$$\int_A f(X_S) \, dP = \int_A P_{S-T} f(X_T) \, dP.$$

This is equivalent to (5). $\qquad\qquad\qquad\qquad\qquad\qquad\qquad\qquad\qquad\qquad\square$

The next theorem demands a thorough understanding of the concepts of "completion" and "augmentation". For the first notion see Theorem 2.2.5 of the *Course*. The second notion will now be discussed in detail.

Let $(\Omega, \mathscr{F}, P)$ be a complete probability space. The class of sets C in $\mathscr{F}$ with $P(C) = 0$ will be denoted by $\mathscr{N}$ and called the P-null sets. Completeness means: a subset of a P-null set is a P-null set. Let $\mathscr{G}$ be a sub-σ-field of $\mathscr{F}$. Put

$$\mathscr{G}^* = \sigma(\mathscr{G} \vee \mathscr{N}),$$

namely $\mathscr{G}^*$ is the minimal σ-field including $\mathscr{G}$ and $\mathscr{N}$. $\mathscr{G}^*$ is called the *augmentation* of $\mathscr{G}$ with respect to $(\Omega, \mathscr{F}, P)$. It is characterized in the following proposition, in which a "function" is from Ω into $[-\infty, +\infty]$. Recall that for a function Y on $\mathbf{E}$, $Y \in \mathscr{G}^*$ is an abbreviation of $Y \in \mathscr{G}^*/\mathscr{E}$ where $\mathscr{E}$ is the Borel field of $\mathbf{E}$.

Lemma. *A subset A of Ω belongs to $\mathscr{G}^*$ if and only if there exists $B \in \mathscr{G}$ such that $A \triangle B \in \mathscr{N}$. A function Y belongs to $\mathscr{G}^*$ if and only if there exists a function $Z \in \mathscr{G}$ such that $\{Y \neq Z\} \in \mathscr{N}$.*

Proof. Let $\mathscr{A}$ denote the class of sets A described in the lemma. Observe that $A \triangle B = C$ is equivalent to $A = B \triangle C$ for arbitrary sets A, B and C. Since $\mathscr{G}^*$ contains all sets of the form $B \triangle C$ where $B \in \mathscr{G}$ and $C \in \mathscr{N}$, it is clear that $\mathscr{G}^* \supset \mathscr{A}$. To show that $\mathscr{A} \supset \mathscr{G}^*$ it is sufficient to verify that $\mathscr{A}$ is a σ-field since it obviously includes both $\mathscr{G}$ and $\mathscr{N}$. Since $A^c \triangle B^c = A \triangle B$, $\mathscr{A}$ is closed under complementation. It is also closed under countable union by the inclusion relation

$$\left(\bigcup_n A_n \right) \triangle \left(\bigcup_n B_n \right) \subset \bigcup_n (A_n \triangle B_n),$$

and the remark that if the right member above is in $\mathscr{N}$, then so is the left member. Thus $\mathscr{A}$ is a σ-field and we have proved the first sentence of the Lemma.

Next, let $Z \in \mathscr{G}$ and $\{Y \neq Z\} \in \mathscr{N}$. Then for any subset S of $[-\infty, +\infty]$, we have the trivial inclusion

$$\{Y \in S\} \triangle \{Z \in S\} \subset \{Y \neq Z\}.$$

For each Borelian S, we have $\{Z \in S\} \in \mathcal{G}$; hence $\{Y \in S\} \in \mathcal{G}^*$ by what we have proved, and so $Y \in \mathcal{G}^*$. Conversely, let $Y \in \mathcal{G}^*$, and Y be countably-valued. Namely Y is of the form $\sum_{j=1}^{\infty} c_j 1_{A_j}$, where the c_j's are points in $[-\infty, +\infty]$, and the A_j's are disjoint sets in $\mathcal{G}^*$. Then for each j there exists $B_j \in \mathcal{G}$ such that $A_j \triangle B_j \in \mathcal{N}$. Since the A_j's are disjoint, $B_i \cap B_j \in \mathcal{N}$ if $i \neq j$. Put $B_1' = B_1$, and $B_j' = B_1^c \cdots B_{j-1}^c B_j$ for $j \geq 2$, then $B_j \triangle B_j' \in \mathcal{N}$ and so $A_j \triangle B_j' \in \mathcal{N}$. The B_j''s are disjoint. Put $Z = \sum_{j=1}^{\infty} c_j 1_{B_j'}$. Then $Z \in \mathcal{G}$ and

$$\{Y \neq Z\} \in \bigcup_{j=1}^{\infty} \{A_j \triangle B_j'\} \in \mathcal{N}.$$

Thus Y is of the form described in the second sentence of the lemma. If $Y \in \mathcal{G}^*$ there exists countably valued $Y^{(k)} \in \mathcal{G}^*$ such that $Y^{(k)} \to Y$. Let $Z^{(k)} \in \mathcal{G}$ such that $\{Y^{(k)} \neq Z^{(k)}\} \in \mathcal{N}$; and $Z = \varliminf_{k \to \infty} Z^{(k)}$. Then $Z \in \mathcal{G}$ and

$$\{Y \neq Z\} \subset \bigcup_{k=1}^{\infty} \{Y^{(k)} \neq Z^{(k)}\} \in \mathcal{N}. \qquad \square$$

Remark. If $(\Omega, \mathcal{F}, P)$ is not complete, the Lemma is true *provided* that we assume $Y \in \mathcal{F}$ in the second assertion. We leave this to the interested reader.

Corollary. $(\mathcal{G}^*)^* = \mathcal{G}^*$. $(\Omega, \mathcal{G}^*, P)$ is complete.

We now review the definition of a conditional expectation, although this has been used hundreds of times above. Let $Y \in \mathcal{F}$ and $E(|Y|) < \infty$. Then the conditional expectations $E(Y|\mathcal{G})$ is the class of functions Z having the following properties:

(a) $Z \in \mathcal{G}$;
(b) for each $W \in b\mathcal{G}$, we have

$$E(WZ) = E(WY). \tag{8}$$

Note that (8) is then also true for $W \in b\mathcal{G}^*$. It follows that if Z is a member of the class $E(Y|\mathcal{G})$, and Z^* is a member of the class $E(Y|\mathcal{G}^*)$, then $\{Z \neq Z^*\} \in \mathcal{N}$. We state this as follows:

$$E(Y|\mathcal{G}) \equiv E(Y|\mathcal{G}^*), \qquad \mathrm{mod}\ \mathcal{N}. \tag{9}$$

We are ready to return to (1) above. First put $T \equiv t$, then shrink $\mathcal{F}_{t+}$ to $\mathcal{F}_{t+}^0$ as we may:

$$E\{f(X_{t+u})|\mathcal{F}_{t+}^0\} = P_u f(X_t). \tag{10}$$

We have seen that there is advantage in augmenting the σ-fields $\mathscr{F}_{t+}^0$ appearing in (10). Since $\mathscr{F}_{t+}^0 \supset \mathscr{F}_t^0 \supset \mathscr{F}_0^0$, it is sufficient to augment $\mathscr{F}_0^0$ to achieve this, for then $\mathscr{N}$ will be included in it, and a fortiori in the others.

Theorem 4. Suppose that $\mathscr{F}_0^0$ is augmented, then the family $\{\mathscr{F}_t^0, t \in \mathbf{T}\}$ is right continuous.

Proof. According to (10), its right member, say Z_1, is a representative of its left member. If Z_2 is any other representative of the latter, then $\{Z_1 \neq Z_2\} \in \mathscr{N}$. This is a consequence of the definition of the conditional expectation which implies its uniqueness mod $\mathscr{N}$. Hence $Z_2 \in \mathscr{F}_t^0$ by the Lemma. Thus the entire class in the left member of (10) belongs to $\mathscr{F}_t^0$. Now the same argument as used in §1.1 to show that (iic) implies (iia) there yields, for any $Y \in b\mathscr{F}_t'$:

$$E\{Y \mid \mathscr{F}_{t+}^0\} \in \mathscr{F}_t^0. \tag{11}$$

Let $Y_0 \in b\mathscr{F}_t^0$, then a representative of $E\{Y_0 \mid \mathscr{F}_{t+}^0\}$ is Y_0 itself because $\mathscr{F}_t^0 \subset \mathscr{F}_{t+}^0$. It follows as before that

$$E\{Y_0 \mid \mathscr{F}_{t+}^0\} \in \mathscr{F}_t^0. \tag{12}$$

Lemma 2 of §1.1 may be used to show that the class of sets Y satisfying (11) constitutes a σ-field. We have just seen that this class includes $\mathscr{F}_t^0$ and $\mathscr{F}_t'$. Since $\mathscr{F}^0 = \sigma(\mathscr{F}_t^0, \mathscr{F}_t')$ for each t, the class must also include $\mathscr{F}^0$. Hence (11) is true for each $Y \in \mathscr{F}^0$. For any $Y \in \mathscr{F}_{t+}^0 \ (\subset \mathscr{F}^0)$, a representative of $E\{Y \mid \mathscr{F}_{t+}^0\}$ is Y itself. Hence $Y \in \mathscr{F}_t^0$ by (11), and this means $\mathscr{F}_{t+}^0 \subset \mathscr{F}_t^0$. Therefore $\mathscr{F}_{t+}^0 = \mathscr{F}_t^0$ as asserted. $\qquad\square$

According to the discussion in §2.2, Theorem 4 may be read as follows. For each probability measure μ on $\mathscr{E}$, let $(\Omega, \mathscr{F}^\mu, P^\mu)$ be the completion of $(\Omega, \mathscr{F}^0, P^\mu)$; and let $\mathscr{F}_t^\mu$ denote the augmentation of $\mathscr{F}_t^0$ with respect to $(\Omega, \mathscr{F}^\mu, P^\mu)$. [Note that if $Y \in \mathscr{F}^\mu$, then $E^\mu(Y)$ is defined as $E^\mu(Z)$ where $Z \in \mathscr{F}^0$ and $P^\mu\{Y \neq Z\} = 0$.] Then $\{\mathscr{F}_t^\mu, t \in \mathbf{T}\}$ is right continuous. Furthermore, we have as a duplicate of (10):

$$E^\mu\{f(X_{t+u}) \mid \mathscr{F}_t^\mu\} = P_u f(X_t),$$

and more generally for $Y \in b\mathscr{F}^0$, and each T optional relative to $\{\mathscr{F}_t^\mu\}$:

$$E^\mu\{Y \circ \theta_T \mid \mathscr{F}_{T+}^\mu\} = E^{X_T}\{Y\}; \tag{13}$$

cf. (3) above. The notation E^μ indicates that the conditional expectation is with respect to $(\Omega, \mathscr{F}^\mu, P^\mu)$; and any two representatives of the left number in

(13) differs by a P^μ-null function. It will be shown below (in the proof of Theorem 5) that (13) is also true for $Y \in b\mathcal{F}^\mu$. Now if $Y \in \mathcal{F}^0$, then $Y \circ \theta_t \in \mathcal{F}^0$. But if $Y \in \mathcal{F}^\mu$, we cannot conclude that $Y \circ \theta_t \in \mathcal{F}^\mu$, even if $Y = f(X_u)$. This difficulty will be resolved by the next step.

Namely, we introduce a smaller σ-field

$$\mathcal{F}^\sim = \bigwedge_\mu \mathcal{F}^\mu \tag{14}$$

and correspondingly for each $t \geq 0$:

$$\mathcal{F}_t^\sim = \bigwedge_\mu \mathcal{F}_t^\mu. \tag{15}$$

In both (14) and (15), μ ranges over all finite measures on $\mathcal{E}$. It is easy to see that the result is the same if μ ranges over all σ-finite measures on $\mathcal{E}$. Leaving aside a string of questions regarding these fields (see Exercises), we state the useful result as follows.

Corollary to Theorem 4. *The family* $\{\mathcal{F}_t^\sim, t \in \mathbf{T}\}$ *is right continuous, as well as the family* $\{\mathcal{F}_t^\mu, t \in \mathbf{T}\}$ *for each μ.*

Proof. We have already proved the right continuity of $\{\mathcal{F}_t^\mu\}$ and will use this to prove that of $\{\mathcal{F}_t^\sim\}$. Let $S(\mu, s)$ be a class of sets in any space, for each μ and s in any index sets. Then the following is a trivial set-theoretic identity:

$$\bigwedge_\mu \bigwedge_s S(\mu, s) = \bigwedge_s \bigwedge_\mu S(\mu, s). \tag{16}$$

Applying this to $S(\mu, s) = \mathcal{F}_s^\mu$, for all finite measures μ on $\mathcal{E}$, and all $s \in (t, \infty)$, we obtain

$$\bigwedge_\mu (\mathcal{F}_t^\mu)_+ = (\mathcal{F}_t^\sim)_+ \tag{17}$$

where

$$(\mathcal{F}_t^\mu)_+ = \bigwedge_{s>t} \mathcal{F}_s^\mu, \qquad (\mathcal{F}_t^\sim)_+ = \bigwedge_{s>t} \mathcal{F}_s^\sim.$$

These are to be distinguished from $(\mathcal{F}_{t+})^\mu$ and $(\mathcal{F}_{t+})^\sim = \bigwedge_\mu (\mathcal{F}_{t+})^\mu$, but see Exercise 6 below. Since $(\mathcal{F}_t^\mu)_+ = \mathcal{F}_t^\mu$, (17) reduces to $\mathcal{F}_t^\sim = (\mathcal{F}_t^\sim)_+$ as asserted. $\qquad\square$

After this travail, we can now extend (13) to Y in $b\mathcal{F}^\sim$. But first we must consider the function $x \to E^x\{Y\}$. We know this is a Borel function if $Y \in b\mathcal{F}^0$; what can we say if $Y \in b\mathcal{F}^\sim$? The reader will do well to hark back to the theory of Lebesgue measure in Euclidean spaces for a clue. There we may begin with the σ-field $\mathcal{B}$ of classical Borel sets, and then complete $\mathcal{B}$ with respect to the Lebesgue measure. But we can also complete $\mathscr{B}$ with respect to other measures defined on $\mathscr{B}$. In an analogous manner, beginning with $\mathscr{F}^0$, we have completed it with respect to P^μ and called the result $\mathscr{F}^\mu$. Then we have taken the intersection of all these completions and called it $\mathscr{F}^\sim$. Exactly the same procedure can be used to complete $\mathscr{E}_\partial$ (the Borel field of $\mathbf{E}_\partial$) with respect to any finite measure μ defined on $\mathscr{E}$. The result will be denoted by $\mathscr{E}^\mu$; we then put

$$\mathscr{E}^\sim = \bigwedge_\mu \mathscr{E}^\mu. \tag{18}$$

As before, we may regard $\mathscr{E}^\mu$ and $\mathscr{E}$ as classes of functions defined on $\mathbf{E}$, as well as subsets of $\mathbf{E}$ which will be identified with their indicators. A function in $\mathscr{E}^\mu$ is called μ-measurable; a function in $\mathscr{E}^\sim$ is called *universally measurable*. A function f is μ-null iff $\mu(\{f \neq 0\}) = 0$. See Exercise 3 below for an instructive example.

The next theorem puts together the various extensions we had in mind. Note that $\mathscr{F}_T^\mu = (\mathscr{F}_T^\mu)_+$ as a consequence of the Corollary to Theorem 4.

Theorem 5. *If $Y \in b\mathscr{F}^\sim$, then the function φ defined on $\mathbf{E}$ by*

$$\varphi(x) = E^x(Y) \tag{19}$$

is in $\mathscr{E}^\sim$. For each μ and each T optional relative to $\{\mathscr{F}_t^\mu\}$, we have

$$Y \circ \theta_T \in \mathscr{F}^\mu, \tag{20}$$

$$E^\mu\{Y \circ \theta_T \,|\, \mathscr{F}_T^\mu\} = \varphi(X_T). \tag{21}$$

Proof. Observe first that for each x, since $Y \in \mathscr{F}^{\varepsilon_x}$, $E^x(Y)$ is defined. Next, it follows easily from the definition of the completion $\mathscr{F}^\mu$ that $Y \in b\mathscr{F}^\mu$ if and only if there exist Y_1 and Y_2 in $b\mathscr{F}^0$ such that $P^\mu\{Y_1 \neq Y_2\} = 0$ and $Y_1 \leq Y \leq Y_2$. Hence if we put

$$\varphi_i(x) = E^x(Y_i), \qquad i = 1, 2;$$

φ_1 and φ_2 are in $\mathscr{E}$, $\varphi_1 \leq \varphi \leq \varphi_2$ and $\mu(\{\varphi_1 \neq \varphi_2\}) = 0$. Thus $\varphi \in \mathscr{E}^\mu$ by definition. This being true for each μ, we have $\varphi \in \mathscr{E}^\sim$. Next, define a measure v on $\mathscr{E}_\partial$ as follows, for each $f \in b\mathscr{E}$:

$$v(f) = \int \mu(dx) E^x\{f(X_T)\} = E^\mu\{f(X_T)\}.$$

This is the distribution of X_T when μ is the distribution of X_0. Thus for any $Z \in b\mathscr{F}^0$, we have

$$E^\mu\{Z \circ \theta_T\} = E^\mu\{E^{X_T}(Z)\} = E^v(Z). \tag{22}$$

If $Y \in \mathscr{F}^{\sim}$, then there exists Z_1 and Z_2 in $\mathscr{F}^0$ such that $P^v\{Z_1 \neq Z_2\} = 0$ and $Z_1 \leq Y \leq Z_2$. We have $Z_i \circ \theta_T \in \mathscr{F}^\mu$ by (22) of §1.3, with $\{\mathscr{F}_t^\mu\}$ for $\{\mathscr{F}_t\}$ (recall $\mathscr{F}^\mu = \mathscr{F}_\infty^\mu$), and

$$P^\mu\{Z_1 \circ \theta_T \neq Z_2 \circ \theta_T\} = P^v\{Z_1 \neq Z_2\} = 0 \tag{23}$$

by (22) above. This and

$$Z_1 \circ \theta_T \leq Y \circ \theta_T \leq Z_2 \circ \theta_T \tag{24}$$

imply (20). Furthermore, we have by (24):

$$E^\mu\{Z_1 \circ \theta_T | \mathscr{F}_T^\mu\} \leq E^\mu\{Y \circ \theta_T | \mathscr{F}_T^\mu\} \leq E^\mu\{Z_2 \circ \theta_T | \mathscr{F}_T^\mu\}. \tag{25}$$

Now put $\psi_i(x) = E^x(Z_i)$, $i = 1, 2$. Since $Z_i \in \mathscr{F}^0$, $\psi_i(X_T)$ is a representative of $E^\mu\{Z_i \circ \theta_T | \mathscr{F}_T^\mu\}$ by (13). On the other hand, we have $\psi_1 \leq \varphi \leq \psi_2$, hence

$$\psi_1(X_T) \leq \varphi(X_T) \leq \psi_2(X_T); \tag{26}$$

and

$$P^\mu\{\psi_2(X_T) \neq \psi_1(X_T)\} = P^\mu\{Z_2 \circ \theta_T \neq Z_1 \circ \theta_T\} = 0. \tag{27}$$

Comparing (25) and (26) we conclude that $\varphi(X_T)$ is a representative of $E\{Y \circ \theta_T | \mathscr{F}_T^\mu\}$. This is the assertion (21).   □

We end this section with a simple but important result which is often called Blumenthal's zero-or-one law.

Theorem 6.   Let $\Lambda \in \mathscr{F}_0^{\sim}$; then for each x we have $P^x(\Lambda) = 0$ or $P^x(\Lambda) = 1$.

Proof. Suppose first that $\Lambda \in \mathscr{F}_0^0$. Then $\Lambda = X_0^{-1}(A)$ for some $A \in \mathscr{E}$. Since $P^x\{X_0 = x\} = 1$, we have

$$P^x(\Lambda) = P^x(X_0^{-1}(A)) = 1_A(x),$$

which can only take the value 0 or 1. If $\Lambda \in \mathscr{F}_0^{\sim}$, then for each x, there exists Λ^x such that $P^x(\Lambda \triangle \Lambda^x) = 0$, so that $P^x(\Lambda) = P^x(\Lambda^x)$ which is either 0 or 1 as just proved.   □

If we think that $\mathscr{F}_0^0$ is a small σ-field, and $\mathscr{F}_0^{\sim}$ is only "trivially" larger, the result appears to be innocuous. Actually it receives its strength from the fact that $\mathscr{F}_0^{\sim} = (\mathscr{F}^{\sim})_{0+}$, as part of the Corollary to Theorem 4. Whether an event will occur "instantly" may be portentous. For instance, let T be optional relative to $\{\mathscr{F}_t^{\sim}, t \in \mathbf{T}\}$, then $\{T = 0\} \in \mathscr{F}_0^{\sim}$ and consequently the event $\{T = 0\}$ has probability 0 or 1 under each P^x. When T is the hitting time of a set, this dichotomy leads to a basic notion in potential theory (see §3.4)—the regularity of a point for a set or the thinness of a set at a point. For a Brownian motion on the line, it yields an instant proof that almost every sample function starting at 0 must change sign infinitely many times in any small time interval.

Exercises

1. Prove that if $\forall t \geq 0$: $\mathscr{F}_t = \mathscr{F}_{t+}$, then for any T which is optional relative to $\{\mathscr{F}_t\}$, we have $\forall t \geq 0$: $\mathscr{F}_{T+t} = \mathscr{F}_{T+t+}$.

2: Let N^* denote the class of all sets in $\mathscr{F}^0$ such that $P^x(\Lambda) = 0$ for every $x \in E$. Let $\mathscr{F}^* = \sigma(\mathscr{F}^0 \vee N^*)$. Show that $\mathscr{F}^* \subset \mathscr{F}^\sim$. The σ-field $\mathscr{F}^*$ is not to be confused with $\mathscr{F}^\sim$; see Problem 3.

3. Let $\Omega = R^1$ and $\mathbf{E} = R^1$. Define $X_t(\omega) = \omega + t$; then $\{X_t, t \geq 0\}$ is the uniform motion (Example 2 of §1.2). Show that $\mathscr{F}_t^0 = \mathscr{B}^1$, the usual Borel field on R^1, and $P^x = \varepsilon_x$, the point mass at x. Show that

$$\bigwedge_{x \in R^1} (\mathscr{F}^0)^{\varepsilon_x} \quad \text{is the class of all subsets of } R^1;$$

$$\bigwedge_{\mu} (\mathscr{F}^0)^{\mu} \quad \text{is the class of universally measurable sets of } R^1$$

where μ ranges over all probability measures on $\mathscr{B}^1$. The class N^* defined in Problem 2 is empty so that $\mathscr{F}^* = \mathscr{B}^1$. (This example is due to Getoor.)

4. Show that

$$\bigvee_{t \in \mathbf{T}} \mathscr{F}_t^{\mu} = \mathscr{F}^{\mu}.$$

Is it true that

$$\bigvee_{t \in \mathbf{T}} \mathscr{F}_t^\sim = \mathscr{F}^\sim ?$$

5. For each n let $\mathscr{G}_n$ be a class of positive (≥ 0) functions closed under the operation of taking the lim sup of a sequence; let $\mathscr{H}$ be a class of positive functions closed under the operation of taking the lim inf of a sequence. Denote by $\mathscr{G}_n + \mathscr{H}$ the class of functions of the form $g_n + h$ where $g_n \in \mathscr{G}_n$ and $h \in \mathscr{H}$. Suppose that $\mathscr{G}_n \supset \mathscr{G}_{n+1}$ for every n and $\mathscr{G} = \bigcap_{n=1}^{\infty} \mathscr{G}_n$. Prove that

$$\bigcap_n (\mathscr{G}_n + \mathscr{H}) = \mathscr{G} + \mathscr{H}.$$

[Hint: if f belongs to the class in the left member, show that $f = \limsup g_n + \liminf h_n$ where $g_n \in \mathscr{G}_n$, $h_n \in \mathscr{H}$; show that $\limsup g_n \in \mathscr{G}$.]

6. Apply Problem 5 to prove that

$$\bigwedge_{s > t} (\mathscr{F}_s^0)^{\mu} = \left(\bigwedge_{s > t} \mathscr{F}_s^0 \right)^{\mu};$$

namely that the two operations of "taking the intersection over (t, ∞)" and "augmentation with respect to P^μ" performed on $\{\mathscr{F}^0_s, s \in \mathbf{T}\}$ are commutative.

7. If both S and T are optional relative to $\{\mathscr{F}^\sim_t\}$, then so is $T + S \circ \theta_T$. [Cf. Theorem 11 of §1.3.]

2.4. Moderate Markov Property and Quasi Left Continuity

From now on we write $\mathscr{F}_t$ for $\mathscr{F}^\sim_t$ and $\mathscr{F}$ for $\mathscr{F}^\sim$, and consider the Feller process $(X_t, \mathscr{F}_t, P^\mu)$ in the probability space $(\Omega, \mathscr{F}, P^\mu)$ for an arbitrary fixed probability measure μ on $\mathscr{E}$. We shall omit the superscript μ in what follows unless the occasion requires the explicit use. For each ω in Ω, the sample function $X(\cdot, \omega)$ is right continuous in $\mathbf{T} = [0, \infty)$ and has left limits in $(0, \infty)$. The family $\{\mathscr{F}_t, t \in \mathbf{T}\}$ is right continuous; hence each optional time T is strictly optional and the pre-T field is $\mathscr{F}_T (= \mathscr{F}_{T+})$. The process has the strong Marov property. Recall that the latter is a consequence of the right continuity of paths without the intervention of left limits. We now proceed to investigate the left limits. We begin with a lemma separated out for its general utility, and valid in any space $(\Omega, \mathscr{F}, P)$.

Lemma 1. *Let $\mathscr{G}$ be a sub-σ-field of $\mathscr{F}$ and $X \in \mathscr{G}$; $Y \in \mathscr{F}$. Suppose that for each $f \in \mathbb{C}_0$ we have*

$$E\{f(Y) | \mathscr{G}\} = f(X). \tag{1}$$

Then $P\{X = Y\} = 1$.

Proof. It follows from (1) that for each open set U:

$$P\{1_U(Y) | \mathscr{G}\} = 1_U(X); \tag{2}$$

see Lemma 1 of §1.1. Take $\Lambda = \{X \notin U\}$, and integrate (2) over Λ:

$$P\{Y \in U; X \notin U\} = P\{X \in U; X \notin U\} = 0. \tag{3}$$

Let $\{U_n\}$ be a countable base of the topology; then we have by (3):

$$P\{X \neq Y\} \leq \sum_n P\{Y \in U_n; X \notin U_n\} = 0. \qquad \square$$

Corollary. *Let X and Y be two random variables such that for any $f \in \mathbb{C}_c$, $g \in \mathbb{C}_c$ we have*

$$E\{f(X)g(X)\} = E\{f(Y)g(X)\}. \tag{4}$$

Then $P\{X = Y\} = 1$.

Proof. As before (4) is true if $g = 1_A$, where A is any open set. Then it is also true if A is any Borel set, by Lemma 2 of §1.1. Thus we have

$$\int_{\{X \in A\}} f(X) \, dP = \int_{\{X \in A\}} f(Y) \, dP;$$

and consequently

$$f(X) = E\{f(X)|\mathscr{G}\} = E\{f(Y)|\mathscr{G}\},$$

where $\mathscr{G} = \sigma(X)$, the σ-field generated by X. Therefore the corollary follows from Lemma 1. $\qquad \square$

The next lemma is also general, and will acquire further significance shortly.

Lemma 2. *Let $\{X_t, t \in R\}$ be an arbitrary stochastic process which is stochastically continuous. If almost every sample function has right and left limits everywhere in R, then for each $t \in R$, X is continuous at t almost surely.*

Proof. The assertion means that for each t,

$$P\left\{\lim_{s \to t} X_s = X_t\right\} = 1.$$

Since X is stochastically continuous at t, there exist $s_n \uparrow\uparrow t$ and $t_n \downarrow\downarrow t$ such that almost surely we have

$$\lim_n X_{s_n} = X_t = \lim_n X_{t_n}.$$

But the limits above are respectively X_{t-} and X_{t+} since the latter are assumed to exist almost surely. Hence $P\{X_{t-} = X_t = X_{t+}\} = 1$ which is the assertion. $\qquad \square$

Remark. It is sufficient to assume that X_{t-} and X_{t+} exist almost surely, for each t.

The property we have just proved is sometimes referred to as follows: "the process has no fixed time of discontinuity." This implies stochastic continuity but is not implied by it. For example, the Markov chain in Example 1 of §1.2, satisfying condition (d) and with a *finite* state space **E**, has this property. But it may not if **E** is infinite (when there are "instantaneous states"). On the other hand, the simple Poisson process (Example 3 of §1.2) has this property even though it is a Markov chain on an infinite **E**. Of course, the said property is weaker than the "almost sure continuity" of sample functions, which means that almost every sample function is a continuous function. The latter is a rather special situation in the theory of Markov processes, but is exemplified by the Brownian motion (Example 4 of §1.2) and will be discussed in §3.1.

The next result is the left-handed companion to Theorem 1 of §2.3. The reader should observe the careful handling of the values 0 and ∞ for the T below—often a nuisance which cannot be shrugged off.

Theorem 3. *For each predictable T, we have for each $f \in \mathbb{C}$, and $u \geq 0$:*

$$E\{f(X_{T+u})1_{\{T<\infty\}}|\mathscr{F}_{T-}\} = P_u f(X_{T-})1_{\{T<\infty\}} \tag{5}$$

where $X_{0-} = X_0$.

Proof. Remember that $\lim_{t\to\infty} X_t$ need not exist at all, so that X_{T-} is undefined on $\{T = \infty\}$. Since $T \in \mathscr{F}_{T-}$, we may erase the two appearances of $1_{\{T<\infty\}}$ in (5) and state the result "on the set $\{T < \infty\}$". However, since T does not necessarily belong to $\mathscr{F}_{T_n}$, we are not allowed to argue blithely below "as if T is finite".

Let $\{T_n\}$ announce T, then we have by Theorem 1 of §2.3:

$$E\{f(X_{T_n+u})|\mathscr{F}_{T_n}\} = P_u f(X_{T_n}). \tag{6}$$

Since $T_n \in \mathscr{F}_{T_n}$, we may multiply both members of (6) by $1_{\{T_n \leq M\}}$, where M is a positive number, to obtain

$$E\{f(X_{T_n+u})1_{\{T_n \leq M\}}|\mathscr{F}_{T_n}\} = P_u f(X_{T_n})1_{\{T_n \leq M\}}. \tag{7}$$

Since $T_n = 0$ on $\{T = 0\}$, $T_n \uparrow\uparrow T$ on $\{T > 0\}$, and f is continuous, we have

$$\lim_{n\to\infty} f(X_{T_n+u}) = \begin{cases} f(X_u) & \text{on } \{T = 0\}, \\ f(X_{T+u-}) & \text{on } \{T > 0\}. \end{cases}$$

Fortunately, $X_u = X_{u-}$ almost surely in view of Lemma 2, and so we can unify the result as $f(X_{T+u-})$ on Ω. On the right side of (7), we have similarly

$$\lim_{n\to\infty} P_u f(X_{T_n}) = P_u f(X_{T-}),$$

because $X_0 = X_{0-}$ by convention on $\{T = 0\}$, and $P_u f$ is continuous. Since also $\mathscr{F}_{T_n}$ increases to $\mathscr{F}_{T-}$ by Theorem 8 of §1.3, and f is bounded, we can take the limit in n simultaneously in the integrand and in the conditioning field in the left side of (7), by a martingale convergence theorem (see Theorem 9.4.8 of *Course*). Thus in the limit (7) becomes

$$E\{f(X_{T+u-})1_{\{T \leq M\}}|\mathscr{F}_{T-}\} = P_u f(X_{T-})1_{\{T \leq M\}}. \tag{8}$$

Letting $M \to \infty$ we obtain (5) with X_{T+u-} replacing X_{T+u} in the left member.

Next, letting $u \downarrow\downarrow 0$, since $\lim_{u \downarrow\downarrow 0} X_{T+u-} = X_T$ by right continuity and $P_u f \to f$ by Feller property, we obtain

$$E\{f(X_T)1_{\{T < \infty\}} | \mathscr{F}_{T-}\} = f(X_{T-})1_{\{T < \infty\}}.$$

Now *define* $X_{\infty-}$ (as well as X_∞) to be ∂ even where $\lim_{t \to \infty} X_t$ exists and is not equal to ∂. Then the factor $1_{\{T < \infty\}}$ in the above may be cancelled. Since $X_{T-} \in \mathscr{F}_{T-}$ by Theorem 10 of §1.3, it now follows from Lemma 1 that we have almost surely:

$$X_T = X_{T-}. \tag{9}$$

This is then true on $\{T < \infty\}$ without the arbitrary fixing of the value of $X_{\infty-}$.

For each optional T and $u > 0$, $T + u$ is predictable. Hence we have just proved that $X_{T+u} = X_{T+u-}$ almost surely on $\{T < \infty\}$. This remark enables us to replace X_{T+u-} by X_{T+u} in the left member of (8), thus completing the proof of Theorem 3. $\qquad \square$

Corollary. *For* $Y \in b\mathscr{F}^\mu$,

$$E\{Y \circ \theta_T | \mathscr{F}_{T-}\} = E^{X_{T-}}\{Y\} \quad on \ \{T < \infty\}. \tag{10}$$

The details of the proof are left as an exercise. The similarity between (10) above and (13) of §2.3 prompts the following definition.

Definition. The Markov process $\{X_t, \mathscr{F}_t\}$ is said to have the moderate Markov property iff almost all sample functions have left limits and (10) is true for each predictable time T.

Thus, this is the case for a Feller process whose sample functions are right continuous and have left limits. The adjective "moderate" is used here in default of a better one; it does not connote "weaker than the strong".

There is an important supplement to Theorem 3 which is very useful in applications because it does not require T to be predictable. Let $\{T_n\}$ be a sequence of optional times, increasing (loosely) to T. Then T is optional but not necessarily predictable. Now for each ω in Ω, the convergence of $T_n(\omega)$ to $T(\omega)$ may happen in two different ways as described below.

Case (i). $\forall n: T_n < T$. In this case $X_{T_n} \to X_{T-}$ if $T < \infty$.

Case (ii). $\exists n_0: T_{n_0} = T$. In this case $X_{T_n} = X_T$ for $n \geq n_0$.

Heuristically speaking, T has the predictable character on the set of ω for which case (i) is true. Hence the result (9) above seems to imply that in both cases we have $X_{T_n} \to X_T$ on $\{T < \infty\}$. The difficulty lies in that we do not know whether we can apply the character of predictability on a part of Ω without regard to the other part. This problem is resolved in the following proof by constructing a truly predictable time, which coincides with the given optional time on the part of Ω in question.

Theorem 4. *Let T_n be optional and increase to T. Then we have almost surely*

$$\lim_n X_{T_n} = X_T \quad on \ \{T < \infty\}. \tag{11}$$

Proof. Put

$$\Lambda = \{\forall n: T_n < T\};$$

and define

$$T'_n = \begin{cases} T_n & \text{on } \{T_n < T\}, \\ \infty & \text{on } \{T_n = T\}; \end{cases} \qquad T' = \begin{cases} T & \text{on } \Lambda, \\ \infty & \text{on } \Omega - \Lambda. \end{cases}$$

Since $\{T_n < T\} \in \mathscr{F}_{T_n}$ and $\Lambda \in \mathscr{F}_T$ by Proposition 6 of §1.3, T'_n and T' are optional by Proposition 4 of §1.3. We have

$$\Lambda \cap \{T < \infty\} \subset \{T' < \infty\} \subset \Lambda. \tag{12}$$

Hence we have

$$\forall n: T'_n = T_n < T = T' \quad \text{on } \Lambda. \tag{13}$$

It follows that $T'_n \wedge n < T'$ and $T'_n \wedge n$ increases to T' on Ω. Hence T' is predictable. (Note: $\{T = 0\} \subset \Omega - \Lambda$, hence $T' > 0$ on Ω.) Applying the result (9) to T', we obtain

$$X_{T'} = X_{T'-} \quad \text{on } \{T' < \infty\};$$

and consequently

$$\lim_n X'_{T_n} = X'_{T'-} = X_{T'} \quad \text{on } \{T' < \infty\}. \tag{14}$$

In view of (12), (13) and (14), Equation (11) is true on $\Lambda \cap \{T < \infty\}$. But (11) is trivially true on $\Omega - \Lambda$. Hence Theorem 4 is proved. $\qquad\square$

The property expressed in Theorem 4 is called the "quasi left continuity" of the process. The preceding proof shows that it is a consequence of the moderate Markov property.

We are now ready to prove the optionality of the hitting time for a closed set as well as an open set. For any set A in $\mathscr{E}_\phi$ define

$$T_A(\omega) = \inf\{t > 0 \,|\, X(t, \omega) \in A\}. \tag{15}$$

This is called the (*first*) *hitting time* of A. Compare with the first entrance time D_A defined in (25) of §1.3. The difference between them can be crucial.

They are related as follows:

$$T_A = \lim_{s \downarrow \downarrow 0} \downarrow (s + D_A \circ \theta_s). \tag{16}$$

This follows from the observation that

$$s + D_A \circ \theta_s = s + \inf\{t \geq 0 \,|\, X(s + t, \omega) \in A\}$$
$$= \inf\{t \geq s \,|\, X(t, \omega) \in A\}.$$

It follows from Theorem 11 of §1.3 that if D_A is optional relative to $\{\mathscr{F}_t^0\}$, then so is $s + D_A \circ \theta_s$ for each $s > 0$, and consequently so is T_A by Proposition 3 of §1.3. It is easy to see that the assertion remains true when $\{\mathscr{F}_t^0\}$ is replaced by $\{\mathscr{F}_t^\mu\}$, for each μ.

Now it is obvious that D_A is the debut of the set

$$H_A = \{(t, \omega) \,|\, X(t, \omega) \in A\} \tag{17}$$

which is progressively measurable relative to $\{\mathscr{F}_t^0\} \subset \{\mathscr{F}_t^\mu\}$, since X is right continuous. For each t, $(\Omega, \mathscr{F}_t^\mu, P^\mu)$ is a complete probability space as a consequence of the definition of $\mathscr{F}_t^\mu$. Hence D_A is optional relative to $\{\mathscr{F}_t^\mu\}$ by Theorem 3 of §1.5. However that theorem relies on another which is not given in this book. The following proof is more direct and also shows why augmentation of $\mathscr{F}_t^0$ is needed.

Theorem 5. *If the Markov process $\{X_t\}$ is right continuous, then for each open set A, D_A and T_A are both optional relative to $\{\mathscr{F}_t^0\}$. If the process is right continuous almost surely and is also quasi left continuous, then for each closed set as well as each open set A, D_A and T_A are both optional relative to $\{\mathscr{\tilde{F}}_t\}$.*

Proof. By a remark above it is sufficient to prove the results for D_A. If A is open, then for each $t > 0$, we have the identity

$$\{D_A < t\} = \bigcup_{r \in Q_t} \{X_r \in A\} \tag{18}$$

where $Q_t = Q \cap [0, t)$, provided all sample functions are right continuous. To see this suppose $D_A(\omega) < t$, then $X(s, \omega) \in A$ for some $s \in [D_A(\omega), t)$, and by right continuity $X(r, \omega) \in A$ for some rational r in $(D_A(\omega), t)$. Thus the left member of (18) is a subset of the right. The converse is trivial and so (18) is true. Since the right member clearly belongs to $\mathscr{F}_t^0$, we conclude that $\{D_A < t\} \in \mathscr{F}_t^0$.

Now suppose that only P-almost all sample functions are right continuous. Then the argument above shows that the two sets in (18) differ by a P-null set. Hence $\{D_A < t\}$ is in the augmentation of $\mathscr{F}_t^0$ with respect to $(\Omega, \mathscr{F}, P)$. Translating this to P^μ, we have $\{D_A < t\} \in \mathscr{F}_t^\mu$ for each μ, hence $\{D_A < t\} \in \mathscr{\tilde{F}}_t$. Since $\{\mathscr{\tilde{F}}_t\}$ is right continuous we have $\{D_A \leq t\} \in \mathscr{\tilde{F}}_t$.

Next, suppose that A is closed. Then there is a sequence of open sets A_n such that $A_n \supset \bar{A}_{n+1}$ for each n, and

$$A = \bigcap_n A_n = \bigcap_n \bar{A}_n. \tag{19}$$

We shall indicate this by $A_n \Downarrow A$. For instance, we may take

$$A_n = \left\{ x \in E_\partial \Big| d(x, A) < \frac{1}{n} \right\}.$$

Clearly D_{A_n} increases and is not greater than D_A; let

$$S = \lim_n \uparrow D_{A_n} \leq D_A. \tag{20}$$

We now make the important observation that for any Borel set B, we have almost surely

$$X(D_B) \in \bar{B} \quad \text{on } \{D_B < \infty\}. \tag{21}$$

For if $D_B(\omega) < \infty$, then for each $\delta > 0$, there exists $t \in [D_B(\omega), D_B(\omega) + \delta)$ such that $X(t, \omega) \in B$. Hence (21) follows by right continuity. Thus we have $X(D_{A_n}) \in \bar{A}_n$ for all n and therefore by quasi left continuity and (19):

$$X(S) = \lim_n X(D_{A_n}) \in \bigcap_n \bar{A}_n = A$$

almost surely on $\{0 \leq S < \infty\}$. The case $S = 0$ is of course trivial. This implies $S \geq D_A$; together with (20) we conclude that $D_A = S$ a.s. on $\{S < \infty\}$. But $D_A \geq S = \infty$ on $\{S = \infty\}$. Hence we have proved a.s.

$$D_A = \lim_n D_{A_n}. \tag{22}$$

Since each D_{A_n} is optional relative to $\{\mathscr{F}_t^\sim\}$; so is D_A by (22).   □

If we replace the D's by T's above, the result (22) is no longer true in general! The reader should find out why. However, since for each Borel set B, $D_B = T_B$ on $\{X_0 \notin B\}$, we have the following consequence of (22) which is so important that we state it explicitly and more generally as follows.

Corollary. Let A_n be closed sets, $A_n \supset A_{n+1}$ for all n, and $\bigcap_n A_n = A$. If $x \notin A$, then P^x-a.s. we have

$$T_A = \lim_n \uparrow T_{A_n} \leq +\infty. \tag{23}$$

Here A may be empty, in which case $T_A = +\infty$.

Exercises

1. Here is a shorter proof of the quasi left continuity of a Feller process (Theorem 4). For $\alpha > 0$, $f \in b\mathbb{C}_+$, $\{e^{-\alpha t}U^\alpha f(X_t)\}$ is a right continuous positive supermartingale because $U^\alpha f$ is continuous. Hence if $T_n \uparrow T$, and $Y = \lim_n X(T_n)$ (which exists on $\{T < \infty\}$), we have a.s.

$$U^\alpha f(Y) = \lim_n U^\alpha f(X(T_n)) = E\left\{U^\alpha f(X(T))\Big| \bigvee_n \mathscr{F}_{T_n}\right\}.$$

Now multiply by α, let $\alpha \to \infty$, and use Lemma 1.

2. Prove the following "separation" property for a Feller process. Let K be compact, G be open such that $K \subset G$. Then for any $\delta > 0$ there exists $t_0 > 0$ such that

$$\inf_{x \in K} P^x\{T_{G^c} \geq t_0\} \geq 1 - \delta.$$

   [Hint: let $0 \leq f \leq 1$, $f = 1$ on K, $f = 0$ in G^c. There exists $t_0 > 0$ such that $\sup_{0 \leq t \leq t_0} \|P_t f - f\| < \delta/2$. Now consider

$$E^x\{f(X(t_0))\} \leq P^x\{T_{G^c} \geq t_0\} + E^x\{f(X(t_0)); T_{G^c} < t_0\}$$

   and apply Theorem 3 of §2.3.]

3. If $Y \in \mathscr{F}^\sim/\mathscr{E}$ and $f \in \mathscr{E}^\sim/\mathscr{B}$ then $f(Y) \in \mathscr{F}^\sim/\mathscr{B}$. Hence $x \to E^x\{f(Y)\}$ is in $\mathscr{E}^\sim$ if the expectation exists. [This will be needed for Lebesgue measurability for important functions associated with the Brownian motion.]

To appreciate the preceding result it is worthwhile to mention that a Lebesgue measurable function of a Borel measurable function need not be Lebesgue measurable (see Exercise 15 on p. 14 of *Course*). Thus if we take $\mathscr{F}^0 = \mathscr{B}$ and $Y \in \mathscr{F}^0$; if $f \in \mathscr{F}^\mu$ we cannot infer that $f(Y) \in \mathscr{F}^\mu$, where μ is the Lebesgue measure on $\mathscr{F}^0$.

NOTES ON CHAPTER 2

§2.1. This chapter serves as an interregnum between the more concrete Feller processes and Hunt's axiomatic theory. It is advantageous to introduce some of the basic tools at an early stage.

§2.2. Feller process is named after William Feller who wrote a series of pioneering papers in the 1950's. His approach is essentially analytic and now rarely cited. The sample function properties of his processes were proved by Kinney, Dynkin, Ray, Knight, among others. Dynkin [1] developed Feller's theory by probabilistic methods. His book is rich in content but difficult to consult owing to excessive codification. Hunt [1] and Meyer [2] both discuss Feller processes before generalizations.

§2.3. It may be difficult for the novice to appreciate the fact that twenty five years ago a formal proof of the strong Markov property was a major event. Who is now interested in an example in which it does not hold?

A full discussion of augmentation is given in Blumenthal and Getoor [1]. This is dry and semi-trivial stuff but inevitable for a rigorous treatment of the fundamental concepts. Instead of beginning the book by these questions it seems advisable to postpone them until their relevance becomes more apparent.

§2.4. There is some novelty in introducing the moderate Markov property before quasi left continuity; see Chung [5]. It serves as an illustration of the general methodology alluded to in §1.3, where both F_{T+} and F_{T-} are considered. Historically, a moderate Markov property was first observed at the "first infinity" of a simple kind of Markov chains, see Chung [3]. It turns out that a strong Markov process becomes moderate when the paths are reversed in time, see Chung and Walsh [1].

A more complete discussion of the measurability of hitting times will be given in §3.3. Hunt practically began his great memoir [1] with this question, fully realizing that the use of hitting times is the principal method in Markov processes.

Chapter 3

Hunt Process

3.1. Defining Properties

Let $\{X_t, \mathscr{F}_t, t \in \mathbf{T}\}$ be a (homogeneous) Markov process with state space $(\mathbf{E}_\partial, \mathscr{E}_\partial)$ and transition function (P_t), as specified in §1.1 and §1.2. Here $\mathscr{F}_t$ is the $\mathscr{F}_t^\sim$ defined in §2.3. Such a process is called a *Hunt process* iff

   (i)   it is right continuous;
   (ii) it has the strong Markov property (embodied in Theorem 1 of §2.3);
   (iii) it is quasi left continuous (as described in Theorem 4 of §2.4).

Among the basic consequences of these hypotheses are the following:

   (iv) $\{\mathscr{F}_t\}$ is right continuous (Corollary to Theorem 4 of §2.3);
   (v) $\{X_t\}$ is progressively measurable relative to $\{\mathscr{F}_t\}$ (Theorem 1 of §1.5);
   (vi) (P_t) is Borelian (Exercise 4 of §1.5).

We have shown in Chapter 2 that given a Feller semigroup (P_t), a Feller process can be constructed having all the preceding properties (and others not implied by the conditions above). Roundly stated: a Feller process is a Hunt process. Whereas a Feller process is constructed from a specific kind of transition function, a Hunt process is prescribed by certain hypotheses regarding the behavior of its sample functions. Thus in the study of a Hunt process we are pursing a deductive development of several fundamental features of a Feller process.

To begin with, we can add the following result to the above list of properties of a Hunt process.

Theorem 1. *Almost surely the sample paths have left limits in* $(0, \infty)$.

Proof. For a fixed $\varepsilon > 0$, define

$$T(\omega) = \inf\{t > 0 \,|\, d(X_t(\omega), X_0(\omega)) > \varepsilon\} \tag{1}$$

where d denotes a metric of the space $\mathbf{E}_\partial$. Our first task is to show that T is optional, indeed relative to $\{\mathscr{F}_t^0\}$. For this purpose let $\{z_k\}$ be a countable dense set in $\mathbf{E}_{\partial}$, and B_{kn} be the closed ball with center z_k and radius n^{-1}: $B_{kn} = \{x \mid d(x, z_k) \leq n^{-1}\}$. Then $\{B_{kn}\}$ forms a countable base of the topology. Put

$$T_{kn}(\omega) = \inf\{t > 0 \mid d(X_t(\omega), B_{kn}) > \varepsilon\}.$$

Then T_{kn} is optional relative to $\{\mathscr{F}_t^0\}$ by Theorem 6 of §2.4, since the set $\{x \mid d(x, B_{kn}) > \varepsilon\}$ is open. We claim that

$$\{T < t\} = \bigcup_{k,n} \{X_0 \in B_{kn}; T_{kn} < t\}. \tag{2}$$

It is clear that right member of (2) is a subset of the left member. To see the converse, suppose $T(\omega) < t$. Then there is $s(\omega) < t$ such that $d(X_s(\omega), X_0(\omega)) > \varepsilon$; hence there are n and k (both depending on ω) such that $d(X_s(\omega), X_0(\omega)) > \varepsilon + 2n^{-1}$ and $X_0(\omega) \in B_{kn}$. Thus $d(X_s(\omega), B_{kn}) > \varepsilon$ and $T_{kn}(\omega) < t$; namely ω belongs to the right member of (2), establishing the identity. Since the set in the right member of (2) belongs to $\mathscr{F}_t^0$, T is optional as claimed.

Next, we define $T_0 \equiv 0$, $T_1 \equiv T$ and inductively for $n \geq 1$:

$$T_{n+1} = T_n + T \circ \theta_{T_n} = \inf\{t > T_n \mid d(X(t), X(T_n)) > \varepsilon\}.$$

Each T_n is optional relative to $\{\mathscr{F}_t^0\}$ by Theorem 11 of §1.3. Since T_n increases with n, the limit $S = \lim_n T_n$ exists and S is optional. On the set $\{S < \infty\}$, we have $\lim_n X(T_n) = X(S)$ by quasi left continuity. On the other hand, right continuity of paths implies that $d(X(T_{n+1}), X(T_n)) \geq \varepsilon$ almost surely for all n, which precludes the existence of $\lim_n X(T_n)$. There would be a contradiction unless $S = \infty$ almost surely. In the latter event, we have $[0, \infty) = \bigcup_{n=0}^{\infty} [T_n, T_{n+1})$. Note that if $T_n = \infty$ then $[T_n, T_{n+1}) = \varnothing$. In each interval $[T_n, T_{n+1})$ the oscillation of $X(\cdot)$ does not exceed 2ε by the definition of T_{n+1}. We have therefore proved that for each ε, there exists Ω_ε with $P(\Omega_\varepsilon) = 1$ such that $X(\cdot)$ does not oscillate by more than 2ε in $[T_n^\varepsilon, T_{n+1}^\varepsilon)$, where

$$[0, \infty) = \bigcup_{n=0}^{\infty} [T_n^\varepsilon, T_{n+1}^\varepsilon). \tag{3}$$

Let $\Omega_* = \bigcap_{m=1}^{\infty} \Omega_{1/m}$; then $P(\Omega_*) = 1$. We assert that if $\omega \in \Omega_*$, then $X(\cdot, \omega)$ must have left limits in $(0, \infty)$. For otherwise there exists $t \in (0, \infty)$ and m such that $X(\cdot, \omega)$ has oscillation $> 2/m$ in $(t - \delta, t)$ for every $\delta > 0$. Thus $t \notin [T_n^{1/m}, T_{n+1}^{1/m})$ for all $n \geq 0$, which is impossible by (3) with $\varepsilon = 1/m$. □

Remark. We prove later in §3.3 that on $\{t < \zeta\}$ we have $X_{t-} \neq \partial$, namely $X_{t-} \in \mathbf{E}$. For a Feller process this is implied by Theorem 7 of §2.2.

Let us observe that quasi left continuity implies that $X(\cdot)$ is left continuous at each fixed $t \in (0, \infty)$, almost surely. For if the t_n's are constant such that $0 \leq t_n \uparrow\uparrow t$, then each t_n is optional and so $X(t_n) \to X(t)$. Coupled with right continuity, this implies the continuity of almost all paths at each fixed t. In other words, the process has no fixed time of discontinuity. For a Feller process, this was remarked in §2.4.

Much stronger conditions are needed to ensure that almost all paths are continuous. One such condition is given below which is particularly adapted to a Hunt process. Another is given in Exercise 1 below.

Theorem 2. *Let* $\{X_t\}$ *be a Markov process with right continuous paths having left limits in* $(0, \infty)$. *Suppose that the transition function satisfies the following condition: for each* $\varepsilon > 0$ *and each compact* $K \subset \mathbf{E}$ *we have*

$$\lim_{t \to 0} \frac{1}{t} \sup_{x \in K} [1 - P_t(x, B(x, \varepsilon))] = 0 \tag{4}$$

where $B(x, \varepsilon) = \{y \in \mathbf{E}_\partial | d(x, y) \le \varepsilon\}$. *Then almost all paths are continuous.*

Proof. The proof depends on the following elementary lemma. □

Lemma. *Let* f *be a function from* $[0,1]$ *to* $\mathbf{E}_\partial$ *which is right continuous in* $[0, 1)$ *and has left limits in* $(0, 1]$. *Then* f *is not continuous in* $[0, 1]$ *(continuity at the endpoints being defined unilaterally) if and only if there exists* $\varepsilon > 0$ *such that for all* $n \ge n_0(\varepsilon)$ *we have*

$$\max_{0 \le k \le n-1} d\left(f\left(\frac{k}{n}\right), f\left(\frac{k+1}{n}\right)\right) > \varepsilon. \tag{5}$$

Proof of the Lemma. If f is not continuous in $[0,1]$, then there exists $t \in (0, 1]$ and $\varepsilon > 0$ such that $d(f(t-), f(t)) > 2\varepsilon$. For each $n \ge 1$ define k by $kn^{-1} < t \le (k + 1)n^{-1}$. Then for $n \ge n_0(\varepsilon)$ we have $d(f(kn^{-1}), f(t-)) < \varepsilon/2$, $d(f((k + 1)n^{-1}), f(t)) < \varepsilon/2$; hence $d(f(kn^{-1}), f((k + 1)n^{-1})) > \varepsilon$ as asserted. Conversely, if f is continuous in $[0,1]$, then f is uniformly continuous there and so for each $\varepsilon > 0$ (5) is false for all sufficiently large n. This is a stronger conclusion than necessary for the occasion.

To prove the theorem, we put for a fixed compact K:

$$M = \{\omega | X(\cdot, \omega) \text{ is not continuous in } [0, 1]; X(s, \omega) \in K \text{ for all } s \in [0, 1]\},$$

$$M_n = \left\{\omega \middle| \sup_{0 \le k \le n-1} d\left(X\left(\frac{k}{n}, \omega\right), X\left(\frac{k+1}{n}, \omega\right)\right) > \varepsilon;\right.$$
$$\left. X(s, \omega) \in K \text{ for all } s \in [0, 1]\right\}.$$

It follows from the lemma that $M \in \mathscr{F}^0$, and

$$M \subset \bigcup_{m=1}^{\infty} \left[\liminf_n M_n^{1/m}\right].$$

If we apply the Markov property at all kn^{-1} for $0 \leq k \leq n-1$, we obtain

$$P(M_n^\varepsilon) \leq \sum_{k=0}^{n-1} P\left\{X\left(\frac{k}{n}\right) \in K; d\left(X\left(\frac{k}{n}\right), X\left(\frac{k+1}{n}\right)\right) > \varepsilon\right\}$$

$$\leq n \sup_{x \in K} P_{1/n}(x, B(x, \varepsilon)^c).$$

Using the condition (4) with $t = n^{-1}$, we see that the last quantity above tends to zero as $n \to \infty$. Hence $P(\liminf_n M_n^\varepsilon) \leq \underline{\lim}_n P(M_n^\varepsilon) = 0$ and $P(M) = 0$. Now replace the interval $[0, 1]$ by $[l/2, l/2 + 1]$ for integer $l \geq 1$, and K by $K_m \cup \{\partial\}$ where K_m is a compact subset of $\mathbf{E}$ and $K_m \uparrow \mathbf{E}$. Denote the resulting M by $M(l, m)$ and observe that we may replace K by $K \cup \{\partial\}$ in (4) because ∂ is an absorbing state. Thus we obtain

$$P\left(\bigcup_{l=1}^\infty \bigcup_{m=1}^\infty M(l, m)\right) = 0$$

which is seen to be equivalent to the assertion of the theorem. $\qquad\square$

EXAMPLE. As we have seen in §2.5, the Brownian motion in R^1 is a Feller process. We have

$$P_t(x, B(x, \varepsilon)^c) = \frac{1}{\sqrt{2\pi t}} \int_{|y-x|>\varepsilon} \exp\left[-\frac{(y-x)^2}{2t}\right] dy$$

$$= \frac{2}{\sqrt{2\pi t}} \int_\varepsilon^\infty \exp\left[-\frac{u^2}{2t}\right] du$$

$$= \sqrt{\frac{2}{\pi t}} \int_\varepsilon^\infty \frac{t}{u}\left(\exp\left[-\frac{u^2}{2t}\right]\frac{u}{t}\right) du \leq \sqrt{\frac{2}{\pi t}} \frac{t}{\varepsilon} \exp\left[-\frac{\varepsilon^2}{2t}\right].$$

Hence (4) is satisfied even if we replace the K there by $\mathbf{E} = R^1$. It follows that almost all paths are continuous. Recall that we are using the fact that the sample paths of a [version of] Feller process are right continuous in $[0, \infty)$ and has left limits in $(0, \infty)$, which was proved in §2.2. Thus the preceding proof is not quite as short as it appears. The result was first proved by Wiener in 1923 in a totally different setting. Indeed it pre-dated the founding of the theory of stochastic processes.

Exercises

1. A stochastic process $\{X(t), t \geq 0\}$ is said to be *separable* iff there exists a countable dense set S in $[0, \infty)$ such that for almost every ω, the sample function $X(\cdot, \omega)$ has the following property. For each $t \geq 0$, there exists $s_n \in S$ such that $s_n \to t$ and $X(s_n, \omega) \to X(t, \omega)$. [The sequence $\{s_n\}$ depends

on ω as well as t!] It is an old theorem of Doob's that every process has a version which is separable; see Doob [1]. If $\{X(t)\}$ is stochastically continuous the proof is easy and the set S may be taken to be any countable dense set in $[0, \infty)$. Now take a sample function $X(\cdot, \omega)$ which has the separability property above. Show that if $X(s, \omega)$ with s restricted S is uniformly continuous in S, then $X(t, \omega)$ is continuous for all $t \geq 0$. Next, suppose that there exist strictly positive numbers δ, α, β and C such that for $t \geq 0$ and $0 < h < \delta$ the following condition is satisfied:

$$E\{|X(t + h) - X(t)|^{\alpha}\} \leq Ch^{1 + \beta}.$$

Take S to be the dyadics and prove that $X(s, \omega)$ with $s \in S$ is continuous on S for almost every ω. Finally, verify the condition above for the Brownian motion in R^1. [Hint: for the last assertion estimate $\sum_{k=0}^{2^n - 1} P\{|X((k + 1)2^{-n}) - X(k2^{-n})| > n^{-2}\}$ and use the Borel-Cantelli lemma. For two dyadics s and s', $X(s) - X(s')$ is a finite sum of terms of the form $X((k + 1)2^{-n}) - X(k2^{-n})$. The stated criterion for continuity is due to Kolmogorov.]

2. Give an example to show that the Lemma in this section becomes false if the condition "f has left limits in $(0, 1]$" is dropped. This does not seem easy, see M. Steele [1].

3. Let X be a homogeneous Markov process. A point x in $\mathbf{E}$ is called a "holding point" iff for some $\delta > 0$ we have $P^x\{X(t) = x \text{ for all } t \in [0, \delta]\} > 0$. Prove that in this case there exists $\lambda \geq 0$ such that $P^x\{T_{\{x\}^c} > t\} = e^{-\lambda t}$ for all $t > 0$. When $\lambda = 0$, x is called an "absorbing point". Prove that if X has the strong Markov property and continuous sample functions, then each holding point must be absorbing.

4. For a homogeneous Markov chain (Example 1 of §1.2), the state i is holding (also called *stable*) if and only if

$$\lim_{t \downarrow 0} \frac{1 - p_{ii}(t)}{t} < \infty.$$

The limit above always exists but may be $+\infty$, when finite it is equal to the λ in Problem 3. In general, a Markov chain does not have a version which is right continuous (even if all states are stable), hence it is not a Hunt process. But we may suppose that it is separable as in Exercise 1 above.

5. For a Hunt process: for each x there exists a countable collection of optional times $\{T_n\}$ such that for P^x-a.e. ω, the set of discontinuities of $X(\cdot, \omega)$ is the union $\bigcup_n T_n(\omega)$. [Hint: for each $\varepsilon > 0$ define $S^{(\varepsilon)} = \inf\{t > 0 \mid d(X_{t-}, X_t) > \varepsilon\}$. Show that each $S^{(\varepsilon)}$ is optional. Let $S_1^{(\varepsilon)} = S^{(\varepsilon)}$, $S_{n+1}^{(\varepsilon)} = S_n^{(\varepsilon)} + S^{(\varepsilon)} \circ \theta(S_n^{(\varepsilon)})$ for $n \geq 1$. The collection $\{S_n^{(1/m)}\}$, $m \geq 1$, $n \geq 1$, is the desired one.]

6. Let T be one of the T_n's in Exercise 5. Suppose that R_n is an increasing sequence of optional times such that $P^x\{\lim_n R_n = T < \infty\} = 1$, then $P^x\{\bigcup_n (R_n = T)\} = 1$. This means intuitively that T cannot be predicted. A stronger property (called "total inaccessibility") holds to the effect that for each x the two probabilities above are equal for any increasing sequence of optional $\{R_n\}$ such that $R_n \le T$ for all n; see Dellacherie and Meyer [1].

7. Prove that for a Hunt process, we have $\mathscr{F}_T = \sigma(\mathscr{F}_{T-}, X_T)$. (See Chung [5].)

3.2. Analysis of Excessive Functions

A basic tool in the study of Hunt processes is a class of functions called by Hunt "excessive". This is a far-reaching extension of the class of super-harmonic functions in classical potential theory. In this section simple basic properties of these functions will be studied which depend only on the semigroup and not on the associated Markov process. Deeper properties derived from the process will be given in the following sections.

Let a Borelian transition semigroup $(P_t, t \ge 0)$ be given where P_0 is the identity. For $\alpha \ge 0$ we write

$$P_t^\alpha = e^{-\alpha t} P_t; \tag{1}$$

thus $P_t^0 = P_t$. For each α, (P_t^α) is also a Borelian semigroup. It is necessarily submarkovian for $\alpha > 0$. The corresponding potential kernel is U^α, where

$$U^\alpha f = \int_0^\infty P_t^\alpha f \, dt. \tag{2}$$

Here $f \in b\mathscr{E}_+$ or $f \in \mathscr{E}_+$. If $f \in b\mathscr{E}_+$ and $\alpha > 0$, then the function $U^\alpha f$ is finite everywhere, whereas this need not be true if $f \in \mathscr{E}_+$. This is one reason why we have to deal with U^α sometimes even when we are interested in $U^0 = U$. But the analogy is so nearly complete that we can save a lot of writing by treating the case $\alpha = 0$ and claiming the result for $\alpha \ge 0$. Only the finiteness of the involved quantities should be watched, and the single forbidden operation is "$\infty - \infty$".

The α-potential has been introduced in §2.1, as are α-superaveraging and α-excessive functions. The class of α-excessive functions will be denoted by S^α, and $S^0 = S$.

We begin with a fundamental lemma from elementary analysis, of which the proof is left to the reader as an essential exercise.

Lemma 1. *Let $\{u_{mn}\}$ be a double array of positive numbers, where m and n are positive integer indices. If u_{mn} increases with n for each fixed m, and increases* *with m for each fixed n, then we have*

$$\lim_m \lim_n u_{mn} = \lim_n \lim_m u_{mn} = \lim_n u_{nn}.$$

Of course these limits may be ∞.

Proposition 2. *For each $\alpha \geq 0$, $\mathbf{S}^\alpha$ is a cone which is closed under increasing sequential limits. The class of α-superaveraging functions is also such a cone which is furthermore closed under the minimum operation "$\wedge$".*

Proof. It is sufficient to treat the case $\alpha = 0$. To say that $\mathbf{S}$ is a cone means: if $f_i \in \mathbf{S}$, and c_i are positive constants for $i = 1, 2$, then $c_1 f_1 + c_2 f_2 \in \mathbf{S}$. This is trivial. Next let $f_n \in \mathbf{S}$ and $f_n \uparrow f$. Then for each n and t

$$f \geq f_n \geq P_t f_n;$$

letting $n \to \infty$ we obtain $f \geq P_t f$ by monotone convergence. If f is superaveraging, then $P_t f$ increases as t decreases. Hence if $f_n \in \mathbf{S}$ and $f_n \uparrow f$, we have

$$f = \lim_n f_n = \lim_n \lim_t P_t f_n = \lim_t \lim_n P_t f_n = \lim_t P_t f,$$

by Lemma 1 applied to n and a sequence of t (>0) decreasing to 0. This proves $f \in \mathbf{S}$. Finally, if f_i is superaveraging for $i = 1, 2$, then $f_i \geq P_t f_i \geq P_t(f_1 \wedge f_2)$; hence $f_1 \wedge f_2 \geq P_t(f_1 \wedge f_2)$; namely that $f_1 \wedge f_2$ is superaveraging. $\quad\square$

It is remarkable that no simple argument exists to show that $\mathbf{S}$ is closed under "$\wedge$". A deep proof will be given in §3.4. The following special situation is very useful.

Proposition 3. *Suppose P_t converges vaguely to P_0 as $t \downarrow 0$. If f is superaveraging and also lower semi-continuous, then f is excessive.*

Proof. By a standard result on vague convergence, we have under the stated hypotheses:

$$f = P_0 f \leq \lim_{t \downarrow 0} P_t f \leq f. \quad\square$$

If (P_t) is Fellerian, then the condition on (P_t) in Proposition 3 is satisfied. More generally, it is satisfied if the sample paths of the associated Markov process are right continuous at $t = 0$. For then if $f \in b\mathbb{C}$, as $t \downarrow 0$:

$$P_t f(x) = E^x\{f(X_t)\} \to E^x\{f(X_0)\} = P_0 f(x) = f(x). \tag{3}$$

Here we have a glimpse of the interplay between the semigroup and the process. Incidentally, speaking logically, we should have said "*an* associated Markov process" in the above.

Proposition 4. *If* $\alpha < \beta$, *then* $\mathbf{S}^\alpha \subset \mathbf{S}^\beta$. *For each* $\alpha \geq 0$, $\mathbf{S}^\alpha = \bigcap_{\beta > \alpha} \mathbf{S}^\beta$.

Proof. It is trivial that if $\alpha < \beta$, then $P_t^\alpha f \geq P_t^\beta f$ for $f \in \mathcal{E}_+$. Now if $f = \lim_{t \downarrow 0} P_t^\beta f$ for any value of $\beta \geq 0$, then the same limit relation holds for all values of $\beta \geq 0$, simply because $\lim_{t \downarrow 0} e^{-\beta t} = 1$ for all $\beta \geq 0$. The proposition follows quickly from these remarks. $\square$

The next result gives a basic connection between superaveraging functions and excessive functions. It is sufficient to treat the case $\alpha = 0$. If f is superaveraging, we define its *regularization* $\hat{f}$ as follows:

$$\hat{f}(x) = \lim_{t \downarrow 0} \uparrow P_t f(x). \tag{4}$$

We have already observed that the limit above is monotone. Hence we may define $P_{t+} f(x) = \lim_{s \downarrow \downarrow t} P_s f$.

Proposition 5. $\hat{f} \in \mathbf{S}$ *and* $\hat{f}$ *is the largest excessive function not exceeding* f. *We have*

$$\forall t \geq 0: P_t \hat{f} = P_{t+} f. \tag{5}$$

Proof. We prove (5) first, as follows:

$$P_t \hat{f} = P_t \left(\lim_{s \downarrow \downarrow 0} \uparrow P_s f \right) = \lim_{s \downarrow \downarrow 0} P_t P_s f = \lim_{s \downarrow \downarrow 0} P_{t+s} f = P_{t+} f.$$

It is immediate from (4) that $\hat{f} \leq f$ and $\hat{f} \geq P_t f \geq P_t \hat{f}$. Furthermore by (5),

$$\lim_{t \downarrow \downarrow 0} P_t \hat{f} = \lim_{t \downarrow \downarrow 0} P_{t+} \hat{f} = \lim_{t \downarrow \downarrow 0} P_t f = \hat{f},$$

where the second equation is by elementary analysis; hence $\hat{f} \in \mathbf{S}$. If $g \in \mathbf{S}$ and $g \leq f$, then $P_t g \leq P_t f$; letting $t \downarrow \downarrow 0$ we obtain $g \leq \hat{f}$. $\square$

The next result contains an essential calculation pertaining to a potential.

Theorem 6. *Suppose* $f \in \mathbf{S}$, $P_t f < \infty$ *for each* $t > 0$, *and*

$$\lim_{t \to \infty} P_t f = 0. \tag{6}$$

Then we have

$$f = \lim_{h \downarrow 0} \uparrow U \left(\frac{f - P_h f}{h} \right). \tag{7}$$

Proof. The hypothesis $P_t f < \infty$ allows us to subtract below. We have for $h > 0$:

$$\int_0^t P_s(f - P_h f)\, ds = \int_0^t P_s f\, ds - \int_h^{t+h} P_s f\, ds = \int_0^h P_s f\, ds - \int_t^{t+h} P_s f\, ds.$$

If we divide though by h above, then let $t \uparrow \infty$, the last term converges to zero by (6) and we obtain

$$\lim_{t\uparrow\infty} \int_0^t P_s \left(\frac{f - P_h f}{h}\right) ds = \frac{1}{h} \int_0^h P_s f\, ds. \tag{8}$$

The integrand on the left being positive because $f \geq P_h f$, this shows that the limit above is the potential shown in the right member of (7). When $h \downarrow 0$, the right member increases to the limit f, because $\lim_{s\downarrow 0} \uparrow P_s f = f$. This establishes (7).   □

Let us observe that there is an obvious analogue of Proposition 6 for S^α. If $\alpha > 0$ and $f \in b\mathscr{E}_+$, then the corresponding conditions in the proposition are satisfied. Hence (7) holds when U and P_h are replaced by U^α and P_h^α for such an f. This is an important case of the theorem.

An alternative approach to excessive functions is through the use of resolvents (U^α) instead of the semigroup (P_t). This yields a somewhat more general theory but less legible formulas, which will now be discussed briefly. We begin with the celebrated *resolvent equation* (9) below.

Proposition 7. *For $\alpha > 0$ and $\beta > 0$, we have*

$$U^\alpha = U^\beta + (\beta - \alpha)U^\alpha U^\beta = U^\beta + (\beta - \alpha)U^\beta U^\alpha. \tag{9}$$

Proof. This is proved by the following calculations, for $f \in b\mathscr{E}_+$ and $\alpha \neq \beta$:

$$U^\alpha U^\beta f = \int_0^\infty e^{-\alpha s} P_s \left[\int_0^\infty e^{-\beta t} P_t f\, dt\right] ds$$

$$= \int_0^\infty \int_0^\infty e^{-\alpha s - \beta t} P_{s+t} f\, dt\, ds$$

$$= \int_0^\infty \int_s^\infty e^{(\beta-\alpha)s} e^{-\beta u} P_u f\, du\, ds, \quad \text{where } u = s + t,$$

$$= \int_0^\infty \left[\int_0^u e^{(\beta-\alpha)s}\, ds\right] e^{-\beta u} P_u f\, du$$

$$= \int_0^\infty \frac{e^{-\alpha u} - e^{-\beta u}}{\beta - \alpha} P_u f\, du = \frac{1}{\beta - \alpha}[U^\alpha f - U^\beta f].$$

Note that the steps above are so organized that it is immaterial where $\beta - \alpha > 0$ or < 0. This remark establishes the second as well as the first equation in (9).   □

If $f \in \mathbf{S}$, then since

$$\alpha U^\alpha f = \int_0^\infty \alpha e^{-\alpha t} P_t f \, dt = \int_0^\infty e^{-u} P_{u/\alpha} f \, du$$

we have

$$\forall \alpha > 0 : \alpha U^\alpha f \leq f, \tag{10}$$

$$\lim_{\alpha \uparrow \infty} \uparrow \alpha U^\alpha f = f. \tag{11}$$

[The case $f(x) = \infty$ in (11) should be scrutinized.] The converse is rather tricky to prove.

Proposition 8. *If* $f \in \mathscr{E}_+$ *and* (10) *is true, then* $\lim_{\alpha \uparrow \infty} \uparrow \alpha U^\alpha f = \hat{f}$. *If* (11) *is also true then* f *is excessive.*

Proof. Let $0 < \beta < \alpha$ and $f \in b\mathscr{E}_+$. Then we have by (9):

$$U^\alpha f = U^\beta [f - (\alpha - \beta) U^\alpha f]. \tag{12}$$

Here subtraction is allowed because $U^\alpha f < \infty$. If (10) is true, then $g_{\alpha\beta} = f - (\alpha - \beta) U^\alpha f \geq 0$. Since $U^\beta g_{\alpha\beta}$ is β-excessive, $U^\alpha f$ is β-excessive for all $\beta \in (0, \alpha)$. Hence $U^\alpha f \in \mathbf{S}$ by Proposition 4. Next, we see from (12) by simple arithmetic that

$$\alpha U^\alpha f = \beta U^\beta f + (\alpha - \beta) U^\beta [f - \alpha U^\alpha f] \geq \beta U^\beta f \tag{13}$$

by (10). Hence if we put

$$f^* = \lim_{\alpha \uparrow \infty} \uparrow \alpha U^\alpha f = \lim_{\alpha \uparrow \infty} \uparrow U^\alpha (\alpha f),$$

we have $f^* \in \mathbf{S}$ by Proposition 2.

For a general $f \in \mathscr{E}_+$ satisfying (10), let $f_n = f \wedge n$. Then $\alpha U^\alpha f_n \leq f_n$ by (10) and the inequality $\alpha U^\alpha n \leq n$. Hence we may apply what has just been proved to each f_n to obtain $f_n^* = \lim_{\alpha \uparrow \infty} \uparrow \alpha U^\alpha f_n$. Also the inequality $\alpha U^\alpha f_n \geq \beta U^\beta f_n$ leads to $\alpha U^\alpha f \geq \beta U^\beta f$, if $\alpha \geq \beta$. Hence we have by Lemma 1:

$$f^* = \lim_{\alpha \uparrow \infty} \uparrow \alpha U^\alpha f = \lim_{\alpha \uparrow \infty} \uparrow \lim_{n \uparrow \infty} \uparrow \alpha U^\alpha f_n$$

$$= \lim_{n \uparrow \infty} \uparrow \lim_{\alpha \uparrow \infty} \uparrow \alpha U^\alpha f_n = \lim_{n \uparrow \infty} \uparrow f_n^*.$$

Since each $f_n^* \in \mathbf{S}$ as shown above, we conclude that $f^* \in \mathbf{S}$. If $g \in \mathbf{S}$ and $g \leq f$, then $\alpha U^\alpha g \leq \alpha U^\alpha f$; letting $\alpha \uparrow \infty$ we obtain $g \leq f^*$. This identifies f^* as the largest excessive function $\leq f$, hence $f^* = \hat{f}$ by Proposition 5. The second assertion of Proposition 8 is now trivial. $\qquad \square$

Proposition 8 has an immediate extension as follows. For each $\beta \geq 0$, $f \in \mathbf{S}^\beta$ if and only if

$$\forall \alpha > 0 : \alpha U^{\alpha+\beta} f \leq f \quad \text{and} \quad \lim_{\alpha \uparrow \infty} \uparrow \alpha U^{\alpha+\beta} f = f. \tag{14}$$

This is left as an exercise.

The next result is extremely useful. It reduces many problems concerning an excessive function to the same ones for potentials, which are often quite easy.

Theorem 9. *If $f \in \mathbf{S}$, then for each $\beta > 0$, there exists $g_n \in b\mathscr{E}_+$ such that*

$$f = \lim_n \uparrow U^\beta g_n. \tag{15}$$

Proof. Suppose first that $f \in b\mathscr{E}_+$. Then as in the preceding proof, we have $U^\alpha(\alpha f) = U^\beta(\alpha g_{\alpha\beta})$. Thus we obtain (15) from (11) with $g_n = n g_{n\beta}$. In general we apply this to obtain for each k, $f \wedge k = \lim_n \uparrow U^\beta g_n^{(k)}$. It follows then by Lemma 1 that

$$f = \lim_k \uparrow \lim_n \uparrow U^\beta g_n^{(k)} = \lim_n \uparrow U^\beta g_n^{(n)}. \qquad \square$$

We remark that this proposition is also a consequence of the extension of Theorem 6 to β-excessive functions, because then $\lim_{t \to \infty} P_t^\beta f = 0$ for $f \in b\mathscr{E}_+$, etc. But the preceding proof is more general in the sense that it makes use only of (10) and (11), and the resolvent equation, without going back to the semigroup.

It is an important observation that Theorem 9 is false with $\beta = 0$ in (15); see Exercise 3 below. An additional assumption, which turns out to be of particular interest for Hunt processes, will now be introduced to permit $\beta = 0$ there. Since we are proceeding in an analytic setting, we must begin by assuming that

$$\forall x \in \mathbf{E} : U(x, \mathbf{E}) > 0. \tag{16}$$

This is trivially true for a process with right continuous paths; for a more general situation see Exercise 2 below. Remember however that $U(\partial, \mathbf{E}) = 0$. To avoid such exceptions we shall agree in what follows that a "point" means a point in $\mathbf{E}$, and a "set" means a subset of $\mathbf{E}$, without specific mention to the contrary. The case of ∂ can always be decided by an extra inspection, when it is worth the pain.

Definition. The semigroup (P_t) or the corresponding U is called *transient* iff there exist a function $h \in \mathscr{E}_+$ such that

$$0 < Uh < \infty \qquad \text{on } \mathbf{E}. \tag{17}$$

An immediate consequence of (17) is as follows. There exists a sequence of $h_n \in b\mathscr{E}_+$ (with $h_n(\partial) = 0$) such that

$$Uh_n > 0 \quad \text{and} \quad Uh_n \uparrow \infty \quad \text{both in } \mathbf{E}. \tag{18}$$

We may take $h_n = nh$ to achieve this.

The probabilistic meaning of transience as defined above, as well as some alternative conditions, will be discussed later in §3.7.

Proposition 10: *If (P_t) is transient, and $f \in \mathbf{S}$, then there exists $g_n \in b\mathscr{E}_+$ such that*

$$f = \lim_n \uparrow Ug_n. \tag{19}$$

In fact, $g_n \leq n^2$ and $Ug_n \leq n$.

Proof. Put

$$f_n = f \wedge Uh_n \wedge n$$

where h_n is as in (18). Then we have

$$P_t f_n \leq P_t Uh_n = \int_t^\infty P_s h_n \, ds \downarrow 0$$

as $t \uparrow \infty$ because $Uh_n < \infty$. By Proposition 2, f_n is superaveraging; let its regularization be $\hat{f}_n$. Since $P_t \hat{f}_n \leq P_t f_n < \infty$, and $\to 0$ as $t \uparrow \infty$, we may apply Theorem 6 to obtain

$$\hat{f}_n = \lim_k \uparrow Ug_{nk}$$

where

$$g_{nk} = k(\hat{f}_n - P_{1/k}\hat{f}_n) \leq kn.$$

From the proof of Theorem 6 we see that

$$Ug_{nk} = k \int_0^{1/k} P_s \hat{f}_n \, dx \leq n.$$

For each n, Ug_{nk} increases with k; for each k, Ug_{nk} increases with n, hence we have by Lemma 1:

$$\lim_{n} \uparrow \hat{f}_n = \lim_{n} \uparrow \lim_{k} \uparrow Ug_{nk} = \lim_{n} \uparrow Ug_{nn}.$$

On the other hand, we have again by Lemma 1:

$$\lim_{n} \uparrow \hat{f}_n = \lim_{n} \uparrow \lim_{t \downarrow 0} \uparrow P_t f_n = \lim_{t \downarrow 0} \uparrow \lim_{n} \uparrow P_t f_n = \lim_{t \downarrow 0} P_t f = f.$$

It follows that (19) is true with $g_n = g_{nn}$. $\square$

Exercises

1. A measure μ on $\mathscr{E}$ is said to be c-finite ("countably finite") iff there is a countable collection of finite measures $\{\mu_n\}$ such that $\mu(A) = \sum_{n=1}^{\infty} \mu_n(A)$ for all $A \in \mathscr{E}$. Show that if μ is σ-finite, then it is c-finite, but not vice versa. For instance, Fubini's theorem holds for c-finite measures. Show that for each x, $U(x, \cdot)$ is c-finite.

2. Show that for each $x \in \mathbf{E}$, $P_t(x, \mathbf{E})$ is a decreasing function of t. The following properties are equivalent:

   (a) $U(x, \mathbf{E}) = 0$;   (b) $\lim_{t \downarrow\downarrow 0} P_t(x, \mathbf{E}) = 0$;   (c) $P^x\{\zeta = 0\} = 1$.

3. For the Brownian motion in R^1, show that there cannot exist positive Borel (or Lebesgue) measurable functions g_n such that $\lim_{n \to \infty} Ug_n = 1$. Thus Theorem 9 is false for $\beta = 0$. [See Theorem 1 of §3.7 below for general elucidation.]

4. If $f \in \mathscr{E}_+$ and $\alpha U^\alpha f \leq f$ for all $\alpha > 0$, then $P_t f \leq f$ for (Lebesgue) almost all $t \geq 0$. [Hint: $\lim_{\alpha \uparrow \infty} \alpha U^\alpha f = f^* \leq f$ and $U^\alpha f^* = U^\alpha f$ for all $\alpha > 0$.]

3.3. Hitting Times

We have proved the optionality of T_A for an open or a closed set A, in Theorem 6 of §2.4. The proof is valid for a Hunt process. It is a major advance to extend this to an arbitrary Borel set and to establish the powerful approximation theorems "from inside" and "from outside". The key is the theory of capacity due to Choquet. We begin by identifying the probability of hitting a set before a fixed time as a Choquet capacity.

   Let $\{X_t, \mathscr{F}_t, t \in \mathbf{T}\}$ be a Hunt process as specified in §3.1. For simplicity of notation, we shall write P for P^μ (for an arbitrary probability measure μ on $\mathscr{E}$); but recall that $\mathscr{F}_t \subset \mathscr{F}_t^\mu$ and $(\mathscr{F}^\mu, P^\mu)$ is complete. Fix $t \geq 0$. For any subset

A of $\mathbf{E}_{\partial}$, we put

$$A' = \{\omega \in \Omega \mid \exists s \in [0,t] : X_s(\omega) \in A\}.$$

This is the projection on Ω of the set of (s, ω) in $[0,t] \times \Omega$ such that $X(s, \omega) \in A$. The mapping $A \to A'$ from the class of all subsets of $\mathbf{E}_{\partial}$ to the class of all subsets of Ω has the following properties:

(i) $A_1 \subset A_2 \Rightarrow A_1' \subset A_2'$;
(ii) $(A_1 \cup A_2)' = A_1' \cup A_2'$;
(iii) $A_n \uparrow A \Rightarrow A_n' \uparrow A'$.

In view of (i) and (ii), (iii) is equivalent to

(iii') $(\bigcup_n A_n)' = \bigcup_n A_n'$ for an arbitrary sequence $\{A_n\}$.

So far these properties are trivial and remain valid when the fundamental internal $[0,t]$ is replaced by $[0, \infty)$. The next proposition will depend essentially on the compactness of $[0,t]$.

We shall write $A_1 \Subset A_2$ to denote $P(A_1 \backslash A_2) = 0$, and $A_1 \doteq A_2$ to denote $P(A_1 \triangle A_2) = 0$.

Proposition 1. *Let G_n be open, K be compact such that*

$$\forall n : G_n \supset \bar{G}_{n+1}, \qquad K = \bigcap_n G_n \left(= \bigcap_n \bar{G}_n \right). \tag{1}$$

Then we have

$$K' \doteq \bigcap_n G_n'. \tag{2}$$

Proof. This is essentially a re-play of Theorem 6 of §2.4. Consider the first entrance times D_{G_n}; these are optional (relative to $\{\mathscr{F}_t^0\}$). If $\omega \in \bigcap_n G_n'$, then $D_{G_n}(\omega) \leq t$ for all n. Let $D(\omega) = \lim_n \uparrow D_{G_n}(\omega)$; clearly $D(\omega) \leq D_K(\omega)$ and $D(\omega) \leq t$. Put

$$N = \left\{ \omega \,\middle|\, \lim_n X(D_{G_n}) \neq X(D) \right\};$$

we have $P(N) = 0$ by quasi left continuity. If $\omega \in \bigcap_n G_n' \backslash N$, then $X(D) \in \bigcap_n \bar{G}_n = K$ since $X(D_{G_n}) \in \bar{G}_n$ by right continuity. Thus $D_K(\omega) \leq D(\omega)$ and so $D_K(\omega) = D(\omega) \leq t$; namely $\omega \in K'$. We have therefore proved $\bigcap_n G_n' \Subset K'$. But $\bigcap_n G_n' \supset K'$ by (i) above, hence (2) is proved. $\qquad\square$

Proposition 2. *Let K_n be compact, $K_n \downarrow$ and $\bigcap_n K_n = K$. Then*

$$K' \doteq \bigcap_n K_n'. \tag{3}$$

Proof. K being given, let $\{G_n\}$ be as in Proposition 1. For each m, $G_m \supset K = \bigcap_n K_n$; hence by a basic property of compactness, there exists n_m such that $G_m \supset K_{n_m}$ and consequently $G'_m \supset K'_{n_m}$ by (i). It follows that

$$K' \doteq \bigcap_m G'_m \supset \bigcap_m K'_{n_m} \supset \bigcap_n K'_n \supset K',$$

where the first equation is just (2). This implies (3). $\qquad\square$

Corollary. *For each compact K, and $\varepsilon > 0$, there exists open G such that $K \subset G$ and*

$$P(K') \leq P(G') \leq P(K') + \varepsilon.$$

In particular

$$P(K') = \inf_{G \supset K} P(G') \tag{4}$$

where the inf *is over all open G containing K.*

From here on in this section, the letters A, B, G, K are reserved for arbitrary, Borel, open, compact sets respectively.

Definition. Define a set function C for all subsets of $\mathbf{E}_\partial$ as follows:

(a) $C(G) = P(G')$;
(b) $C(A) = \inf_{G \supset A} C(G)$.

Clearly C is "monotone", namely $A_1 \subset A_2$ implies $C(A_1) \leq C(A_2)$. Note also that it follows from (4) that

$$C(K) = P(K'). \tag{5}$$

We proceed to derive properties of this function C. Part (b) of its definition is reminiscent of an outer measure which is indeed a particular case; on the other hand, it should be clear from part (a) that C is not additive on disjoint sets. What replaces additivity is a strong kind of subadditivity, as given below.

Proposition 3. *C is "strongly subadditive" over open sets, namely;*

$$C(G_1 \cup G_2) + C(G_1 \cap G_2) \leq C(G_1) + C(G_2). \tag{6}$$

Proof. We have, as a superb example of the facility (felicity) of reasoning with sample functions:

$$(G_1 \cup G_2)' - G'_1 = \{\omega \mid \exists s \in [0,t]: X_s(\omega) \in G_1 \cup G_2; \forall s \in [0,t]: X_s(\omega) \notin G_1\}$$
$$\subset \{\omega \mid \exists s \in [0,t]: X_s(\omega) \in G_2; \forall s \in [0,t]: X_s(\omega) \notin G_1 \cap G_2\}$$
$$= G'_2 - (G_1 \cap G_2)'.$$

Taking probabilities, we obtain

$$P((G_1 \cup G_2)') - P(G_1') \le P(G_2') - P((G_1 \cap G_2)').$$

Up to here G_1 and G_2 may be arbitrary sets. Now we use definition (a) above to convert the preceding inequality into (6). □

The same argument shows C is also strongly additive over all compact sets, because of (5). Later we shall see that C is strongly additive over all capacitable sets by the same token.

Lemma 4. *Let* $A_n\uparrow$, $A_n \subset G_n$, $\varepsilon_n > 0$ *such that*

$$\forall n: C(G_n) \le C(A_n) + \varepsilon_n. \tag{7}$$

Then we have for each finite m:

$$C\left(\bigcup_{n=1}^{m} G_n\right) \le C(A_m) + \sum_{n=1}^{m} \varepsilon_n. \tag{8}$$

Proof. For $m = 1$, (8) reduces to (7). Assume (8) is true as shown and observe that

$$A_m \subset \left(\bigcup_{n=1}^{m} G_n\right) \cap G_{m+1}. \tag{9}$$

We now apply the strong subadditivity of C over the two open sets $\bigcup_{n=1}^{m} G_n$ and G_{m+1}, its monotonicity, and (9) to obtain

$$C\left(\bigcup_{n=1}^{m+1} G_n\right) + C(A_m) \le C\left(\left(\bigcup_{n=1}^{m} G_n\right) \cup G_{m+1}\right) + C\left(\left(\bigcup_{n=1}^{m} G_n\right) \cap G_{m+1}\right)$$

$$\le C\left(\bigcup_{n=1}^{m} G_n\right) + C(G_{m+1}) \le C(A_m) + \sum_{n=1}^{m} \varepsilon_n + C(A_{m+1}) + \varepsilon_{m+1}.$$

This completes the induction on m. □

We can now summarize the properties of C in the theorem below.

Theorem 5. *We have*

(i) $A_1 \subset A_2 \Rightarrow C(A_1) \le C(A_2)$;

(ii) $A_n \uparrow A \Rightarrow C(A_n) \uparrow C(A)$;

(iii) $K_n \downarrow K \Rightarrow C(K_n) \downarrow C(K)$.

Proof. We have already mentioned (i) above. Next, using the notation of Lemma 4, we have

$$C(A) \le C\left(\bigcup_{n=1}^{\infty} G_n\right) = \lim_m C\left(\bigcup_{n=1}^{m} G_n\right)$$

by definition (a) of C, and the monotone property of the probability measure P. Applying Lemma 4 with $\varepsilon_n = \varepsilon 2^{-n}$, we see from (8) that

$$C(A) \le \lim_m C(A_m) + \varepsilon.$$

Since ε is arbitrary, (ii) follows from this inequality and (i). Finally, it follows from (5), Proposition 2, and the monotone property of P that

$$C(K) = P(K') = P\left(\bigcap_n K'_n\right) = \lim_n P(K'_n) = \lim_n C(K_n).$$

This proves (iii). $\qquad\qquad\qquad\qquad\qquad\qquad\qquad\qquad\qquad\qquad\quad\square$

Definition. A function defined on all the subsets of $\mathbf{E}_\partial$ and taking values in $[-\infty, +\infty]$ is called a *Choquet capacity* iff it has the three properties (i), (ii) and (iii) in Theorem 5. A subset A of $\mathbf{E}_\partial$ is called *capacitable* iff given any $\varepsilon > 0$, there exist an open G and a compact K such that $K \subset A \subset G$ and

$$C(G) \le C(K) + \varepsilon. \tag{10}$$

This definition (which can be further generalized) is more general than we need here, since the C defined above takes values in $[0,1]$ only. Here is the principal theorem of capacitability.

Choquet's Theorem. *In a locally compact Hausdorff space with countable base, each analytic set is capacitable.*

A set $A \subset \mathbf{E}_\partial$ is *analytic* iff there exists a separable complete metric space (alias "Polish space") M and a continuous mapping φ of M into $\mathbf{E}_\partial$ such that $A = \varphi(M)$.

Such a definition needs much study to make sense, but we must forego it since we shall not use analytic sets below. Suffice it to say that each Borel set, namely each A in $\mathscr{E}$ in our general notation, is analytic and therefore capacitable by Choquet's theorem. We are ready for the principal result.

Theorem 6. *For each Borel set B (i.e., $B \in \mathscr{E}_\partial$) we have*

$$C(B) = P(B'). \tag{11}$$

Proof. Since B is capacitable by Choquet's theorem, for each $n \geq 1$ there exist $K_n \subset B \subset G_n$ such that

$$C(K_n) \leq C(B) \leq C(G_n) \leq C(K_n) + \frac{1}{n}. \tag{12}$$

We have $K_n' \subset B' \subset G_n'$. Let

$$\Lambda_1 = \bigcup_n K_n' \subset B' \subset \bigcap_n G_n' = \Lambda_2. \tag{13}$$

Then for each $n \geq 1$:

$$P(\Lambda_2) - P(\Lambda_1) \leq P(G_n') - P(K_n') = C(G_n) - C(K_n) \leq \frac{1}{n},$$

and consequently

$$\Lambda_1 \doteq \Lambda_2 \doteq B'. \tag{14}$$

It follows that

$$P(B') = P(\Lambda_1) = \lim_n P(K_n') = \lim_n C(K_n) = C(B)$$

where the last equation is by (12). $\square$

Corollary. $P(B') = \lim_n P(K_n') = \lim_n P(G_n')$.

The optionality of D_B and T_B follows quickly.

Theorem 7. *For each $B \in \mathscr{E}_\partial$, D_B and T_B are both optional relative to $\{\mathscr{F}_t\}$.*

Proof. Recall that t is fixed in the foregoing discussion. Let us now denote the B' above by $B'(t)$. For the sake of explicitness we will replace $(\Omega, \mathscr{F}, P)$ above by $(\Omega, \mathscr{F}_t, P^\mu)$ for each probability measure μ on $\mathscr{E}$, as fully discussed in §2.3. Put also $Q^t = (Q \cap [0,t)) \cup \{t\}$. Since G is open, the right continuity of $s \to X_s$ implies that

$$G'(t) = \{\omega \mid \exists s \in Q^t : X_s(\omega) \in G\}.$$

Hence $G'(t) \in \mathscr{F}_t^0 \subset \mathscr{F}_t^\mu$. Next, we have $P^\mu\{B'(t) \bigtriangleup \bigcap_n G_n'(t)\} = 0$ by (13), hence $B'(t) \in \mathscr{F}_t^\mu$ because $\mathscr{F}_t^\mu$ contains all P^μ-null sets. Finally, a careful scrutiny shows that for each $B \in \mathscr{E}_\partial$ and $t > 0$:

$$\{\omega \mid D_B(\omega) < t\} = \bigcup_{r \in Q \cap [0,t)} \{\omega \mid \exists s \in [0,r] : X_s(\omega) \in B\}$$

$$= \bigcup_{r \in Q \cap [0,t)} B'(r) \in \mathscr{F}_t^\mu.$$

Hence D_B is optional relative to $\{\mathscr{F}_t^\mu\}$. The same is true of T_B by the argument in Theorem 6 of §2.4.   □

It should be recalled that the preceding theorem is also an immediate consequence of Theorem 3 of §1.5. The more laborious proof just given "tells more" about the approximation of a Borel set by compact subsets and open supersets, but the key to the argument which is concealed in Choquet's theorem, is the projection from $\mathbf{T} \times \Omega$ to Ω, just as in the case of Theorem 3 of §1.5. This is also where analytic sets enter the picture; see Dellacherie and Meyer [1].

The approximation theorem will be given first for D_B, then for T_B, in two parts. The probability measure μ is defined on $\mathscr{E}_\partial$, though the point ∂ plays only a nuisance role for what follows and may be ignored.

Theorem 8(a). *For each μ and $B \in \mathscr{E}_\partial$, there exist $K_n \subset B$, $K_n\uparrow$ and $G_n \supset B$, $G_n\downarrow$ such that*

$$D_{K_n} \downarrow D_B, \qquad D_{G_n} \uparrow D_B, \qquad P^\mu\text{-a.s.} \tag{15}$$

Proof. The basic idea is to apply the Corollary to Theorem 6 to $B'(r)$ for all $r \in Q$. Thus for each $r \in Q$, we have

$$K_{rn} \subset B \subset G_{rn} \tag{16}$$

such that

$$\lim_n P^\mu\{G'_{rn}(r) - K'_{rn}(r)\} = 0. \tag{17}$$

Let $\{r_j\}$ be an enumeration of Q, and put

$$K_n = \bigcup_{j=1}^n K_{r_jn}, \qquad G_n = \bigcap_{j=1}^n G_{r_jn}.$$

Then $K_n \subset B$, $K_n\uparrow$; $G_n \supset B$, $G_n\downarrow$; let

$$D_{K_n} \downarrow D', \qquad D_{G_n} \uparrow D''.$$

Then $D' \geq D_B \geq D''$. We have for each j:

$$\{D' > r_j > D''\} \subset \bigcap_n \{D_{K_n} > r_j > D_{G_n}\} \subset \bigcap_n \{G'_n(r_j) - K'_n(r_j)\}. \tag{18}$$

For $n \geq j$, we have

$$K_n \supset K_{r_jn}, \qquad G_n \subset G_{r_jn}$$

hence

$$K'_n(r_j) \supset K'_{r_jn}(r_j), \qquad G'_n(r_j) \subset G'_{r_jn}(r_j).$$

Hence by (17),

$$\lim_n P^\mu \{G'_n(r_j) - K'_n(r_j)\} = 0$$

and consequently by (18), $P^\mu \{D' > r_j > D''\} = 0$. This being true for all r_j, we conclude that $P^\mu \{D' = D_B = D''\} = 1$, which is the assertion in (15). □

Theorem 8(b). *For each μ and $B \in \mathcal{E}_\partial$, there exists $K_n \subset B, K_n\uparrow$ such that*

$$T_{K_n} \downarrow T_B, \qquad P^\mu\text{-a.s.} \tag{19}$$

If μ is such that $\mu(B) = 0$, then there exist $G_n \supset B, G_n\downarrow$ such that

$$T_{G_n} \uparrow T_B, \qquad P^\mu\text{-a.s.} \tag{20}$$

Proof. The basic formula here is (16) of §2.4:

$$T_B = \lim_{s\downarrow\downarrow 0} \downarrow (s + D_B \circ \theta_s), \tag{21}$$

applied to a sequence $s_k \downarrow\downarrow 0$. Let $\mu_k = \mu P_{s_k}$. Then for each $k \geq 1$, we have by part (a): there exist $K_{kn} \subset B$, $K_{kn}\uparrow$ as $n\uparrow$ such that

$$D_{K_{kn}} \downarrow D_B, \qquad P^{\mu_k}\text{-a.s.}$$

which means

$$D_{K_{kn}} \circ \theta_{s_k} \downarrow D_B \circ \theta_{s_k}, \qquad P^\mu\text{-a.s.} \tag{22}$$

Let $K_n = \bigcup_{k=1}^n K_{kn}$: then $K_n \subset B$, $K_n\uparrow$, and it is clear from (22) that

$$\forall k: D_{K_n} \circ \theta_{s_k} \downarrow D_B \circ \theta_{s_k}, \qquad P^\mu\text{-a.s.}$$

Therefore, we have by (an analogue of) Lemma 1 in §3.2:

$$T_B = \lim_k \downarrow (s_k + D_B \circ \theta_{s_k}) = \lim_k \downarrow \lim_n \downarrow (s_k + D_{K_n} \circ \theta_{s_k})$$

$$= \lim_n \downarrow \lim_k \downarrow (s_k + D_{K_n} \circ \theta_{s_k}) = \lim_n \downarrow T_{K_n}, \qquad P^\mu\text{-a.s.}$$

This proves (19) and it is important to see that the additional condition $\mu(B) = 0$ is needed for (20). Observe that $T_B = D_B$ unless $X_0 \in B$; hence under the said condition we have $P^\mu \{T_B = D_B\} = 1$. It follows from this remark that $T_G = D_G$ for an open G, because if $X_0 \in G$ we have $T_G = 0$ (which is not the case for an arbitrary B). Thus we have by part (a):

$$T_{G_n} = D_{G_n} \uparrow D_B = T_B, \qquad P^\mu\text{-a.s.} \qquad \square$$

Remark. The most trivial counterexample to (20) when $\mu(B) \neq 0$ is the case of uniform motion with $B = \{0\}$, $\mu = \varepsilon_0$. Under P^μ, it is obvious that $T_B = \infty$ but $T_G = 0$ for each open $G \supset B$. Another example is the case of Brownian motion in the plane, if $B = \{x\}$ and $\mu = \varepsilon_x$; see the Example in §3.6.

As a first application of Theorem 8, we consider the "left entrance time" and "left hitting time" of a Borel set B, defined as follows:

$$
\begin{aligned}
D_B^-(\omega) &= \inf\{t \geq 0 \,|\, X_{t-}(\omega) \in B\}, \\
T_B^-(\omega) &= \inf\{t > 0 \,|\, X_{t-}(\omega) \in B\}.
\end{aligned}
\tag{23}
$$

We must make the convention $X_{0-} = X_0$ to give a meaning to D_B.

Theorem 9. *We have almost surely*:

$$
\begin{aligned}
D_G^- &= D_G; & T_G^- &= T_G; \\
D_B^- &\geq D_B; & T_B^- &\geq T_B.
\end{aligned}
$$

Proof. Since the relation (21) holds for T_B^- and D_B^- as well, it is sufficient to consider the D's.

If $X_t \in G$ and G is open, then there is an open set G_1 such that $\bar{G}_1 \subset G$ and $X_t \in G_1$. The right continuity of paths then implies that for each ω there exists $\delta_0(\omega) > 0$ such that $X_{t+\delta} \in \bar{G}_1$ for $0 < \delta \leq \delta_0$ and so $X_{t+\delta-} \in \bar{G}_1$. This observation shows $D_G^- \leq D_G$. Conversely if $t > 0$ and $X_{t-} \in G$, then there exist $t_n \upuparrows t$ such that $X_{t_n} \in G$; whereas if $X_{0-} \in G$ then $X_0 \in G$. This shows $D_G \leq D_G^-$ and consequently $D_G = D_G^-$.

For a general B, we apply the second part of (15) to obtain $G_n \supset B$ such that $D_{G_n} \uparrow D_B$, P^μ-a.s. Since $D_B^- \geq D_{G_n}^- = D_{G_n}$, it follows that $D_B^- \geq D_B$, P^μ-a.s. $\qquad\square$

We can now settle a point remarked in §3.1, by applying Theorem 9 with $B = \{\partial\}$.

Corollary. *For a Hunt process, we have almost surely*:

$$
\forall t < \zeta: X_{t-} \in \mathbf{E}.
\tag{24}
$$

This implies: on the set $\{t < \zeta\}$, the closure of the set of values $\bigcup_{s \in [0,t]} X(s, \omega)$ is a compact subset of $\mathbf{E}$.

We close this section by making a "facile generalization" which turns out to be useful, in view of later developments in §3.4.

Definition. A set $A \subset \mathbf{E}_\partial$ is called *nearly Borel* iff for each finite measure μ on $\mathscr{E}_\partial$, there exist two Borel sets B_1 and B_2, depending on μ, such that $B_1 \subset A \subset B_2$ and

$$
P^\mu\{D_{B_2 - B_1} < \infty\} = 0.
\tag{25}
$$

This is equivalent to:

$$P^{\mu}\{\exists t \in [0, \infty): X_t \in B_2 - B_1\} = 0. \tag{26}$$

Of course, this definition depends on the process $\{X_t\}$.

The class of nearly Borel sets will be denoted by $\mathcal{E}^*$. It is easy to verify that this is a σ-field. It is included in $\mathcal{E}^{\sim}$, the universally measurable σ-field, because (26) implies

$$\mu(B_2 - B_1) = P^{\mu}\{X_0 \in B_2 - B_1\} = 0.$$

One can describe $\mathcal{E}^*$ in a folksy way by saying that the "poor" Hunt process $\{X_t\}$ cannot distinguish a set in $\mathcal{E}^*$ from a set in $\mathcal{E}_{\partial}$. It should be obvious that the hitting time of a nearly Borel set is optional, and the approximation theorems above hold for it as well. [We do not need "nearly compact" or "nearly open" sets!] Indeed, if $A \in \mathcal{E}^*$ then for each μ there exists $B \in \mathcal{E}_{\partial}$ such that $D_B = D_A$, P^{μ}-a.s. Needless to say, this B depends on μ.

A function f on $\mathbf{E}_{\partial}$ to $[-\infty, +\infty]$ is nearly Borel when $f \in \mathcal{E}^*$. This is the case if and only if for each μ, there exist two Borel functions f_1 and f_2, depending on μ, such that $f_1 \leq f \leq f_2$ and

$$P^{\mu}\{\exists t \in [0, \infty): f_1(X_t) \neq f_2(X_t)\} = 0. \tag{27}$$

It follows that we may replace f_1 or f_2 in the relation above by f. We shall see in the next section that all excessive functions are nearly Borel.

3.4. Balayage and Fundamental Structure

Let $\{X_t, \mathcal{F}_t\}$ be a Hunt process, where $\{\mathcal{F}_t\}$ is as specified in §3.1.

For $f \in \mathcal{E}_+$, $\alpha \geq 0$, and optional T, we define the operator P_T^{α} as follows:

$$P_T^{\alpha} f(x) = E^x\{e^{-\alpha T}f(X_T); T < \infty\}. \tag{1}$$

We write P_T for P_T^0. If we use the convention that $X_{\infty} = \partial$ and $f(\partial) = 0$ for each f, then we may omit "$T < \infty$" in the expression above. We shall frequently do so without repeating this remark.

The following composition property is fundamental. Recall that if S and T are optional relative to $\{\mathcal{F}_t\}$, so is $S + T \circ \theta_S$: see Exercise 7 of §2.3.

Proposition 1. *We have*

$$P_S^{\alpha} P_T^{\alpha} = P_{S + T \circ \theta_S}^{\alpha}. \tag{2}$$

Proof. Since $X_T(\omega) = X(T(\omega), \omega)$, we have

$$X_T \circ \theta_S(\omega) = X(T(\theta_S\omega), \theta_S(\omega)) = X(S(\omega) + T \circ \theta_S(\omega), \omega) = X_{S+T\circ\theta_S}(\omega). \quad (3)$$

It follows that for $f \in b\mathcal{E}_+$:

$$\begin{aligned}
P_S^\alpha P_T^\alpha f &= E^\cdot\{e^{-\alpha S} P_T^\alpha f(X_S)\} = E^\cdot\{e^{-\alpha S} E^{X_S}[e^{-\alpha T} f(X_T)]\} \\
&= E^\cdot\{e^{-\alpha S}[e^{-\alpha T} f(X_T)] \circ \theta_S\} \\
&= E^\cdot\{e^{-\alpha(S+T\circ\theta_S)} f(X_{S+T\circ\theta_S})\}.
\end{aligned}$$ □

Here and henceforth we will adopt the notation $E^\cdot(\cdots)$ to indicate the function $x \to E^x(\cdots)$.

Two important instances of (2) are: when $S = t$, a constant; and when S and T are hitting times of a nearly Borel set A. In the latter case we write P_A^α for $P_{T_A}^\alpha$. We shall reserve the letter A for a nearly Borel set below.

Definition. A point x is said to be *regular* for the set A iff

$$P^x\{T_A = 0\} = 1. \quad (4)$$

The set of all points in $\mathbf{E}_\partial$ which are regular for A will be denoted by A^r; and the union $A \cup A^r$ is called the *fine closure* of A and denoted by A^*. The nomenclature will be justified in §3.5.

According to the zero-or-one law (Theorem 6 of §2.3), x is not regular for A iff $P^x\{T_A = 0\} = 0$ or $P^x\{T_A > 0\} = 1$. In this case we say also that the set A is *thin* at the point x. Let $\bar{A}$ denote the topological closure of A, then if $x \notin \bar{A}$, a path starting at x must remain for some time in an open neighborhood of x which is disjoint from A, hence x cannot be regular for A. Thus $A^* \subset \bar{A}$. Since in general the set $\{T_A = 0\}$ belongs to $\mathcal{F}_0 = \mathcal{F}_{\tilde{0}}$ and not to $\mathcal{F}_0^0$, the function $x \to P^x\{T_A = 0\}$ belongs to $\mathcal{E}^\sim$ rather than $\mathcal{E}$. This is a nuisance which will be ameliorated. Observe that (4) is equivalent to

$$P_A^\alpha 1(x) \equiv E^x\{e^{-\alpha T_A}\} = 1 \quad (5)$$

for each $\alpha > 0$; indeed $P^x\{T_A = 0\}$ may be regarded as $\lim_{\alpha \uparrow \infty} P_A^\alpha 1(x)$. Finally if $x \in A^r$, then for any f we have

$$P_A f(x) = E^x\{T_A < \infty; f(X_{T_A})\} = E^x\{T_A = 0; f(X_0)\} = f(x).$$

Theorem 2. *For each x and A, the measure $P_A(x, \cdot)$ is concentrated on A^*. In other words,*

$$\forall x: P^x\{X(T_A) \in A^*; T_A < \infty\} = P^x\{T_A < \infty\}.$$

Proof. We have by definition $(A^*)^c = A^c \cap (A^r)^c$. If $y \in (A^r)^c$, then $P^y\{T_A > 0\}$ $= 1$. It follows that

$$P^x\{T_A < \infty; X(T_A) \in (A^*)^c\}$$
$$\leq E^x\{T_A < \infty; X(T_A) \in A^c; P^{X(T_A)}[T_A > 0]\}. \qquad (6)$$

Applying the strong Markov property at T_A, we see that the right member of (6) is equal to the P^x-probability of the set

$$\{T_A < \infty; X(T_A) \in A^c; T_A \circ \theta_{T_A} > 0\}$$

where $T_A \circ \theta_{T_A}$ means the "time lapse between the first hitting time of A and the first hitting time of A thereafter." If $X_{T_A}(\omega) \notin A$ and $T_A \circ \theta_{T_A}(\omega) > 0$ then the sample function $X(\cdot, \omega)$ is not in A for $T \in [T_A(\omega), T_A(\omega) + T_A \circ \theta_{T_A}(\omega))$, a nonempty interval. This is impossible by the definition of T_A. Hence the right member of (6) must be equal to zero, proving the assertion of the theorem. $\square$

The following corollary for $\alpha = 0$ is just (21) of §2.4, which is true for T_B as well as D_B.

Corollary. *If A is a closed set, then $P_A^\alpha(x, \cdot)$ is concentrated in A for each $\alpha \geq 0$ and each x.*

The operator $P_A(x, \cdot)$ corresponds to what is known as "balayage" or "sweeping out" in potential theory. A unit charge placed at the point x is supposed to be swept onto the set A. The general notion is due to Poincaré, for a modern analytical definition see e.g. Helms [1]. Hunt was able to identify it with the definition given above, and that apparently convinced the potential theorists that "he's got something there"!

We interrupt the thrust of our preceding discussion by an obligatory extension of measurability. According to Theorem 5 of §2.3, for $Y \in b\mathscr{F}^\sim$ hence also for $Y \in \mathscr{F}_+^\sim$, the function $x \to E^x(Y)$ is universally measurable, namely in $\mathscr{E}^\sim$ defined in (18) of §2.3. In particular $P_A^\alpha 1$ is so measurable for $A \in \mathscr{E}$. We have therefore no choice but to enlarge the class of functions considered from $\mathscr{E}$ to $\mathscr{E}^\sim$. This could of course have been done from the outset but it is more convincing to introduce the extension when the need has arisen.

From here on, the transition probability measure $P_t(x, \cdot)$ is extended to $\mathscr{E}^\sim$. It is easy to see that for each $A \in \mathscr{E}^\sim$,

$$(t, x) \to P_t(x, A)$$

belongs to $\mathscr{B} \times \mathscr{E}^\sim$; alternately, $(t, x) \to P_t f(x)$ is in $\mathscr{B} \times \mathscr{E}^\sim$ for each $f \in b\mathscr{E}^\sim$ or $\mathscr{E}_+^\sim$. It follows that $U^\alpha f$ is in $\mathscr{E}^\sim$ for such an f and $\alpha \geq 0$. Finally, an α-excessive function is defined as before except that $f \in \mathscr{E}^\sim$, rather than $f \in \mathscr{E}$

as previously supposed. A universally measurable function is sandwiched between two Borel measurable functions (for each finite measure μ on $\mathscr{E}_\partial$) and this furnishes the key to its handling. See the proof of Theorem 5 of §2.3 for a typical example of this remark.

In what follows we fix the notation as follows: f and g are functions in $\mathscr{E}^\sim_+$; $x \in \mathbf{E}_\partial$; $\alpha \geq 0$; T is optional relative to $\{\mathscr{F}^\sim_t\}$; A is a nearly Borel set; K is a compact set; G is an open set. These symbols may appear with subscripts.

We begin with the formula:

$$P_T^\alpha U^\alpha f = E^\cdot \left\{ \int_T^\infty e^{-\alpha t} f(X_t)\, dt \right\}, \tag{7}$$

which is derived as follows. Making the substitution $t = T + u$, we transform the right member of (7) by the strong Markov property into

$$E^\cdot \left\{ e^{-\alpha T} \left[\int_0^\infty e^{-\alpha u} f(X_u)\, du \right] \circ \theta_T \right\} = E^\cdot \left\{ e^{-\alpha T} E^{X_T} \left[\int_0^\infty e^{-\alpha u} f(X_u)\, du \right] \right\}$$

$$= E^\cdot \{ e^{-\alpha T} U^\alpha f(X_T) \},$$

which is the left member of (7). It follows at once that

$$U^\alpha f = E^\cdot \left\{ \int_0^T e^{-\alpha t} f(X_t)\, dt \right\} + P_T^\alpha U^\alpha f. \tag{8}$$

On $\{T = \infty\}$ we have $P_T^\alpha U^\alpha f = 0$ by our convention. Note also that when T is a constant (8) has already been given in §2.1.

We use the notation $\lfloor f$ for the support of f, namely the set

$$\{ x \in \mathbf{E}_\partial \mid f(x) > 0 \}.$$

Theorem 3.

   (a) $P_A^\alpha U^\alpha f \leq U^\alpha f$;
   (b) $P_A^\alpha U^\alpha f = U^\alpha f$, if $\lfloor f \subset A$;
   (c) If we have

$$U^\alpha f \leq U^\alpha g \tag{9}$$

   on $\lfloor f$, then (9) is true everywhere.

Proof. Assertion (a) is obvious from (8) with $T = T_A$; so is (b) if we observe that $f(X_t) = 0$ for $t < T_A$ so that the first term on the right side of (8) vanishes. To prove (c) we need the approximation in Theorem 8(b) of §3.3. Let $A = \lfloor f$; given x, let $K_n \subset A$ such that $T_{K_n} \downarrow T_A$, P^x-a.s. We have then by (7), as $n \to \infty$:

$$P_{K_n}^\alpha U^\alpha f(x) = E^x \left\{ \int_{T_{K_n}}^\infty e^{-\alpha t} f(X_t)\, dt \right\} \uparrow E^x \left\{ \int_{T_A}^\infty e^{-\alpha t} f(X_t)\, dt \right\}$$

$$= P_A^\alpha U^\alpha f(x). \tag{10}$$

This is a fundamental limit relation which makes potentials easy to handle. By the Corollary to Theorem 2, $P_{K_n}(x, \cdot)$ is concentrated on K_n, hence $U^\alpha f \leq U^\alpha g$ on K_n by the hypothesis of (c), yielding

$$P^\alpha_{K_n} U^\alpha f(x) \leq P^\alpha_{K_n} U^\alpha g(x).$$

Letting $n \to \infty$, using (a) and (b), and (10) for both f and g, we obtain

$$U^\alpha f(x) = P^\alpha_A U^\alpha f(x) \leq P^\alpha_A U^\alpha g(x) \leq U^\alpha g(x). \qquad \square$$

Assertion (c) is known as the "domination principle" in potential theory, which will be amplified after Theorem 4. Note however it deals with the potentials of functions rather than potentials of measures, which are much more difficult to deal with.

We state the next few results for $\mathbf{S}$, but they are true for $\mathbf{S}^\alpha$, *mutatis mutandis*.

Theorem 4. *Let* $f \in \mathbf{S}$. *Then* $P_A f \in \mathbf{S}$ *and* $P_A f \leq f$. *If* $A_1 \subset A_2$, *then*

$$P_{A_1} f \leq P_{A_2} f. \tag{11}$$

For each x, *there exist compacts* $K_n \subset A$ *such that*

$$P_{K_n} f(x) \uparrow P_A f(x). \tag{12}$$

Proof. Let us begin by recording an essential property of a hitting time which is not shared by a general optional time:

$$\forall t \geq 0: T_A \leq t + T_A \circ \theta_t; \quad \text{and} \quad T_A = \lim_{t \downarrow 0} \downarrow (t + T_A \circ \theta_t). \tag{13}$$

This is verified exactly like (16) of §2.4, and the intuitive meaning is equally obvious. We have from (7):

$$P^\alpha_t P^\alpha_A U^\alpha g = E^\cdot \left\{ \int_{t + T_A \circ \theta_t}^\infty e^{-\alpha s} g(X_s) \, ds \right\}.$$

It follows from this and (13) that $P^\alpha_A U^\alpha g \in \mathbf{S}$. For $f \in \mathbf{S}$, we have by Theorem 9 of §3.2, $f = \lim_k \uparrow U^\alpha g_k$ where $\alpha > 0$. Hence $P^\alpha_A f = \lim_k \uparrow P^\alpha_A U^\alpha g_k$ and so $P^\alpha_A f \in \mathbf{S}^\alpha$ by Proposition 2 of §3.2. Finally $P_A f = \lim_{\alpha \downarrow 0} \uparrow P^\alpha_A f$; hence $P_A f \in \mathbf{S}$ by Proposition 4 of §3.2. Next if $A_1 \subset A_2$, then it follows at once from (7) that $P^\alpha_{A_1} U^\alpha g \leq P^\alpha_{A_2} U^\alpha g$. Now the same approximations just shown establish (11), which includes $P_A f \leq f$ as a particular case.

Next, given x and A, let $\{K_n\}$ be as in the preceding proof. As in (10), we have

$$P^\alpha_{K_n} U^\alpha g_k(x) \uparrow P^\alpha_A U^\alpha g_k(x).$$

Hence by Lemma 1 of §3.2,

$$P_A^\alpha f = \lim_k \uparrow P_A^\alpha U^\alpha g_k = \lim_k \uparrow \lim_n \uparrow P_{K_n}^\alpha U^\alpha g_k$$

$$= \lim_n \uparrow \lim_k \uparrow P_{K_n}^\alpha U^\alpha g_k(x) = \lim_n \uparrow P_{K_n}^\alpha f(x).$$

Letting $\alpha \downarrow 0$ and using Lemma 1 once again we obtain (12). $\qquad \square$

Corollary. Part (c) of Theorem 3 is true if $U^\alpha g$ is replaced by any excessive function f in (9).

The next result is crucial, though it will soon be absorbed into the bigger Theorem 6 below.

Theorem 5. *Let* $f \in S$, *and* $A \in \mathscr{E}^*$. *If* $x \in A^r$, *then*

$$\inf_A f \le f(x) \le \sup_A f.$$

Proof. Since $x \in A^r$, $f(x) = P_A f(x)$ as noted above. It follows from Theorem 8(b) of §3.3 that for each $\varepsilon > 0$, there exists $K \subset A$ such that

$$P^x\{T_K > \varepsilon\} = P^x\{T_K > T_A + \varepsilon\} \le \varepsilon \qquad (14)$$

since $P^x\{T_A = 0\} = 1$. At the same time, we may choose K so that

$$P_K f(x) \le P_A f(x) \le P_K f(x) + \varepsilon$$

by (12). Hence by the Corollary to Theorem 2, we have

$$P_K(x, K) \inf_{y \in K} f(y) \le P_K f(x) \le \sup_{y \in K} f(y).$$

Since $P_K(x, K) \ge P^x\{T_K \le \varepsilon\} \ge 1 - \varepsilon$ by (14), it follows from the inequalities above that

$$(1 - \varepsilon) \inf_{y \in A} f(y) \le f(x) \le \sup_{y \in A} f(y) + \varepsilon.$$

Since ε is arbitrary this is the assertion of the theorem. $\qquad \square$

We have arrived at the following key result. In its proof we shall be more circumspect than elsewhere about pertinent measurability questions.

Theorem 6. *Let* $f \in S$. *Then almost surely* $t \to f(X_t)$ *is right continuous on* $[0, \infty)$ *and has left limits (possibly* $+\infty$*) in* $(0, \infty]$. *Moreover,* $f \in \mathscr{E}^*$.

Proof. We begin by assuming that $f \in \mathbf{S} \cap \mathscr{E}^*$. Introduce the metric ρ on $[0, \infty]$ which is compatible with the extended Euclidean topology: $\rho(a, b) = |(a/1 + a) - (b/1 + b)|$, where $\infty/(1 + \infty) = 1$. For each $\varepsilon > 0$ define

$$S_\varepsilon(\omega) = \inf\{t > 0 \,|\, \rho(f(X_0(\omega)), f(X_t(\omega))) > \varepsilon\}.$$

An argument similar to that for the T in (2) of §3.1 shows that S_ε is optional (relative to $(\mathscr{F}_t)$). It is clear that we have

$$\bigcap_{k=1}^{\infty} \{S_{1/k} > 0\} \subset \left\{\omega \,\middle|\, \lim_{t \downarrow 0} f(X_t(\omega)) = f(X_0(\omega))\right\} = \Lambda, \tag{15}$$

say. For each x put $A = \{y \in \mathbf{E}_\partial \,|\, \rho(f(x), f(y)) > \varepsilon\}$. Then $A \in \mathscr{E}^*$; and S_ε reduces to the hitting time T_A under P^x. It is a consequence of Theorem 5 that $x \notin A^r$; for obviously the value $f(x)$ does not lie between the bounds of f on A. Hence for each x, $P^x\{S_\varepsilon > 0\} = 1$ by Blumenthal's zero-one law. Now $\{S_\varepsilon > 0\} \in \mathscr{F}$, hence $P^\cdot\{S_\varepsilon > 0\} \in \mathscr{E}^\sim$ by Theorem 5 of §2.3, and this implies $P^\mu\{S_\varepsilon > 0\} = 1$ for each probability measure μ on $\mathscr{E}_\partial$. It now follows from (15) that $\Lambda \in \mathscr{F}^\mu$ since $(\mathscr{F}^\mu, P^\mu)$ is complete. This being true for each μ, we have proved that $\Lambda \in \mathscr{F}$ and Λ is an almost sure set. For later need let us remark that for each optional S, $\Lambda \circ \theta_S \in \mathscr{F}$ by Theorem 5 of §2.3.

Our next step is to define inductively a family of optional times T_α such that the sample function $t \to f(X_t)$ oscillates no more than 2ε in each interval $[T_\alpha, T_{\alpha+1})$. This is analogous to our procedure in the proof of Theorem 1 of §3.1, but it will be necessary to use transfinite induction here. The reader will find it instructive to compare the two proofs and scrutinize the difference. Let us denote by $\mathbb{O}$ the well-ordered set of ordinal numbers before the first uncountable ordinal (see e.g., Kelley [1; p. 29]). Fix ε, put $T_0 \equiv 0$, $T_1 \equiv S_\varepsilon$, and for each $\alpha \in \mathbb{O}$, define T_α as follows:

(i)   if α has the immediate predecessor $\alpha - 1$, put

$$T_\alpha = T_{\alpha-1} + T_1 \circ \theta_{T_{\alpha-1}};$$

(ii) if α is a limit ordinal without an immediate predecessor, put

$$T_\alpha = \sup_{\beta < \alpha} T_\beta \quad \text{where } \beta \text{ ranges over all ordinals} < \alpha.$$

Then each T_α is optional by induction. [Note that the sup in case (ii) may be replaced by the sup of a countable sequence.] Now for each $\alpha \in \mathbb{O}$, we have for each μ:

$$P^\mu\left\{T_\alpha < \infty; \lim_{t \downarrow 0} f(X_{T_\alpha + t}) = f(X_{T_\alpha})\right\}$$

$$= E^\mu\left\{T_\alpha < \infty; P^{X(T_\alpha)}\left[\lim_{t \downarrow 0} f(X_t) = f(X_0)\right]\right\} = P^\mu\{T_\alpha < \infty\} \tag{16}$$

by the strong Markov property and the fact that $P^x(\Lambda) = 1$ for each x, proved above. It follows that P^μ-a.e. on the set $\{T_\alpha < \infty\}$, the limit relation in the first member of (16) holds; hence $T_1 \circ \theta_{T_\alpha} > 0$ by the definition of T_1, hence $T_{\alpha+1} > T_\alpha$. Consider now the numbers

$$b_\alpha = P^\mu\{T_\alpha < \infty\}, \qquad c_\alpha = E^\mu\{e^{-T_\alpha}\}, \qquad \alpha \in \mathbb{O}.$$

We have just shown that if $b_\alpha > 0$, then

$$c_{\alpha+1} = E^\mu\{e^{-T_{\alpha+1}}; T_{\alpha+1} < \infty\} < E^\mu\{e^{-T_\alpha}; T_\alpha < \infty\} = c_\alpha.$$

Now the set $\mathbb{O}$ has uncountable cardinality. If $b_\alpha > 0$ for all $\alpha \in \mathbb{O}$, then the uncountably many intervals $(c_{\alpha+1}, c_\alpha)$, $\alpha \in \mathbb{O}$, would be all nonempty and disjoint, which is impossible because each must contain a rational number. Therefore there exists $\alpha^* \in \mathbb{O}$ for which $b_{\alpha^*} = 0$; namely that

$$P^\mu\{T_{\alpha^*} = \infty\} = 1.$$

For each $\alpha < \alpha^*$, if s and t both belong to $[T_\alpha, T_{\alpha+1})$, then $\rho(f(X_s), f(X_t)) \leq 2\varepsilon$. This means: there exists Ω_ε^μ with $P^\mu(\Omega_\varepsilon) = 1$ such that if $\omega \in \Omega_\varepsilon^\mu$, then the sample function $t \to f(X(t, \omega))$ does not oscillate more than 2ε *to the right of any* $t \in [0, \infty)$. It follows that the set Ω_ε^* of ω satisfying the latter requirement belongs to $\mathscr{F}$ and is an almost sure set. Since $\bigcap_{k=1}^\infty \Omega_{1/k}^*$ is contained in the set of ω for which $t \to f(X_t(\omega))$ is right continuous in $[0, \infty)$, we have proved that the last-mentioned set belongs to $\mathscr{F}$ and is an almost sure set. This is the main assertion of the theorem.

   In order to fully appreciate the preceding proof, it is necessary to reflect that no conclusion can be drawn as to the existence of left limits of the sample function in $(0, \infty)$. The latter result will now follows from an earlier theorem on supermartingales. In fact, if we write $f_n = f \wedge n$ then $\{f_n(X_t), \mathscr{F}_t\}$ is a bounded supermartingale under each P^x, by Proposition 1 of §2.1, since f_n is superaveraging. Almost every sample functions $t \to f_n(X_t)$ is right continuous by what has just been proved, hence it has left limits in $(0, \infty]$ by Corollaries 1 and 2 to Theorem 1 of §1.4. Since n is arbitrary, it follows (why?) that $t \to f(X_t)$ has left limits (possibly $+\infty$) as asserted.

   It remains to prove that $f \in \mathbf{S}$ implies $f \in \mathscr{E}^\cdot$. In view of Theorem 9 of §3.2, it is sufficient to prove this for $U^\alpha g$ where $\alpha > 0$ and $g \in b\mathscr{E}_+^\sim$. For each μ, there exist g_1 and g_2 in $\mathscr{E}_\partial$ such that $g_1 \leq g \leq g_2$ and $\mu U^\alpha(g_2 - g_1) = 0$. For each $t \geq 0$, we have

$$P_t U^\alpha(g_2 - g_1) = e^{\alpha t} P_t^\alpha U^\alpha(g_2 - g_1) \leq e^{\alpha t} U^\alpha(g_2 - g_1).$$

It follows that

$$E^\mu\{U^\alpha(g_2 - g_1)(X_t)\} = \mu P_t U^\alpha(g_2 - g_1) \leq e^{\alpha t} \mu U^\alpha(g_2 - g_1) = 0.$$

Thus we have for each $t \geq 0$:

$$U^\alpha(g_2 - g_1)(X_t) = 0, \qquad P^\mu\text{-a.s.} \tag{17}$$

Therefore (17) is also true simultaneously for all rational $t \geq 0$. But $U^\alpha(g_2 - g_1) \in \mathbf{S}^\alpha \cap \mathscr{E}$, hence the first part of our proof extended to $\mathbf{S}^\alpha$ shows that the function of t appearing in (17) is right continuous, hence must vanish identically in t, P^μ-a.s. This means $U^\alpha g \in \mathscr{E}^*$ as asserted. Theorem 6 is completely proved.   □

In the proof of Theorem 6, we have spelled out the $\mathscr{F}$-measurability of "the set of all ω such that $f(X(\cdot, \omega))$ is right continuous". Such questions can be handled by a general theory based on $\mathbf{T} \times \Omega$-analytic sets and projection, as described in §1.5; see Dellacherie and Meyer [1], p. 165.

If $f \in \mathbf{S}$, when can $f(X_t) = \infty$? The answer is given below. We introduce the notation

$$\forall t \in (0, \infty]: f(X_t)_- = \lim_{s \uparrow\uparrow t} f(X_s). \tag{18}$$

This limit exists by Theorem 6, and is to be carefully distinguished from $f(X_{t-})$. Note also that in a similar notation $f(X_t)_+ = f(X_t) = f(X_{t+})$.

Theorem 7. *Let $F = \{f < \infty\}$. Then we have almost surely:*

$$f(X_t)_- < \infty \quad \text{and} \quad f(X_t) < \infty \qquad \text{for all } t > D_F, \tag{19}$$

where D_F is the first entrance time into F.

Proof. If $x \in F$, then under P^x, $\{f(X_t), \mathscr{F}_t\}$ is a right continuous positive supermartingale by Theorem 6. Hence by Corollary 1 to Theorem 1 of §1.4, $f(X_t)$ is bounded in each finite t-interval, and by Corollary 2 to the same theorem, converges to a finite limit as $t \to \infty$. Therefore (19) is true with $D_F = 0$. In general, let $F_n = \{f \leq n\}$, then $F_n \in \mathscr{E}^*$ by Theorem 6 and so D_{F_n} is optional; also $f(X(D_{F_n})) \leq n$ by right continuity. It follows by the strong Markov property that (19) is true when the D_F there is replaced by D_{F_n}. Now an easy inspection shows that $\lim_n \downarrow D_{F_n} = D_F$, hence (19) holds as written.   □

It is obvious that on the set $\{D_F > 0\}$, $f(X_t) = \infty$ for $t < D_F$, and $f(X_t)_- = \infty$ for $t \leq D_F$; but $f(X(D_F))$ may be finite or infinite.

There are numerous consequences of Theorem 6. One of the most important is the following basic property of excessive functions which can now be easily established.

Theorem 8. *The class of excessive functions is closed under the minimum operation.*

Proof. Let $f_1 \in \mathbf{S}$, $f_2 \in \mathbf{S}$, then we know $f_1 \wedge f_2$ is superaveraging. On the other hand,

$$\lim_{t \downarrow 0} P_t(f_1 \wedge f_2)(x) = \lim_{t \downarrow 0} E^x\{f_1(X_t) \wedge f_2(X_t)\}$$

$$\geq E^x\left\{\lim_{t \downarrow 0} [f_1(X_t) \wedge f_2(X_t)]\right\}$$

$$= E^x\{f_1(X_0) \wedge f_2(X_0)\} = (f_1 \wedge f_2)(x);$$

by Fatou's lemma followed by the right continuity of $f_i(X_t)$ at $t = 0$, $i = 1, 2$. Hence there is equality above and the theorem is proved. $\qquad\square$

In particular, if $f \in \mathbf{S}$ then $f \wedge n \in \mathbf{S}$ for any n. This truncation allows us to treat bounded excessive functions before a passage to the limit for the general case, which often simplifies the argument.

As another application of Theorem 6, let us note that $P_A^\alpha 1$ is α-excessive for each $\alpha \geq 0$ and each $A \in \mathscr{E}^*$, hence $P_A^\alpha 1 \in \mathscr{E}^*$. Since $A^r = \{x \mid P_A^\alpha 1 = 1\}$ as remarked in (5), we have $A^r \in \mathscr{E}^*$, $A^* \in \mathscr{E}^*$. The significance of these sets will be studied in the next section.

Exercises

1. Let $f \in b\mathscr{E}_+$, $T_n \uparrow T$. Prove that for a Hunt process

$$\lim_{n \to \infty} U^\alpha f(X(T_n)) = E\left\{U^\alpha f(X(T)) \Big| \bigvee_n \mathscr{F}_{T_n}\right\}.$$

Hence if we assume that

$$X(T) \in \bigvee_n \mathscr{F}_{T_n}, \tag{20}$$

then we have

$$\lim_n U^\alpha f(X(T_n)) = U^\alpha f(X(T)).$$

Remark. Under the hypothesis that (20) holds for any optional $T_n \uparrow T$, the function $t \to U^\alpha f(X(t))$ is left continuous. This is a complement to Theorem 6 due to P. A. Meyer: the proof requires projection onto the predictable field, see Chung [5].

2. Let $A \in \mathscr{E}^*$ and suppose $T_n \uparrow T_A$. Then we have for each $\alpha > 0$, a.s. on the set $\bigcap_n \{T_n < T_A < \infty\}$:

$$\lim_{n \to \infty} E^{X(T_n)}\{e^{-\alpha T_A}\} = 1.$$

[Hint: consider $E^x\{e^{-\alpha T_A}; T_n < T_A | \mathscr{F}_{T_n}\}$ and observe that $T_A \in \bigvee_n \mathscr{F}_{T_n}$.]

3. Let x_0 be a holding but not absorbing point (see Problem 3 of §3.1). Define

$$T = \inf\{t > 0 : X(t) \neq x_0\}.$$

For an increasing sequence of optional T_n such that $T_n \leq T$ and $T_n \uparrow T$ we have

$$P^{x_0}\left\{\bigcap_{n=1}^{\infty} [T_n < T]\right\} = 0.$$

Indeed, it follows that for any optional S, $P^{x_0}\{0 < S < T\} = 0$. [Hint: the second assertion follows from the first, by contraposition and transfinite induction, as suggested by J. B. Walsh.]

4. Let $f \in \mathscr{E}_+$ and $f \geq P_K f$ for every compact K. Let $g \in \mathscr{E}$ such that $U(g^+) \wedge U(g^-) < \infty$ (namely $U(g)$ is defined). Then

$$f \geq Ug \quad \text{on } \lfloor g^+$$

implies $f \geq Ug$ (everywhere). [Hint: $f + U(g^-) \geq U(g^+)$; for each $K \subset \lfloor g^+$ we have $f + U(g^-) \geq P_K(f + U(g^-)) \geq P_K U(g^+)$, now approximate $\overline{\lfloor g^+}$ by compacts.]

5. Suppose U is transient (see (17) of §3.1). For the f in Problem 4 we have $f \geq \alpha U^{\alpha} f$ for every $\alpha > 0$. [Hint: we may suppose f bounded with compact support; put $g = f - \alpha U^{\alpha} f$, then $f \geq U(\alpha g) = U^{\alpha}(\alpha f)$ on $\{g \geq 0\}$. This is a true *leger-demain* due to E. B. Dynkin [1].]

6. Prove Theorem 6 for a Feller process by using Theorem 5 of §1.4. [Hint: show first $U^{\alpha} f(X_t)$ is right continuous for $f \in b\mathscr{E}_+$, which is trivially true for $f \in \cdot C_0$.]

7. Let f be excessive and $A = \{f = \infty\}$. If $f(x_0) < \infty$, then $P_A 1(x_0) = 0$.

8. The optional random variable T is called *terminal* iff for every $t \geq 0$ we have

$$T = t + T \circ \theta_t \quad \text{on } \{T > t\}$$

and

$$T = \lim_{t \downarrow 0} (t + T \circ \theta_t).$$

Thus the notion generalizes that of a hitting time. Prove that if T is terminal, then $x \to P^x\{T < \infty\}$ is excessive; and $x \to E^x\{e^{-\alpha T}\}$ is α-excessive for each $\alpha > 0$.

3.5. Fine Properties

Theorem 5 of §3.4 may be stated as follows: if $f \in S$, then the bounds of f on the fine closure of A are the same as those on A itself. Now if f is continuous in any topology, then this is the case when the said closure is taken in that topology. We proceed to define such a topology.

Definition. An arbitrary set A is called *finely open* iff for each $x \in A$, there exists $B \in \mathscr{E}$ such that $x \in B \subset A$, and

$$P^x\{T_{B^c} > 0\} = 1. \tag{1}$$

Namely, almost every path starting at a point in A will remain in a nearly Borel subset of A for a nonempty initial interval of time. Clearly an open set is finely open. It is immediately verified that the class of finely open sets meets the two requirements below to generate a topology:

(a) the union of any collection of finely open sets is finely open;
(b) the intersection of any finite collection of finely open sets is finely open.

We shall denote this class by **O**. Observe that if we had required a finely open set to belong to $\mathscr{E}^*$, condition (b) would fail. On the other hand, there is a fine base consisting of set in $\mathscr{E}^*$; see Exercise 5 below.

Since ∂ is absorbing, the singleton $\{\partial\}$ forms a finely open set so that ∂ is an isolated point in the fine topology. If $\varphi(x) = E^x\{e^{-\alpha\zeta}\}$, where $\zeta = T_{\{\partial\}}$, then $\varphi(\partial) = 1$. We must therefore *abandon the convention made earlier that a function defined on $\mathbf{E}_\partial$ vanishes at ∂*. There may well be other absorbing points besides ∂. A point x is called holding iff x is not regular for the set $\mathbf{E}_\partial - \{x\}$; in other words iff P^x-almost every path starting at x will remain at x for a nonempty initial period of time. This is the case if and only if x is isolated in the fine topology, but it need not be absorbing. A *stable* state in a Markov chain is such an example; indeed under a mild condition all states are stable (see Chung [2]).

A set is finely closed iff its complement in $\mathbf{E}_\partial$ is finely open. It is immediate that if $B \in \mathscr{E}^*$ then B is finely closed iff $B^r \subset B$, which is in turn equivalent to $B^* = B$. Thus the operation "*" is indeed the closure operation in the fine topology, justifying the term "fine closure".

The next result yields a vital link between two kinds of continuity.

Theorem 1. *Let $f \in \mathscr{E}^*$. Then f is finely continuous if and only if $t \to f(X_t)$ is right continuous on $[0, \infty)$, almost surely.*

Proof. Assume the right continuity. For any real constant c, put

$$A = \{f < c\}, \qquad A' = \{f > c\}.$$

For each $x \in A$, under P^x we have $f(X_0) = f(x) < c$, hence $P^x\{\overline{\lim}_{t \downarrow 0} f(X_t) < c\} = 1$ by right continuity at $t = 0$. This implies (1) with $B = A$. Hence A is finely open; similarly A' is finely open. Consequently $f^{-1}(U)$ is finely open, first for $U = (c_1, c_2)$ where $c_1 < c_2$, and then for each open set U of $(-\infty, \infty)$. Therefore f is finely continuous, namely continuous in the fine topology, by a general characterization of continuity.

Conversely, let $f \in \mathscr{E}^*$ and f be finely continuous. For each $q \in Q$, put

$$A = \{f > q\}, \qquad \varphi_q(x) = E^x\{e^{-T_A}\}.$$

Then A is finely open; also $A \in \mathscr{E}^*$ implies that T_A is optional, and φ_q is 1-excessive. It follows from Theorem 6 of §3.4 that there exists Ω_* with $P^x(\Omega_*) = 1$ such that if $\omega \in \Omega_*$ then

$$\forall q \in Q : t \to \varphi_q(X_t) \quad \text{is right continuous.} \tag{2}$$

We claim that for such an ω, $\overline{\lim}_{s \downarrow\downarrow t} f(X(s, \omega)) \le f(X(t, \omega))$ for all $t \ge 0$. Otherwise there exist $t \ge 0$, $t_n \downarrow\downarrow t$, and a rational q such that

$$f(X(t, \omega)) < q, \qquad f(X(t_n, \omega)) > q.$$

Since $X(t_n, \omega) \in A$ and A is finely open, the point $X(t_n, \omega)$ is certainly regular for A, hence $\varphi_q(X(t_n, \omega)) = 1$ for all n. Therefore we have $\varphi_q(X(t, \omega)) = 1$ by (2). But $B = \{f < q\}$ is also finely open and $X(t, \omega) \in B$, hence by definition the point $X(t, \omega)$ is not regular for B^c, a fortiori not regular for A since $B^c \supset A$. Thus $\varphi_q(X(t, \omega)) < 1$. This contradiction proves the claim. A similar argument shows that $\lim_{s \downarrow\downarrow t} f(X(s, \omega)) \ge f(X(t, \omega))$ for all $t \ge 0$, P^x-a.s. Hence $t \to f(X(t, \omega))$ is right continuous in $[0, \infty)$, a.s. $\square$

Corollary. *If $f \in \mathbf{S}^\alpha$, then f is finely continuous.*

Theorem 2. *If A_1 and A_2 are in $\mathscr{E}^*$, then*

$$(A_1 \cup A_2)^r = A_1^r \cup A_2^r. \tag{3}$$

For each $A \in \mathscr{E}^*$, we have

$$(A^r)^r \subset A^r; \tag{4}$$

$$T_{A^r} \ge T_A \quad \text{almost surely.} \tag{5}$$

Proof. (3) is easy. To prove (4) let $f = \mathscr{E}^*\{e^{-T_A}\}$. Then $f(x) = 1$ if and only if $x \in A^r$. If $x \in (A^r)^r$ then $f(x) \ge \inf_{A^r} f$ by theorem 5 of §3.4. Hence $f(x) = 1$. To prove (5), note first that $A^r \in \mathscr{E}^*$ so that T_{A^r} is optional. We have by Theorem 2 of §3.4, almost surely:

$$X(T_{A^r}) \in (A^r)^* = A^r \cup (A^r)^r = A^r$$

where the last equation is by (4). Hence $T_A \circ \theta_{T_{A^r}} = 0$ by the strong Markov property applied at T_{A^r}, and this says that T_A cannot be strictly greater than T_{A^r}. $\square$

We shall consider the amount of time that the sample paths of a Hunt process spend in a set. Define for $A \subset \mathbf{E}_\partial$:

$$J_A(\omega) = \{t \geq 0 \mid X(t, \omega) \in A\}. \tag{6}$$

Thus for each $t \geq 0$, "$t \in J_A(\omega)$" is equivalent to "$X(t, \omega) \in A$". We leave it as an exercise to show that if $A \in \mathscr{E}^{\,*}$, then $J_A(\omega) \in \mathscr{B}$, a.s. Thus if m denotes the Borel-Lebesgue measure on $\mathbf{T}$, then $m(J_A(\omega))$ is the total amount of time that the sample function $X(\cdot, \omega)$ spends in A, called sometimes the "occupation time" of A. It follows that

$$m(J_A(\omega)) = \int_0^\infty 1_A(X(t, \omega)) \, dt; \tag{7}$$

and consequently by Fubini's theorem for each x:

$$E^x\{m(J_A)\} = \int_0^\infty P_t(x, A) \, dt = U(x, A). \tag{8}$$

Thus the potential of A is the expected occupation time.

Definition. A set A in $\mathscr{E}^{\,*}$ is said to be of zero potential iff $U(\cdot, A) \equiv 0$.

Proposition 3. *If* $U^\alpha(\cdot, A) \equiv 0$ *for some* $\alpha \geq 0$, *then* $U^\alpha(\cdot, A) \equiv 0$ *for all* $\alpha \geq 0$.

Proof. We have by the resolvent equation, for any β:

$$U^\beta(x, A) = U^\alpha(x, A) + (\alpha - \beta)U^\beta U^\alpha 1_A(x).$$

If $U^\alpha(\cdot, A) \equiv U^\alpha 1_A(\cdot) \equiv 0$, then $U^\beta U^\alpha 1_A(\cdot) \equiv 0$, hence $U^\beta(\cdot, A) \equiv 0$. $\square$

It follows from (8) that if A is of zero potential, then $m(J_A) = 0$ almost surely; namely almost every path spends "almost no time" (in the sense of Borel-Lebesgue measure) in A.

Proposition 4. *If* A *is of zero potential, then* $A^c = \mathbf{E}_\partial - A$ *is finely dense.*

Proof. A set is dense in any topology iff its closure in that topology is the whole space; equivalently iff each nonempty open set in that topology contains at least one point of the set. Let O be a finely open set and $x \in O$. Then almost every path starting at x spends a nonempty initial interval of time in a nearly Borel subset B of O, hence we have $E^x\{m(J_B)\} > 0$. Since $E^x\{m(J_A)\} = 0$ we have $E^x\{m(J_{B \cap A^c})\} > 0$. Thus $O \cap A^c$ is not empty for each finely open O, and so A^c is finely dense. $\square$

The following corollary is an illustration of the utility of fine concepts. It can be deduced from (11) of §3.2, but the language is more pleasant here.

Corollary. *If two excessive functions agree except on a set of zero potential, then they are identical.*

For they are both finely continuous and agree on a finely dense set. The assertion is therefore reduced to a familiar topological one.

From certain points of view a set of zero potential is not so "small" and scarcely "negligible". Of course, what is negligible depends on the context. For example a set of Lebesgue measure zero can be ignored in integration but not in differentiation or in questions of continuity. We are going to define certain small sets which play great roles in potential theory.

Definition. A set A in $\mathscr{E}^*$ is called *thin* iff $A^r = \varnothing$; namely iff it is thin at every point of $\mathbf{E}_\partial$. A set is *semipolar* iff it is the union of countably many thin sets. A set A in $\mathscr{E}^*$ is called *polar* iff

$$\forall x \in \mathbf{E}_\partial: \ P^x\{T_A < \infty\} = 0. \tag{9}$$

The last condition is equivalent to: $E^x\{e^{-\alpha T_A}\} \equiv P_A^\alpha 1(x) \equiv 0$ for each $\alpha > 0$. For comparison, A is thin if and only if

$$\forall x \in E_\partial: P^x\{T_A = 0\} = 0, \quad \text{or} \quad E^x\{e^{-\alpha T_A}\} < 1. \tag{10}$$

It is possible to extend the preceding definitions to all subsets of $\mathbf{E}_\partial$, without requiring them to be nearly Borel. For instance, a set is polar iff it is contained in a nearly Borel set which satisfies (9). We shall not consider such an extension.

The case of uniform motion on R^1 yields facile examples. Each singleton $\{x_0\}$ is thin but not polar. Next, let $x_n \downdownarrows x_\infty > -\infty$, then the set $A = \bigcup_{n=1}^\infty \{x_n\}$ is semipolar but not thin. For a later discussion we note also that each compact subset of A is thin.

From the point of view of the process, the smallness of a set A should be reflected in the rarity of the incidence set J_A. We proceed to study the relations between the space sets and time sets. For simplicity we write φ_A for $P_A^\alpha 1$ below for a fixed α which may be taken as 1. A set A will be called *very thin* iff $A \in \mathscr{E}^*$ and

$$\sup_{x \in A} \varphi_A(x) < 1. \tag{11}$$

Note that (11) does not imply that $\sup_{x \in \mathbf{E}} \varphi_A(x) < 1$. An example is furnished by any singleton $\{x_0\}$ in the uniform motion, since $\varphi_{\{x_0\}}(x_0) = 0$ and $\lim_{x \uparrow\uparrow x_0} \varphi_{\{x_0\}}(x) = 1$.

Proposition 5. *If A is very thin, then it is thin. Furthermore almost surely J_A is a discrete set (namely, finite in each finite time interval).*

Proof. Recall that φ_A is α-excessive; hence by Theorem 5 of §3.4, the sup in (11) is the same if it is taken over A^* instead of A. But if $x \in A^r$, then $\varphi_A(x) = 1$. Hence under (11) $A^r = \varnothing$ and A is thin.

Next, denote the sup in (11) by θ so that $\theta < 1$. Define $T_1 \equiv T_A$ and for $n \geq 1$: $T_{n+1} = T_n + T_A \circ \theta_{T_n}$. Thus $\{T_n, n \geq 1\}$ are the successive hitting times of A. These are actually the successive entrance times into A, because $X(T_n) \in A^* = A$ here. The adjective "successive" is justified as follows. On $\{T_n < \infty\}$, $X(T_n)$ is a point which is not regular for A; hence by the strong Markov property we must have $T_A \circ \theta_{T_n} > 0$, namely $T_{n+1} > T_n$. Now we have, taking $\alpha = 1$ in φ_A:

$$E\{e^{-T_{n+1}}\} = E\{e^{-T_n}E^{X(T_n)}[e^{-T_A}]; T_n < \infty\} \leq \theta E\{e^{-T_n}\}.$$

It follows that $E\{e^{-T_n}\} \leq \theta^n \downarrow 0$ as $n \to \infty$. Let $T_n \uparrow T_\infty$. Then $E\{e^{-T_c}\} = 0$; consequently $P\{T_\infty = \infty\} = 1$. Therefore, we have almost surely

$$J_A = \bigcup_{n=1}^{\infty} \{T_n\} 1_{\{T_n < \infty\}}; \tag{12}$$

Since $T_n \uparrow \infty$, this is a discrete set in $[0, \infty)$. $\qquad\square$

Theorem 6. *Each semipolar set is the union of countably many very thin sets. For each $A \in \mathscr{E}^*$, $A \backslash A^r$ is semipolar.*

Proof. For any $A \in \mathscr{E}^*$, we define for $n \geq 1$:

$$A^{(n)} = \left\{ x \in \mathbf{E}_\partial \,\middle|\, \varphi_A(x) \leq 1 - \frac{1}{n} \right\}. \tag{13}$$

Then we have

$$A = (A \cap A^r) \cup \left(\bigcup_{n=1}^{\infty} A \cap A^{(n)} \right). \tag{14}$$

For each n, it is obvious that

$$\sup_{x \in A \cap A^{(n)}} \varphi_{A \cap A^{(n)}}(x) \leq \sup_{x \in A^{(n)}} \varphi_A(x) \leq 1 - \frac{1}{n}.$$

Hence each $A \cap A^{(n)}$ is a very thin set. If A is thin, then (14) exhibits A as a countable union of very thin sets. Therefore each semipolar set by its definition is also such a union. Finally, (14) also shows that $A - (A \cap A^r)$ is such a union, hence semipolar. $\qquad\square$

Corollary. *If A is semipolar then there is a countable collection of optional times $\{T_n\}$ such that (12) holds, where each $T_n > 0$, a.s.*

This follows at once from the first assertion of the theorem and (12). Note however that the collection $\{T_n\}$ is not necessarily well-ordered (by the increasing order) as in the case of a very thin set A.

We summarize the hierarchy of small sets as follows.

Proposition 7. *A polar set is very thin; a very thin set is thin; a thin set is semipolar; a semipolar set is of zero potential.*

Proof. The first three implications are immediate from the definitions. If A is semipolar, then J_A is countable and so $m(J_A) = 0$ almost surely by Theorem 6. Hence $U(\cdot, A) \equiv 0$ by (8). This proves the last implication.   □

For Brownian motion in R^1, each Borel set of measure zero is of zero potential, and vice versa. Each point x is regular for $\{x\}$, hence the only thin or semipolar set is the empty set. In particular, it is trivial that semipolar and polar are equivalent notions in this case. The last statement is also true for Brownian motion in any dimension, but it is a deep result tantamount to Kellogg-Evans's theorem in classical potential theory, see §4.5 and §5.2.

Let us ponder a little more on a thin set. Such a set is finely separated in the sense that each of its points has a fine neighborhood containing no other point of the set. If the fine topology has a countable base (namely, satifies the second axiom of countability), such a set is necessarily countable. In general the fine topology does not even satisfy the first axiom of countability, so a finely separated set may not be so sparse. Nevertheless Theorem 6 asserts that almost every path meets the set only countably often, and so only on a countable subset (depending on the path). The question arises if the converse is also true. Namely, if A is a nearly Borel set such that almost every path meets it only countably often, or more generally only on a countable subset, must A be semipolar? Another related question is: if every compact subset of A is semipolar, must A be semipolar? [Observe that if every compact subset of A is polar, then A is polar by Theorem 8 of §3.3.]

These questions are deep but have been answered in the affirmative by Dellacherie under an additional hypothesis which is widely used, as follows.

Hypothesis (L). *There exists a measure ξ_0 on $\mathscr{E}^*$, which is the sum of countably many finite measures, such that if $\xi_0(A) = 0$ then A is of zero potential. In other words, $U(x, \cdot) \ll \xi_0$ for all x.*

It follows easily that if we put $\xi = \xi_0 U^\alpha$ for any $\alpha > 0$, then A is of potential zero *if and only if* $\xi(A) = 0$. Hence under Hypothesis (L) such a measure exists. It will be called a *reference measure* and denoted by ξ below. For example, for Brownian motion in any dimension, the corresponding Borel-Lebesgue measure is a reference measure. It is trivial that there exists a probability measure which is equivalent to ξ. We can then use P^ξ with ξ as the initial distribution.

We now state without proof one of Dellacherie's results; see Dellacherie [1].

Theorem 8. *Assume Hypothesis (L). Let $A \in \mathcal{E}^{*}$ and suppose that almost surely the set J_A in (6) is countable. Then A is semipolar.*

The expediency of Hypothesis (L) will now be illustrated.

Proposition 9. *If f_1 and f_2 are two excessive functions such that $f_1 \leq f_2$ ξ-a.e., then $f_1 \leq f_2$ everywhere. In particular the result is true with equalities replacing the inequalities. If f is excessive and $\int_{\mathbf{E}_\partial} f \, d\xi = 0$, then $f \equiv 0$.*

This is just the Corollary to Proposition 4 with a facile extension.

Proposition 10. *Let $A \in \mathcal{E}^{*}$. Under (L) there exists a sequence of compact subsets K_n of A such that $K_n \uparrow$ and for each x:*

$$P^x \left\{ \lim_n T_{K_n} = T_A \right\} = 1. \tag{15}$$

Proof. This is an improvement of Theorem 8 of §3.3 in that the sequence $\{K_n\}$ does not depend on x. Write $\xi(f)$ for $\int_{\mathbf{E}_\partial} f \, d\xi$ below; and put for a fixed $\alpha > 0$:

$$c = \sup_{K \subset A} \xi(P_K^\alpha 1).$$

There exists $K_n \subset A$, $K_n\uparrow$ such that

$$c = \lim_n \xi(P_{K_n}^\alpha 1).$$

Let $f = \lim_n \uparrow P_{K_n}^\alpha 1$; then $c = \xi(f)$ by monotone convergence. For any compact $K \subset A$, let $g = \lim_n \uparrow P_{K_n \cup K}^\alpha 1$. Both f and g are α-excessive and $g \geq f$ by Theorem 4 of §3.4 extended to $\mathbf{S}^\alpha$. But $\xi(g) \leq c = \xi(f)$, hence $g = f$ by Proposition 9 (extended to $\mathbf{S}^\alpha$). Thus $P_K^\alpha 1 \leq f$ for each compact $K \subset A$, and together with the definition of f we conclude that

$$\forall x: f(x) = \sup_{K \subset A} P_K^\alpha 1(x) \leq P_A^\alpha 1(x). \tag{16}$$

On the other hand, for each x it follows from Theorem 4 of §3.4 that there exists a sequence of compact subsets $L_n(x)$ such that

$$\lim_n P_{L_n}^\alpha(x) \, 1(x) = P_A^\alpha 1(x).$$

Therefore (16) also holds with the inequality reversed, hence it holds with the inequality strengthened to an equality. Recalling the definition of f we have proved that $\lim_n P^\alpha_{K_n} 1 = P^\alpha_A 1$. Let $T_{K_n} \downarrow S \geq T_A$. We have $E^x\{e^{-\alpha S}\} = E^x\{e^{-\alpha T_A}\}$, hence $P^x\{S = T_A\} = 1$ which is (15).   □

The next curious result is a lemma in the proof of Dellacherie's Theorem 8.

Proposition 11. *Let A be a semipolar set. There exists a finite measure v on $\mathscr{E}^*$ such that if $B \in \mathscr{E}^*$ and $B \subset A$, then B is polar if and only if $v(B) = 0$.*

Proof. Using the Corollary to Theorem 6, we define v as follows:

$$v(B) = \sum_{n=1}^{\infty} \frac{1}{2^n} P^\xi\{X(T_n) \in B; T_n < \infty\}. \tag{17}$$

If $B \in \mathscr{E}^*$ and $v(B) = 0$, then

$$P^\xi\left\{\bigcup_{n=1}^{\infty} \{X(T_n) \in B; T_n < \infty\}\right\} = 0.$$

If $B \subset A$, then $J_B \subset J_A = \bigcup_n \{T_n\} 1_{\{T_n < \infty\}}$ by the definition of $\{T_n\}$, hence the above implies that

$$P^\xi\{T_B < \infty\} = P^\xi\{J_B \neq \varnothing\} = 0.$$

Hence B is polar by Proposition 9 applied to $f = P_B 1$. The converse is trivial.   □

The following analogous result, without Hypothesis (L), is of interest.

Proposition 12. *Let A be a semipolar set. Then for each x, there exists $F = \bigcup_{n=1}^{\infty} K_n$ where the K_n's are compact subsets of A, such that P^x-a.s. we have $T_{A-F} = \infty$.*

Proof. Replace ξ by x in (17) and call the resulting measure v^x. Since this is a finite measure on a locally compact separable metric space, it is regular. Hence

$$v^x(A) = \sup_{K \subset A} v^x(K)$$

where K ranges over compact sets. Let K_n be a sequence of compact subsets of A such that $v^x(K_n)$ increases to $v^x(A)$. Then its union F has the property that $v^x(A - F) = 0$. The argument in the proof of Proposition 11 then shows that J_{A-F} is empty P^x-almost surely.   □

Exercises

1. Prove that $\mathscr{E}^*$ is a σ-field and that $\mathscr{E}^* \subset \mathscr{E}^\sim$.

2. Prove that if A is finely dense, then $A^r = E_\partial$. Furthermore for almost every ω, the set $J_A(\omega)$ is dense in $\mathbf{T}$. [Hint: consider $E^x\{e^{-T_A}\}$.]

3. Suppose that Hypothesis (L) holds. Then for $\alpha \geq 0$, each α-excessive function is Borelian. If $A \in \mathscr{E}$, then $A^r \in \mathscr{E}$. For each $t \geq 0$, $x \to P^x\{T_B \leq t\}$ is Borelian. [Hint: the last assertion follows from the first through the Stone-Weierstrass theorem; this is due to Getoor.]

4. Suppose that for some $\alpha > 0$, all α-excessive functions are lower semicontinuous. Then Hypothesis (L) holds. [Hint: consider λU^α where $\lambda = \sum_n 2^{-n} \varepsilon_{x_n}$ and $\{x_n\}$ is a dense set in $\mathbf{E}$.]

5. Let $A \subset \mathbf{E}$ and A be a fine neighborhood of x. Then for $\alpha > 0$ there exists an α-excessive function φ and a compact set K such that $\{\varphi < 1\} \subset K \subset A$, and $\{\varphi < 1\}$ is a fine neighborhood of x. [Hint: Let V be open with compact closure, $B = A^c \cup V^c$. By Theorem 8(a) of §3.3, there exists open $G \supset B$ such that $\varphi(x) = E^x\{e^{-\alpha T_G}\} < 1$. Take $K = G^c$.]

6. Fix $\alpha > 0$. If $\mathscr{T}$ is a topology on $\mathbf{E}_\partial$ which renders all α-excessive functions continuous, then $\mathscr{T}$ is a finer topology than the fine topology, namely each finely open set is open in $\mathscr{T}$. Thus, the fine topology is the least fine topology rendering all α-excessive functions continuous. [Hint: consider the φ in Problem 5.]

7. If f is bounded and finely continuous, then for each $\beta \geq 0$, $\lim_{\alpha \to \infty} \alpha U^{\alpha+\beta} f = f$. As a consequence, if the function f in Problem 4 of §3.2 is finely continuous, then it is excessive.

8. Let $A \in \mathscr{E}^*$ and $x \in A^r$ and consider the following statement: "for almost every ω, $X(t, \omega) = x$ implies that there exists $t_n \downarrow\downarrow t$ such that $X(t_n, \omega) \in A$." This statement is true for a *fixed* t (namely when t is a constant independent of ω), or when $t = T(\omega)$ where T is optional; but it is false when t is *generic* (namely, when t is regarded as the running variable in $t \to X(t, \omega)$). [Hint: take $x = 0$, $A = \{0\}$ in the Brownian motion in R^1. See §4.2 for a proof that $0 \in \{0\}^r$.]

9. Let A be a finely open set. Then for almost every ω, $X(t, \omega) \in A$ implies that there exists $\delta(\omega) > 0$ such that $X(u, \omega) \in A$ for all $u \in [t, t + \delta(\omega))$. Here t is generic. The contrast with Problem 8 is interesting. [Hint: suppose $A \in \mathscr{E}^*$ and let $\varphi(x) = E^x\{e^{-T_{A^c}}\}$. There exists $\delta'(\omega) \leq 1$ such that $\varphi(X(u, \omega)) \leq \lambda < 1$ for $u \in [t, t + \delta'(\omega))$. The set $B = A^c \cap \{\varphi \leq \lambda\}$ is very thin. By Proposition 5 of §3.5, $J_B(\omega) \cap [t, t + \delta'(\omega))$ is a finite set. Take $\delta(\omega)$ to be the minimum element of this set.]

10. If A is finely closed, then almost surely $J_A(\omega)$ is closed from the right, namely, $t_n \downarrow\downarrow t$ and $t_n \in J_A(\omega)$ implies $t \in J_A(\omega)$, for a generic t.

3.6. Decreasing Limits

The limit of an increasing sequence of excessive functions is excessive, as proved in Proposition 2 of §3.2. What can one say about a decreasing sequence? The following theorem is due to H. Cartan in the Newtonian case and is actually valid for any convergent sequence of excessive functions. It was proved by a martingale method by Doob, see Meyer [2]. The proof given below is simpler, see Chung [4].

Theorem 1. Let $f_n \in S$ and $\lim f_n = f$. Then f is superaveraging and the set $\{f > \hat{f}\}$ is semipolar. In case $\hat{f} < \infty$ everywhere, then

$$\{f > \hat{f} + \varepsilon\} \quad is \ thin \tag{1}$$

for every $\varepsilon > 0$.

Proof. For each $t \geq 0$, $f_n \geq P_t f_n$; letting $n \to \infty$ we obtain $f \geq P_t f$ by Fatou's lemma. Hence f is superaveraging. For each compact K, $f_n \geq P_K f_n$ by Theorem 4 of §3.4; letting $n \to \infty$ we obtain $f \geq P_K f$ as before. Now let A denote the set in (1) and let K be a compact subset of A; then by the Corollary to Theorem 2 of §3.4 we have $f > \hat{f} + \varepsilon$ on the support of $P_K(x, \cdot)$ for every x. Therefore, since $\hat{f} < \infty$ on A,

$$f \geq P_K f \geq P_K(\hat{f} + \varepsilon) = P_K \hat{f} + \varepsilon P_K 1$$

and consequently

$$P_t f \geq P_t P_K \hat{f} + \varepsilon P_t P_K 1. \tag{2}$$

Both $P_K \hat{f}$ and $P_K 1$ are excessively by Theorem 4 of §3.4. Letting $t \downarrow 0$ in (2), we obtain

$$\hat{f} \geq P_K \hat{f} + \varepsilon P_K 1. \tag{3}$$

For a fixed x there exists a sequence of compact subsets K'_n of A such that $P_{K'_n} \hat{f}(x) \uparrow P_A \hat{f}$ and another sequence K''_n such that $P_{K''_n} 1(x) \uparrow P_A 1(x)$; by (12) of §3.4. Taking $K_n = K'_n \cup K''_n$ we see that

$$P_{K_n} \hat{f}(x) \uparrow P_A \hat{f}(x), \quad and \quad P_{K_n} 1(x) \uparrow P_A 1(x).$$

Using this sequence of K_n in (3) we obtain

$$\hat{f}(x) \geq P_A \hat{f}(x) + \varepsilon P_A 1(x). \tag{4}$$

If $\hat{f} < \infty$ everywhere this relation implies that $A^r = \varnothing$; for otherwise if $x \in A^r$ it would read $\hat{f}(x) \geq \hat{f}(x) + \varepsilon$ which is impossible. Hence in this case

A is thin. Letting $\varepsilon \downarrow 0$ through a sequence we conclude that $\{f > \hat{f}\}$ is semipolar.

In the general case, let $f_n^{(m)} = f_n \wedge m$, $f^{(m)} = f \wedge m$. We have just shown that $\{f^{(m)} > \hat{f}^{(m)} + \varepsilon\}$ is thin for each $\varepsilon > 0$. Notice that $f^{(m)}$ is superaveraging and $\hat{f} \wedge m \geq \hat{f}^{(m)}$. It follows that

$$\{f > \hat{f} + \varepsilon\} = \bigcup_{m=1}^{\infty} \{f \wedge m > (\hat{f} \wedge m) + \varepsilon\}$$

$$\subset \bigcup_{m=1}^{\infty} \{f^{(m)} > \hat{f}^{(m)} + \varepsilon\}$$

and therefore $\{f > \hat{f} + \varepsilon\}$ is semipolar. Hence so is $\{f > \hat{f}\}$. $\qquad\square$

Corollary. *If the limit function f above vanishes except on a set of zero potential, then it vanishes except on a polar set.*

Proof. Let $\varepsilon > 0$, $A = \{f \geq \varepsilon\}$ and K be a compact subset of A. Then we have by the Corollary to Theorem 2 of §3.4:

$$f \geq P_K f \geq \varepsilon P_K 1. \tag{5}$$

Hence $P_K 1$ vanishes except on a set of zero potential, and so it vanishes on a finely dense set by Proposition 4 of §3.5. But $P_K 1$ being excessive is finely continuous, hence it vanishes identically by Corollary to Proposition 4 of §3.5. Thus K is polar. This being true for each compact $K \subset A$, A is polar by a previous remark. $\qquad\square$

The corollary has a facile generalization which will be stated below together with an analogue. Observe that the condition (6) below holds for an excessive f.

Proposition 2. *Let $f \in \mathscr{E}_+^{\sim}$ and suppose that we have for each compact K:*

$$f \geq P_K f. \tag{6}$$

If $\{f \neq 0\}$ is of zero potential, then it is polar. If $\{f = \infty\}$ is of zero potential, then it is polar.

Proof. The first assertion was proved above. Let $K \subset \{f = \infty\}$, then we have as in (5):

$$f \geq P_K f \geq \infty \, P_K 1.$$

Hence $P_K 1$ vanishes on $\{f < \infty\}$. The rest goes as before. $\qquad\square$

An alternative proof of the second assertion in Proposition 2, for an excessive function, may be gleaned from Theorem 7 of §3.4. For according to that theorem, almost surely the set $\{t: f(X(t)) = \infty\}$ is either empty or of the form $[0, D_F)$ or $[0, D_F]$, where D_F is the first entrance time of $\{f < \infty\}$. If $\{f = \infty\}$ is of zero potential then $\{f < \infty\}$ is finely dense by Proposition 4 of §3.5, which implies that $D_F = 0$ because $\{t \mid f(X(t)) = \infty\}$ cannot contain any nonempty interval (see Exercise 2 of §3.5).

A cute illustration of the power of these general theorems is given below.

EXAMPLE. Consider Brownian motion in R^2. We have $P_t(x, dy) = p_t(x - y)\,dy$ where

$$p_t(x) = \frac{1}{2\pi t} \exp\left(-\frac{\|x\|^2}{2t}\right), \qquad \|x\|^2 = x_1^2 + x_2^2.$$

Well known convolution property yields

$$P_s p_t(x) = \int_{R^2} p_s(x - y) p_t(y)\,dy = p_{s+t}(x).$$

Define for $\alpha > 0$:

$$f(x) = \int_0^\infty e^{-\alpha t} p_t(x)\,dt;$$

this is just $u^\alpha(x)$ in (10) of §3.7 below. For each $s \geq 0$, we have

$$e^{-\alpha s} P_s f(x) = \int_0^\infty e^{-\alpha(s+t)} P_s p_t(x)\,dt = \int_s^\infty e^{-\alpha t} p_t(x)\,dt,$$

from which it follows that f is α-excessive. It is obvious from the form of $p_t(x)$ that $f(x)$ is finite for $x \neq o$, and $f(o) = +\infty$. Thus $\{f = \infty\} = \{o\}$. Since $P_t(\cdot, \{o\}) = 0$ for each t, $\{o\}$ is of zero potential. By the second assertion of Proposition 2 applied to the semigroup $(e^{-\alpha t} P_t)$, we conclude that $\{o\}$ is a polar set. Therefore every singleton is polar for Brownian motion in R^2, and consequently also in R^d, $d \geq 2$, by considering the projection on R^2. We shall mention this striking fact later on more than one occasion.

For $\alpha = 0$ we have $f \equiv \infty$ in the above. It may be asked whether we can find an excessive function to serve in the preceding example. The answer will be given in Theorem 1 and Example 2 of the next section.

The next theorem deals with a more special situation than Theorem 1. It is an important part of the Riesz representation theorem for an excessive function (see Blumenthal and Getoor [1], p. 84).

Theorem 3. *Let $B_n \in \mathcal{E}^*$ and $B_n \downarrow$; let $T_n = T_{B_n}$. Let $f \in S$ and $g = \lim_n P_{T_n} f$. Then g is superaveraging and $g = \hat{g}$ on $\{g < \infty\} \cap (\bigcap_n B_n^r)^c$.*

Proof. We have by Fatou's lemma,

$$P_t g = P_t \left(\lim_n P_{T_n} f \right) \leq \lim_n P_t P_{T_n} f \leq \lim_n P_{T_n} f = g$$

since $P_{T_n} f$ is excessive by Theorem 4 of §3.4. Thus g is superaveraging. If $g(x) < \infty$, then there exists n_0 such that $P_{T_n} f(x) < \infty$ for $n \geq n_0$. We have by Proposition 1 of §3.4,

$$\begin{aligned} P_t P_{T_n} f(x) &= E^x \{ f(X(t + T_n \circ \theta_t); T_n < \infty \} \\ &\geq E^x \{ f(X(t + T_n \circ \theta_t); t < T_n < \infty \}. \end{aligned} \tag{7}$$

But for any $A \in \mathcal{E}^{\cdot}$

$$t + T_A \circ \theta_t = T_A \quad \text{on } \{ t < T_A \}. \tag{8}$$

This relation says that the first hitting time on A after t *is* the first hitting time of A after 0, provided that A has not yet been hit at time t. It follows from (7) that

$$P_t P_{T_n} f(x) \geq E^x \{ f(X(T_n)); t < T_n \},$$

and consequently by subtraction:

$$P_{T_n} f(x) - P_t P_{T_n} f(x) \leq E^x \{ f(X(T_n)); T_n \leq t \}. \tag{9}$$

Now under P^x, $\{ f(X(T_n)), \mathscr{F}_{T_n}, n \geq 1 \}$ is a supermartingale, and $\{ T_n \leq t \} \in \mathscr{F}_{T_n}$. Therefore, by the supermartingale inequality we have

$$\begin{aligned} E^x \{ f(X(T_n)); T_n \leq t \} &\geq E^x \{ f(X(T_{n+1})); T_n \leq t \} \\ &\geq E^x \{ f(X(T_{n+1})); T_{n+1} \leq t \}. \end{aligned} \tag{10}$$

Thus the right member of (9) decreases as n increases. Hence for each $k \geq n_0$ we have for all $n \geq k$:

$$P_{T_n} f(x) - P_t P_{T_n} f(x) \leq E^x \{ f(X(T_k)); T_k \leq t \}. \tag{11}$$

If $n \to \infty$, then $P_t P_{T_n} f(x) \to P_t g(x)$ by dominated convergence, because $P_{T_n} f(x) \leq P_{T_k} f(x)$, and $P_t P_{T_k} f(x) \leq P_{T_k} f(x) < \infty$. It follows from (11) that

$$g(x) - P_t g(x) \leq E^x \{ f(X(T_k)); T_k \leq t \}. \tag{12}$$

If $x \notin B_k^r$, then $P^x \{ T_k > 0 \} = 1$. Letting $t \downarrow 0$ in (12) we obtain $g(x) - \hat{g}(x) = 0$. This conclusion is therefore true for each x such that $g(x) < \infty$ and $x \notin \bigcap_{k=1}^{\infty} B_k^r$, as asserted.   □

Two cases of Theorem 3 are of particular interest.

Case 1. Let G_n be open, $\bar{G}_{n-1} \subset G_n$ and $\bigcup_n G_n = \mathbf{E}$. For each $x \in \mathbf{E}$, there exists n such that $x \in G_{n-1}$ so that $x \notin (G_n^c)^r$. Hence $\bigcap_n (G_n^c)^r = \{\partial\}$. If we put $f(\partial) = 0$ as by usual convention, then

$$\lim_n P_{T_{G_n^c}} f = g = \hat{g} \quad \text{on } \{f < \infty\}. \tag{13}$$

We can also take K_n compact, $K_{n-1} \subset K_n^0$ and $\bigcup_n K_n = \mathbf{E}$. Then $\bigcap_n (K_n^c)^r = \{\partial\}$ and (13) is true when G_n is replaced by K_n.

Case 2. For a fixed $x_0 \neq \partial$ let G_n be open, $G_n \supset \bar{G}_{n+1}$ and $\bigcap_n G_n = \{x_0\}$. For each $x \neq x_0$, there is an n such that $x \notin G_n^r$. Hence for any such x for which $f(x) < \infty$, we have

$$\lim_n P_{T_{G_n}} f(x) = g(x) = \hat{g}(x).$$

On closer examination the two cases are really the same!

In a sense the next result is an analogue of Theorem 3 when the optional times T_n are replaced by the constant times n.

Theorem 4. *Let $f \in \mathbf{S}$ and suppose that*

$$h(x) = \lim_{t \to \infty} \downarrow P_t f(x) < \infty \tag{14}$$

for all x. Then $h = P_t h$ for each $t \geq 0$; and

$$f = h + f_0 \tag{15}$$

where f_0 is excessive with the property that

$$\lim_{t \to \infty} P_t f_0 \equiv 0. \tag{16}$$

Proof. For each x, the assumption (14) implies that there exists $t_0 = t_0(x)$ such that $P_{t_0} f(x) < \infty$. Since $P_t f \leq P_{t_0} f$ for $t \geq t_0$, and $P_s(P_{t_0} f)(x) \leq P_{t_0} f(x) < \infty$ for $s \geq 0$, the following limit relation holds by dominated convergence, for each $s \geq 0$:

$$P_s h = P_s \left(\lim_{t \to \infty} P_t f \right) \quad \lim_{t \to \infty} P_s P_t f = \lim_{t \to \infty} P_{s+t} f = h. \tag{17}$$

Put $f_0 = f - h$, so that $f_0(x) = \infty$ if $f(x) = \infty$. Then we have for $t \geq t_0$:

$$P_t f_0 = P_t f - P_t h = P_t f - h, \tag{18}$$

and it follows that f_0 is excessive. Letting $t \to \infty$ in (18) we obtain (16). $\square$

This theorem gives a decomposition of an excessive function into an "invariant" part h and a "purely excessive" part f_0 which satisfies the important condition (6) in Theorem 6 of §3.2. It is an easy and rather emasculated version of the Riesz decomposition in potential theory. The reader is welcome to investigate what happens when the condition (14) is dropped.

A noteworthy instance of Theorem 4 is when $f = P_A 1$ where $A \in \mathscr{E}^*$. In this case we have

$$h(x) = \lim_{t \to \infty} P_t P_A 1(x) = \lim_{t \to \infty} P^x \{t + T_A \circ \theta_t < \infty\}. \tag{19}$$

Since $t + T_A \circ \theta_t$ increases with t, we may write (19) as

$$h(x) = P^x \left\{ \bigcap_{n=1}^{\infty} T_A \circ \theta_n < \infty \right\}. \tag{20}$$

A little reflection shows that the set within the braces in (20) is exactly the set of ω for which $J_A(\omega)$ is an unbounded subset of $\mathbf{T}$, where J_A is defined in (6) of §3.5.

Definition. The set A is *recurrent* iff the probability in (19) is equal to one for every $x \in \mathbf{E}$; and *transient* iff it is equal to zero for every $x \in \mathbf{E}$. Of course this is not a dichotomy in general, but for an important class of processes it indeed is (see Exercise 6 of §4.1).

Clearly if A is recurrent then $P_A 1 \equiv 1$ (on $\mathbf{E}$). The converse is true if and only if (P_t) is strictly Markovian. The phenomenon of recurrence may be described in a more vital way, as follows. For each $A \in \mathscr{E}^*$ we introduce a new function:

$$L_A(\omega) = \sup\{t \geq 0 \mid X(t, \omega) \in A\} \tag{21}$$

where sup $\varnothing = 0$ as by standard and meaningful convention. We call L_A the "quitting time of A", or "the last exit time from A". Its contrast to the hitting time T_A should be obvious. To see that $L_A \in \mathscr{F}^{\sim}$, note that for each $\omega \in \Omega$:

see pp. 61,62

$$\{L_A(\omega) \leq t\} \equiv \{X(s, \omega) \notin A \text{ for } s > t\} \equiv \{t + T_A \circ \theta_t(\omega) = \infty\}. \tag{22}$$

It follows that

$$\{L_A < \infty\} = \bigcup_{n=1}^{\infty} \{T_A \circ \theta_n = \infty\}; \tag{23}$$

$$\{L_A = \infty\} = \bigcap_{n=1}^{\infty} \{T_A \circ \theta_n < \infty\}. \tag{24}$$

Thus A is recurrent [transient] if and only if $L_A = \infty \; [< \infty]$ almost surely. This may be used as an alternative definition.

It is clear from the first equivalence in (22) that L_A is not optional; indeed $\{L_A \leq t\} \in \sigma(X_s, s \in (t, \infty))$, augmented. Such a random time has special properties and is a kind of dual to a hitting time. We shall see much of it in §5.1.

Exercises

1. Let $f \in \mathscr{E}_+$ and $T = T_A$ where $A \in \mathscr{E}$. Suppose that for each x and $t > 0$, we have

$$f(x) \geq E^x\{f(X_t)1_{\{t < T\}}\}.$$

   Show that if $f(x) < \infty$, then $\{f(X_t)1_{\{t < T\}}, \mathscr{F}_t, P^x\}$ is a supermartingale.

2. In the notation of Case 1 of Theorem 3, show that for each compact $K \subset \mathbf{E}$, we have $g = P_{K^c}g$ on $\{g < \infty\}$.

3. Under Hypothesis (L), let F be a class of excessive functions. Then there exists a decreasing sequence $\{f_n\}$ in F such that if $u = \lim_n \downarrow f_n$ we have $u \geq \inf_{f \in F} f \geq \hat{u}$. [Hint: choose f_n decreasing and $\lim_n \xi(f_n/(1 + f_n)) = \inf_{f \in F} \xi(f/(1 + f))$.]

3.7. Recurrence and Transience

A Hunt process will be called *recurrent* iff any, hence every, one of the equivalent conditions in the following theorem is satisfied. We use the convention below that an unspecified point, set, function is of, in, on $\mathbf{E}$, not $\mathbf{E}_\partial$. Thus a function is said to be a constant if it is a constant on $\mathbf{E}$; its value at ∂, when defined, is not at issue. Nevertheless, the proof of Theorem 1 is complicated by the fact that it is not obvious that assertion (1) holds *prima facie* under each of the conditions (i) to (iv) there. The additional care needed to avoid being trapped by ∂ is well worth the pain from the methodological point of view. Recall that $\zeta = T_{\{\partial\}}$.

Theorem 1. *The following four propositions are equivalent for a Hunt process, provided that* $\mathbf{E}$ *contains more than one point.*

   (i)   *Each excessive function is a constant.*
   (ii)   *For each* $f \in \mathscr{E}_+^*$, *either* $Uf \equiv 0$ *or* $Uf \equiv \infty$.
   (iii) *For each* B *in* $\mathscr{E}^*$ *which is not thin, we have* $P_B1 \equiv 1$.
   (iv)   *For each* B *in* $\mathscr{E}^*$, *either* $P_B1 \equiv 0$ *or* $P_B1 \equiv 1$. *Namely, each nearly Borel set is either polar or recurrent as defined at the end of §3.6.*

Moreover, any of these conditions implies the following:

$$\forall x: P^x\{\zeta = \infty\} = 1. \tag{1}$$

In other words, (P_t) *is a strictly Markovian semigroup on E.*

Proof (i) $\Rightarrow$ (ii). Since Uf is excessive, it must be a constant, say c, by (i). Suppose $c \neq \infty$. We have for $t \geq 0$:

$$P_t Uf(x) = E^x \left\{ \int_t^\infty f(X_s)\, ds \right\}. \tag{2}$$

Thus $\int_0^\infty f(X_s)\, ds$ is integrable under E^x, hence finite P^x-almost surely. We now prove that (i) implies (1), as follows.

Let x_1 and x_2 be two distinct points. Then there exist open sets O_1 and O_2 with disjoint closures such that $x_1 \in O_1$, $x_2 \in O_2$. Both $P_{O_1} 1$ and $P_{O_2} 1$ are excessive functions which take the value one, hence both must be identically equal to 1 by (i). Put

$$T_1 = T_{O_1}, \qquad T_2 = T_1 + T_{O_2} \circ \theta_{T_1},$$

and for $n \geq 1$:

$$T_{2n+1} = T_{2n} + T_{O_1} \circ \theta_{T_{2n}}, \qquad T_{2n+2} = T_{2n+1} + T_{O_2} \circ \theta_{T_{2n+1}}.$$

Since $X(T_{O_1}) \in \bar{O}_1 \subset (\bar{O}_2)^c$, we have $T_1 < T_2$ on $\{T_1 < \infty\}$. The same argument then shows that the sequence $\{T_n\}$ is strictly increasing. Let $\lim_n T_n = T$; then on $\{T < \infty\}$, $X(T-)$ does not exist because $X(T_n)$ oscillates between $\bar{O}_1$ and $\bar{O}_2$ at a strictly positive distance apart. This is impossible for a Hunt process. It follows that almost surely we have $T = \infty$, and this implies (1) because ∂ is absorbing. Hence $P_t c = c$ for all t.

We now have by (2) and dominated convergence:

$$c = \lim_{t \to \infty} P_t c = \lim_{t \to \infty} P_t Uf = 0.$$

Remark. It is instructive to see why (i) does not imply (ii) when $\mathbf{E}$ reduces to one point.

(ii) $\Rightarrow$ (iii). Fix $B \in \mathscr{E}^*$ and write φ for $P_B 1$. By Theorem 4 of §3.6, we have $\varphi = h + \varphi_0$ where h is invariant and φ_0 is purely excessive. Hence by Theorem 6 of §3.2, there exists $f_n \in \mathscr{E}_+^*$ such that $\varphi_0 = \lim_n \uparrow Uf_n$. Since $Uf_n \leq \varphi_0 \leq 1$, condition (ii) implies that $Uf_n \equiv 0$ and so $\varphi_0 \equiv 0$. Therefore $\varphi = P_t \varphi$ for every $t \geq 0$. If B is not thin, let $x_0 \in B^r$ so that $\varphi(x_0) = 1$. Put $A = \{x \in \mathbf{E} \mid \varphi(x) < 1\}$. We have for each $t \geq 0$:

$$1 = \varphi(x_0) = \int_{\mathbf{E}_\partial} P_t(x_0, dy)\varphi(y).$$

Since $P_t(x_0, \mathbf{E}_\partial) = 1$, the equation above entails that $P_t(x_0, A) = 0$. This being true for each t, we obtain $U1_A(x_0) = 0$. Hence condition (ii) implies that $U1_A \equiv 0$. But A is finely open since φ is finely continuous. If $x \in A$, then $U1_A(x) > 0$ by fine openness. Hence A must be empty and so $\varphi \equiv 1$.

(iii) $\Rightarrow$ (iv). Under (iii), $P_B 1 \equiv 1$ for any nonempty open set B. Hence (1) holds as under (i). If B is not a polar set, there exists x_0 such that $P_B 1(x_0) = \delta > 0$. Put

$$A = \left\{ x \,\middle|\, P_B 1(x) > \frac{\delta}{2} \right\};\tag{3}$$

then A is finely open. Since $x_0 \in A$, $x_0 \in A^r$ by fine openness. Hence $P_A 1 \equiv 1$ by (iii). Now we have

$$P_B 1 \geq P_A P_B 1 \geq \inf_{x \in A^*} P_B 1(x) \geq \frac{\delta}{2},\tag{4}$$

where the first inequality is by Theorem 4 of §3.4, the second by Theorem 2 of §3.4, and the third by (3) and the fine continuity of $P_B 1$. Next, we have for $t \geq 0$, P^x-almost surely:

$$P^x\{T_B < \infty \,|\, \mathscr{F}_t\} = P^x\{T_B \leq t \,|\, \mathscr{F}_t\} + P^x\{t < T_B < \infty \,|\, \mathscr{F}_t\}$$
$$= 1_{\{T_B \leq t\}} + 1_{\{t < T_B\}} P^{X(t)}[T_B < \infty].\tag{5}$$

This is because $\{T_B \leq t\} \in \mathscr{F}_{t+} = \mathscr{F}_t$; and $T_B = t + T_B \circ \theta_t$ on $\{t < T_B\}$ (see (8) of §3.6), so that we have by the Markov property:

$$P^x\{t < T_B < \infty \,|\, \mathscr{F}_t\} = P^x\{t < T_B; t + T_B \circ \theta_t < \infty \,|\, \mathscr{F}_t\}$$
$$= 1_{\{t < T_B\}} P^{X(t)}[T_B < \infty].$$

As $t \to \infty$, the first member in (5) as well as the first term in the third member converges to $1_{\{T_B < \infty\}}$. Hence the second term in the third member must converge to zero. Now we have $X(t) \in \mathbf{E}$ for $t < \infty$ and therefore $P^{X(t)}[T_B < \infty] \geq \delta/2$ by (4). It follows that $\lim_{t \to \infty} 1_{\{t < T_B\}} = 0$, P^x-almost surely. This is just $P_B 1(x) = 1$, and x is arbitrary.

(iv) $\Rightarrow$ (i). Let f be excessive. If f is not a constant, then there are real numbers a and b such that $a < b$, and the two sets $A = \{f < a\}$, $B = \{f > b\}$ are both nonempty. Let $x \in A$, then we have by Theorems 4 and 2 of §3.4:

$$a > f(x) \geq P_B f(x) = \int_{B^*} P_B(x, dy) f(y) \geq b P_B(x, B^*) = b P_B 1(x),\tag{6}$$

since $f \geq b$ on B^* by fine continuity. Thus $P_B 1(x) < 1$. But B being finely open and nonempty is not polar, hence $P_B 1 \equiv 1$ by (iv). This contradiction proves that f is a constant. $\qquad\square$

Finally, we have proved above that (i) implies (1), hence each of the other three conditions also implies (1), which is equivalent to $P_t(x, \mathbf{E}) = 1$ for every $t \geq 0$ and $x \in \mathbf{E}$.

Remark. For the argument using (5) above, cf. *Course*, p. 344.

EXAMPLE 1. It is easy to construct Markov chains (Example 1 of §1.2) which are recurrent Hunt processes. In particular this is the case if there are only a finite number of states which communicate with each other, namely for any two states i and j we have $p_{ij} \neq 0$ and $p_{ji} \neq 0$. Condition (ii) is reduced to the following: for some (hence every) state i we have

$$\int_0^\infty p_{ii}(t)\, dt = \infty. \tag{7}$$

Since each i is regular for $\{i\}$, the only thin or polar set is the empty set. Hence condition (iv) reduces to: $P^i\{T_{\{j\}} < \infty\} = 1$ for every i and j. Indeed,

$$P^i\{L_{\{j\}} = \infty\} = 1 \tag{8}$$

in the notation of (21) of §3.6. For a general Markov chain (7) and (8) are equivalent provided that all states communicate, even when the process is not a Hunt process because it does not satisfy the hypotheses (i), (ii) and (iii); see Chung [2].

EXAMPLE 2. For the Brownian motion in R^d, the transition probability kernel $P_t(x, dy) = p_t(x - y)\, dy$ where

$$p_t(z) = \frac{1}{(2\pi t)^{d/2}} \exp\left[-\frac{\|z\|^2}{2t}\right], \quad \text{where } \|z\|^2 = \sum_{j=1}^d z_j^2. \tag{9}$$

For $\alpha \geq 0$, define

$$u^\alpha(z) = \int_0^\infty e^{-\alpha t} p_t(z)\, dt; \qquad u(z) = u^0(z). \tag{10}$$

Then the resolvent kernel $U^\alpha(x, dy) = u^\alpha(x - y)\, dy$.

For $d = 1$ or $d = 2$, we have $u(z) = \infty$. Hence by the Fubini-Tonelli theorem,

$$U(x, B) = \int_B u(x - y)\, dy = 0 \quad \text{or} \quad \infty$$

according as $m(B) = 0$ or $m(B) > 0$. It follows that condition (ii) of Theorem 1 holds. Therefore, Brownian motion in R^1 or R^2 is a recurrent Hunt process. Since each nonempty ball is not thin, the sample path will almost surely enter into it and returns to it after any given time, by condition (iv). Hence it will almost surely enter into a sequence of concentric balls shrinking to a given point, and yet it will not go through the point because a singleton is a polar set. Such a phenomenon can happen only if the entrance times into the successive balls increase to infinity almost surely, for otherwise at a finite limit time the path must hit the center by continuity. These statements can be made quantitative by a more detailed study of the process.

Note: Whereas under recurrence $U1_B \not\equiv 0$ implies $P_B1 \equiv 1$, the converse is not true in general. For Brownian motion in R^2, the boundary circle C of a nenempty disk is recurrent by the discussion above, but $U1_C \equiv 0$ because $m(C) = 0$.

Turning in the opposite direction, we call a Hunt process "transient" iff its semigroup is transient as defined at the end of §3.2. However, there are several variants of the condition (17) used there. It is equivalent to the condition that $\mathbf{E}$ is "σ-transient", namely it is the union of a sequence of transient sets. Hence the condition is satisfied if each compact subset of $\mathbf{E}$ is transient, because $\mathbf{E}$ is σ-compact. However, to relate the topological notion of compactness to the stochastic notion of transience some analytic hypothesis is needed. Consider the following two conditions on the potentials (resolvents) of a Hunt process:

$$\forall f \in \mathscr{E}^*_+ : Uf \text{ is lower semi-continuous}; \tag{11}$$

$$\exists \alpha > 0 \text{ such that } \forall f \in \mathscr{E}^*_+ : U^\alpha f \text{ is lower semi-continuous}. \tag{12}$$

Observe that there is no loss of generality if we use only bounded f, or bounded f with compact supports in (11) and (12), because lower semi-continuity is preserved by increasing limits. It is easy to see that (12) implies (11) by the resolvent equation.

Recall that condition (16) of §3.2 holds for a Hunt process and will be used below tacitly. Also, the letter K will be reserved for a compact set.

Theorem 2. *In general, either* (ii) *or* (iv) *below implies* (iii). *Under the condition* (11), (iii) *implies* (iv). *Under the condition* (12), (iv) *implies* (i).

(i) $\forall K : U1_K$ is bounded.
(ii) $\forall K : U1_K$ is finite everywhere.
(iii) $\exists h \in \mathscr{E}_+$ such that $0 < Uh < \infty$ on $\mathbf{E}$.
(iv) $\forall K; \lim_{t \to \infty} P_t P_K 1 = 0.$ $\tag{13}$

Proof. (ii) $\Rightarrow$ (iii) under (11). Let K_n be compact and increase to $\mathbf{E}$, and put

$$A_{nk} = \{x \in K_n \mid U(x, K_n) \le k\}.$$

The function $U1_{A_{nk}}$ is bounded above by k on the support of $1_{A_{nk}}$, hence by the domination principle (Corollary to Theorem 4 of §3.4), we have

$$U1_{A_{nk}} \le k \tag{14}$$

everywhere in **E**. Define h as follows:

$$h = \sum_{n=1}^{\infty} \sum_{k=1}^{\infty} \frac{1_{A_{nk}}}{k2^{k+n}}. \tag{15}$$

Clearly $h \leq 1$. For each x, there exist n and k such that $x \in A_{nk}$ by assumption (ii). Hence $h > 0$ everywhere and so also $Uh > 0$. It follows from (14) that $Uh \leq 1$. Thus h has all the properties (and more!) asserted in (ii).

(iii) $\Rightarrow$ (iv) under (11). Since Uh is finite, we have

$$\lim_{t \to \infty} P_t Uh = 0 \tag{16}$$

as in (2). Since $Uh > 0$ and Uh is lower semi-continuous by (11), for each K we have $\inf_{x \in K} Uh(x) = C(K) > 0$. Hence by the corollary to Theorem 2 of §3.4 we have

$$P_K 1 \leq P_K \left(\frac{Uh}{C(K)} \right) = \frac{1}{C(K)} P_K Uh \leq \frac{1}{C(K)} Uh. \tag{17}$$

The assertion (13) follows from (16) and (17).

(iv) $\Rightarrow$ (iii) and under (12)(iv) $\Rightarrow$ (i). Let $K_n \subset K_{n+1}^0$ and $\bigcup_n K_n = \mathbf{E}$. Applying Theorem 6 of §3.2 to each $P_{K_n} 1$ we have

$$P_{K_n} 1 = \lim_k \uparrow Ug_{nk} \tag{18}$$

where $g_{nk} \leq k$ by the proof of Theorem 6 of §3.2. It follows by Lemma 1 of §3.2 that

$$1 = \lim_n \uparrow P_{K_n} 1 = \lim_n \uparrow Ug_{nn}. \tag{19}$$

Define g as follows:

$$g = \sum_{n=1}^{\infty} \frac{g_{nn}}{n2^n}.$$

Then $g \leq 1$ and $0 < Ug \leq 1$ by (19). Now put $h = U^{\alpha}g$; then $h \leq Ug \leq 1$. It follows from the resolvent equation that

$$Uh = UU^{\alpha}g \leq \frac{1}{\alpha} Ug \leq \frac{1}{\alpha}. \tag{20}$$

The fact that for each x we have

$$Ug(x) = \int_0^{\infty} P_t g(x)\, dt > 0$$

implies that

$$h(x) = U^\alpha g(x) = \int_0^\infty e^{-\alpha t} P_t g(x)\, dt > 0.$$

Hence $Uh > 0$ and h satisfies the conditions in (iii). Finally, under (12) h is lower semi-continuous. Since $h > 0$, for each K we have $\inf_{x \in K} h(x) = C(K) > 0$; hence by (20):

$$U1_K \le \frac{Uh}{C(K)} \le \frac{1}{\alpha C(K)},$$

namely (i) is true. Theorem 2 is completely proved.   □

Let us recall that (iv) is equivalent to "every compact subset of $\mathbf{E}$ is transient", i.e.,

$$\forall K: L_K < \infty \quad \text{almost surely.} \tag{21}$$

In this form the notion of transience is probabilistically most vivid.

EXAMPLE 3. For a Markov chain as discussed in Example 1 above, in which all states communicate, transience is equivalent to the condition $\int_0^\infty p_{ii}(t)\, dt < \infty$ for some (hence every) state. A simple case of this is the Poisson process (Example 3 of §1.2).

EXAMPLE 4. Brownian motion R^d, when $d \ge 3$, is transient. A simple computation using the gamma function yields

$$u(z) = \frac{\Gamma\left(\dfrac{d}{2} - 1\right)}{2\pi^{d/2}|z|^{d-2}}; \tag{22}$$

in particular for $d = 3$:

$$U(x, dy) = \frac{dy}{2\pi|x - y|}. \tag{23}$$

If B is a ball with center at the origin and radius r, we have by calculus:

$$U(o, B) = \frac{1}{2\pi} \int_B \frac{dy}{|y|} = \frac{\pi r^2}{2}.$$

Let f be bounded (Lebesgue) measurable with compact support K, then

$$Uf(x) = \frac{1}{2\pi} \int_K \frac{f(y)}{|x - y|}\, dy = \frac{1}{2\pi} \int_{K \cap B} \frac{f(y)}{|x - y|}\, dy + \frac{1}{2\pi} \int_{K \setminus B} \frac{f(y)}{|x - y|}\, dy \tag{24}$$

where B is the ball with center x and radius δ. The first term in the last member of (24) is bounded uniformly in x by $(2\pi)^{-1}\|f\|U(o,\delta) = 4^{-1}\|f\|\delta^2$. The second term there is continuous at x by bounded convergence. It is also bounded by $\delta^{-1}\|f\|m(K)$. Thus Uf is bounded continuous. In particular if $f = 1_K$, this implies condition (i) of Theorem 2. The result also implies condition (11) as remarked before. Hence all the conditions in Theorem 2 are satisfied. The case $d > 3$ is similar.

We can also compute for $d = 3$ and $\alpha \geq 0$:

$$u^\alpha(z) = \frac{e^{-|z|\sqrt{2\alpha}}}{2\pi|z|} \tag{25}$$

by using the following formula, valid for any real c:

$$\int_0^\infty \exp\left[-\alpha\left(t - \frac{c}{t}\right)^2\right] dt = \frac{1}{2}\sqrt{\frac{\pi}{\alpha}}.$$

Hence for any bounded measurable f, we have

$$U^\alpha f(x) = \int_{R^3} \frac{\exp[-|x - y|\sqrt{2\alpha}]}{2\pi|x - y|} f(y)\, dy. \tag{26}$$

For $\alpha > 0$, an argument similar to that indicated above shows that $U^\alpha f$ belongs to $\mathbb{C}_0$. This is stronger than the condition (12). For $d \geq 3$, $\alpha \geq 0$, u^α can be expressed in terms of Bessel functions. For odd values of d, these reduce to elementary functions.

Let us take this occasion to introduce a new property for the semigroup (P_t) which is often applicable. It will be said to have the "strong Feller property" iff for each $t > 0$ and $f \in b\mathscr{E}_+$, we have $P_t f \in \mathbb{C}$, namely continuous. [The reader is warned that variants of this definition are used by other authors.] This condition is satisfied by the Brownian motion in any dimension. To see this, we write

$$P_t f(x) = \int_B p_t(x - y)f(y)\, dy + \int_{R^d - B} p_t(x - y)f(y)\, dy. \tag{27}$$

For any x_0 and a neighborhood V of x_0, we can choose B to be a large ball containing V to make the second term in (27) less than ε for all x in V. The first term is continuous in x by dominated convergence. Hence $P_t f$ is continuous at x_0.

We can now show that $U^\alpha f \in \mathbb{C}$ for $\alpha > 0$, and $f \in b\mathscr{E}_+$ without explicit computation of the kernel U^α in (26). Write

$$U^\alpha f(x) = \int_0^\infty e^{-\alpha t} P_t f(x)\, dt$$

and observe that the integrand is dominated by $\|f\|e^{-\alpha t}$, and is continuous in x for each $t > 0$. Hence $U^{\alpha}f(x_n) \to U^{\alpha}f(x)$ as $x_n \to x$ by dominated convergence. Since $Uf = \lim_{\alpha \to 0} \uparrow U^{\alpha}f$ it now follows that Uf is lower semicontinuous, without the explicit form of U. This is a good example where the semigroup (P_t) is used to advantage—one which classical potential theory lacks. Of course, a Hunt process is a lot more than its semigroup.

Let us remark that for any (P_t), the semi-group (P_t^{α}) where $P_t^{\alpha} = e^{-\alpha t}P_t$ is clearly transient because $U^{\alpha}(x, \mathbf{E}) \le 1/\alpha$. This simple device makes available in a general context the results which are consequences of the hypothesis of transience. Theorem 9 of §3.2 is such an example by comparison with Proposition 10 there.

Exercises

1. For a Hunt process, not necessarily transient, suppose $A \in \mathscr{E}$, $B \in \mathscr{E}$ such that $\bar{B} \subset A$ and $U(x_0, A) < \infty$. Then if $\inf_{x \in \bar{B}} U(x, A) > 0$, we have $P^{x_0}\{L_{\bar{B}} < \infty\} = 1$, where $L_{\bar{B}}$ is the last exist time from $\bar{B}$. [Hint: define $T_n = n + T_{\bar{B}} \circ \theta_n$, then $U(x_0, A) \ge E^{x_0}\{U(X(T_n), A); T_n < \infty\}$.]

2. Let X be a transient Hunt process. Let $\bar{A}_1$ be compact, $A_n\downarrow$ and $\bigcap_n \bar{A}_n = B$. Put $f = \lim_n P_{A_n}1$. Prove that f is superaveraging and its excessive regularization $\hat{f}$ is equal to P_B1. [Hint: $f \ge \hat{f} \ge P_B1$. If $x \notin B$, then

$$P^x\left\{\bigcap_n [T_{A_n} < \infty]\right\} = P^x\{T_B < \infty\}$$

by transience. Hence $\hat{f} = P_B1$ on $B^c \cup B^r$ which is a finely dense set.]

3. Let $h \in \mathscr{E}_+$ and $Uh < \infty$. In the same setting as Problem 2, prove that if $f = \lim_n P_{A_n}Uh$ then $\hat{f} = P_BUh$.

4. Let X be a transient Hunt process and suppose Hypothesis (L) holds with ξ a reference probability measure. Let B a compact polar set. Prove that there exists a function $f \in \mathscr{E}_+$ such that $\xi(f) \le 1$ and $f = \infty$ on B. [Hint: in Problem 2 take A_n to be open, and $f_n = P_{A_n}1$. On B^c, $f_n \downarrow P_B1$, hence $\xi(f_n) \downarrow 0$; on B, $f_n = 1$. Take $\{n_k\}$ so that $\xi(\sum_k f_{n_k}) \le 1$. This is used in classical potential theory to define a polar set as the set of "poles" of a superharmonic function.]

5. Prove that (iii) in Theorem 2 is equivalent to $\mathbf{E}$ being σ-transient. [Hint: under (iii), $\{Uh > \varepsilon\}$ is transient.]

3.8. Hypothesis (B)

This section is devoted to a new assumption for a Hunt process which Hunt listed as Hypothesis (B). What is Hypothesis (A)? This is more or less the set of underlying assumptions for a Hunt process given in §3.1. Unlike some other assumptions such as the conditions of transience, or the Feller property and its generalization or strengthening mentioned in §3.7, Hypothesis (B) is not an analytic condition but restricts the discontinuity of the sample paths. It turns out to be essential for certain questions of potential theory and Hunt used it to fit his results into the classical grain.

Hypothesis (B).   *For some $\alpha \geq 0$, any $A \in \mathscr{E}^*$, and open G such that $A \subset G$, we have*

$$P_A^\alpha = P_G^\alpha P_A^\alpha. \tag{1}$$

This means of course that the two measures are identical. Recalling Proposition 1 of §3.4, for $S = T_G$ and $T = T_A$, we see that (1) is true if

$$T_A = T_G + T_A \circ \theta_{T_G} \quad \text{almost surely on } \{T_A < \infty\}. \tag{2}$$

On the other hand, if $\alpha > 0$, then the following particular case of (1)

$$P_A^\alpha 1 = P_G^\alpha P_A^\alpha 1 \tag{3}$$

already implies the truth of (2). For (3) asserts that

$$E\{e^{-\alpha T_A}\} = E\{\exp[-\alpha(T_G + T_A \circ \theta_{T_G})]\},$$

whereas the left member in (2) never exceeds the right member which is $\inf\{t > T_G : X(t) \in A\}$. We will postpone a discussion of the case $\alpha = 0$ until the end of this section.

Next, the two members of (2) are equal on the set $\{T_G < T_A\}$. This is an extension of the "terminal" property of T_A expressed in (8) of §3.6. Since $G \supset A$, we have always $T_G \leq T_A$. Can $T_G = T_A$? [If $\bar{A} \subset G$, this requires the sample path to jump from G^c into $\bar{A}$.] There are two cases to consider.

Case (i). $T_G = T_A < \infty$ and $X(T_G) = X(T_A) \in A^r$. In this case, the strong Markov property at T_G entails that $T_A \circ \theta_{T_G} = 0$. Hence the two members of (2) are manifestly equal.

Case (ii). $T_G = T_A < \infty$ and $X(T_G) = X(T_A) \notin A^r$. Then since $X(T_A) \in A \cup A^r$, we must have $X(T_G) \in A \backslash A^r$. Consequently $T_A \circ \theta_{T_G} > 0$ and the left member of (2) is strictly less than the right.

We conclude that (2) holds if and only if case (ii) does not occur, which may be expressed as follows:

$$\forall x: P_G(x, A \backslash A^r) = 0. \tag{4}$$

Now suppose that (2) holds and A is thin. Then neither case (i) nor case (ii) above can occur. Hence for any open $G_n \supset A$ we have almost surely

$$T_{G_n} < T_A, \quad \text{on } \{T_A < \infty\}. \tag{5}$$

Suppose first that $x \notin A$; then by Theorem 8(b) of §3.3 there exist open sets $G_n \supset A$, $G_n \downarrow$ such that $T_{G_n} \uparrow T_A$, P^x-a.s. Since left limits exist everywhere for a Hunt process, we have $\lim_{n \to \infty} X(T_{G_n}) = X(T_A-)$; on the other hand, the limit is equal to $X(T_A)$ by quasi left continuity. Therefore, $X(\cdot)$ is continuous at T_A, P^x-a.s. For an arbitrary x we have

$$P^x\{T_A < \infty; X(T_A-) = X(T_A)\} = \lim_{t \downarrow 0} P^x\{t < T_A < \infty; X(T_A-) = X(T_A)\}$$

$$= \lim_{t \downarrow 0} P^x\{t < T_A; P^{X(t)}[T_A < \infty;$$

$$X(T_A-) = X(T_A)]\}$$

$$= \lim_{t \downarrow 0} P^x\{t < T_A; P^{X(t)}[T_A < \infty]\}$$

$$= \lim_{t \downarrow 0} P^x\{t < T_A; P_A 1(X_t)\}$$

$$= P_A 1(x).$$

In the above, the first equation follows from $P^x\{T_A > 0\} = 1$; the second by Markov property; the third by what has just been proved because $X(t) \notin A$ for $t < T_A$; and the fifth by the right continuity of $t \to P_A 1(X_t)$. The result is as follows, almost surely:

$$X(T_A-) = X(T_A) \quad \text{on } \{T_A < \infty\}. \tag{6}$$

Equation (6) expresses the fact that the path is almost surely continuous at the first hitting time of A. This will now be strengthened to assert continuity whenever the path is in A. To do so suppose first that A is very thin. Then by (12) of §3.5, the incidence set J_A consists of a sequence $\{T_n, n \geq 1\}$ of successive hitting times of A. By the strong Markov property and (6), we have for $n \geq 2$, almost surely on $\{T_{n-1} < \infty\}$:

$$P^x\{T_n < \infty; X(T_n-) = X(T_n)\} = P^{X(T_{n-1})}\{T_A < \infty; X(T_A-) = X(T_A)\}$$
$$= P^{X(T_{n-1})}\{T_A < \infty\} = P^x\{T_n < \infty\}.$$

It follows that $X(T_n-) = X(T_n)$ on $\{T_n < \infty\}$, and consequently

$$\forall t \geq 0: X(t-) = X(t) \quad \text{almost surely on } \{X_t \in A\}. \tag{7}$$

Since (7) is true for each very thin set A, it is also true for each semi-polar set by Theorem 6 of §3.5.

Thus we have proved that (2) implies the truth of (7) for each thin set $A \subset G$. The converse will now be proved. First let A be thin and contained in a compact K which in turn is contained in an open set G. We will show that (4) is true. Otherwise there is an x such that $P^x\{T_G < \infty; X(T_G) \in A\} > 0$. Since A is at a strictly positive distance from G^c, this is ruled out by (7). Hence

$P_G(x, A) = 0$. Next, let $K \subset G$ as before. Then $K \backslash K^r = \bigcup_n C_n$ where each C_n is (very) thin by Theorem 6 of §3.5. Since $C_n \subset K$, the preceding argument yields $P_G(x, C_n) = 0$ and consequently $P_G(x, K \backslash K^r) = 0$. By the discussion leading to (4), this is equivalent to:

$$T_K = T_G + T_K \circ \theta_{T_G} \quad \text{almost surely on } \{T_K < \infty\}. \tag{8}$$

Finally let $A \in \mathscr{E}^*$, $A \subset G$. By Theorem 8(b) of §3.5, for each x there exist compact sets K_n, $K_n \subset A$, $K_n \uparrow$ such that $T_{K_n} \downarrow T_A$, P^x-almost surely. On $\{T_A < \infty\}$ we have $T_{K_n} < \infty$ for all large n. Hence (8) with $K = K_n$ yields (2) as $n \to \infty$.

We summarize the results in the theorem below.

Theorem 1. *For a Hunt process the following four propositions are equivalent*:

(i) *Hypothesis (B), namely (1), is true for some $\alpha > 0$.*
(ii) *Equation (3) is true for some $\alpha > 0$.*
(iii) *For each semi-polar A and open G such that $A \subset G$, we have $P_G(x, A) = 0$ for each x.*
(iv) *For each semi-polar A, (7) is true.*

Moreover, if in (1) or (3) we restrict A to be a compact set, the resulting condition is also equivalent to the above.

It should also be obvious that if Hypothesis (B) holds for some $\alpha > 0$, it holds for every $\alpha > 0$. We now treat the case $\alpha = 0$.

Theorem 2. *If the Hunt process is transient in the sense that it satisfies condition (iii) of Theorem 2 of §3.7, then the following condition is also equivalent to the conditions in Theorem 1.*

(v) *Hypothesis (B) is true for $\alpha = 0$.*

Proof. Let $h > 0$ and $Uh \leq 1$. Then applying (1) with $\alpha = 0$ to the function Uh, we obtain

$$0 = P_A Uh - P_G P_A Uh = E^{\cdot} \left\{ \int_{T_A}^{T_G + T_A \circ \theta_{T_G}} h(X_t) \, dt \right\}.$$

On the set $\{T_A < T_G + T_A \circ \theta_{T_G}; T_A < \infty\}$, the integral above is strictly positive. Hence the preceding equation forces this set to be almost surely empty. Thus (2) is true, which implies directly (1) for every $\alpha \geq 0$. $\square$

Needless to say, we did not use the full strength of (v) above, since it is applied only to a special kind of function. It is not apparent to what extent the transience assumption can be relaxed.

In Theorem 9 of §3.3, we proved that for a Hunt process, $T_B \leq T_B^-$ a.s. for each $B \in \mathscr{E}^*$. It is trivial that $T_B = T_B^-$ if all paths are continuous. The next theorem is a deep generalization of this and will play an important role in §5.1. It was proved by Meyer [3], p. 112, under Hypothesis (L), and by Azema [1] without (L). The following proof due to Walsh [1] is shorter but uses Dellacherie's theorem.

Theorem 3. *Under Hypotheses* (L) *and* (B), *we have* $T_B = T_B^-$ a.s. *for each* $B \in \mathscr{E}^*$.

Proof. The key idea is the following *constrained* hitting time, for each $B \in \mathscr{E}^*$:

$$S_B = \inf\{t > 0 \,|\, X(t-) = X(t) \in B\}. \tag{9}$$

To see that S_B is optional, let $\{J_n\}$ be the countable collection of all jump times of the process (Exercise 5 of §3.1), and $\Lambda = \bigcup_{n=1}^{\infty}\{T_B \neq J_n\}$. Then $\Lambda \in \mathscr{F}(T_B)$ by Theorem 6 of §1.3, hence $S_B = (T_B)_\Lambda$ is optional by Proposition 4 of §1.3. It is easy to verify that S_B is a terminal time. Define $\varphi(x) = E^x\{e^{-S_B}\}$. Then φ is 1-excessive by Exercise 8 of §3.4. Now suppose first that B is compact. For $0 < \varepsilon < 1$ put

$$A = B \cap \{x \,|\, \varphi(x) \leq 1 - \varepsilon\}.$$

Define a sequence of optional times as follows: $R_1 = S_A$, and for $n \geq 1$:

$$R_{n+1} = R_n + S_A \circ \theta_{R_n}.$$

On $\{R_n < \infty\}$ we have $X(R_n) \in A$ because A is finely closed. Hence $\varphi(X(R_n)) \leq 1 - \varepsilon$ and so $P^{X(R_n)}\{S_A > 0\} = 1$ by the zero-one law. It follows that $R_n < R_{n+1}$ by the strong Markov property. On the other hand, we have as in the proof of Proposition 5 of §3.5, for each x and $n \geq 1$:

$$E^x\{e^{-R_{n+1}}\} \leq (1 - \varepsilon)E^x\{e^{-R_n}\} \leq (1 - \varepsilon)^n.$$

Hence $R_n \uparrow \infty$ a.s. Put $I(\omega) = \{t > 0 \,|\, X(t, \omega) \in A\}$. By the definition of R_n, for a.e. ω such that $R_n(\omega) < \infty$, the set $I(\omega) \cap (R_n(\omega), R_{n+1}(\omega))$ is contained in the set of t where $X(t-, \omega) \neq X(t, \omega)$, and is therefore a countable set. Since $R_n(\omega) \uparrow \infty$, it follows that $I(\omega)$ itself is countable. Hence by Theorem 8 of §3.5, A is a semi-polar set. This being true for each ε, we conclude that the set $B' = B \cap \{x \,|\, \varphi(x) < 1\}$ is semi-polar.

Now we consider three cases on $\{T_B < \infty\}$, and omit the ubiquitous "a.s."

(i) $X(T_B-) = X(T_B)$. Since B is compact, in this case $X(T_B-) \in B$ and so $T_B^- \leq T_B$.

(ii) $X(T_B-) \neq X(T_B)$ and $X(T_B) \in B - B'$. In this case $\varphi(X(T_B)) = 1$ and so $S_B \circ \theta_{T_B} = 0$. Since $T_B^- \leq S_B$ we have $T_B^- \leq T_B$.

(iii) $X(T_B-) \neq X(T_B)$ and $X(T_B) \in B'$. Since B' is semi-polar this is ruled out by Theorem 1 under Hypothesis (B).

We have thus proved for each compact B that $T_B^- \le T_B$ on $\{T_B < \infty\}$; hence $T_B^- = T_B$ by Theorem 9 of §3.3. For an arbitrary $B \in \mathscr{E}^*$, we have by Theorem 8(b) of §3.3, for each x a sequence of compacts $K_n \subset B$, $K_n \uparrow$ such that $P^x\{T_{K_n} \downarrow T_B\} = 1$. Since $T_{K_n}^- = T_{K_n}$ as just proved and $T_B^- \le T_{K_n}^-$ for $n \ge 1$, it follows that $P^x\{T_B^- \le T_B\} = 1$. As before this implies the assertion of Theorem 3. $\square$

Exercises

1. Give an example such that $P_A 1 = P_G P_A 1$ but $P_A \ge P_G P_A$ is false. [Hint: starting at the center of a circle, after holding the path jumps to a fixed point on the circle and then execute uniform motion around the circle. This is due to Dellacherie. It is a recurrent Hunt process. Can you make it transient?]

2. Under the conditions of Theorem 1, show that (7) remains true if the set $\{X_t \in A\}$ is replaced by $\{X_{t-} \in A\}$. [Hint: use Theorem 9 of §3.3.]

3. If (7) is true for each compact semi-polar set A, then it is true for every semi-polar set. [Hint: Proposition 11 of §3.5 yields a direct proof without use of the equivalent conditions in Theorem 1.]

4. Let $\mathbf{E} = \{0\} \cup [1, \infty)$; let 0 be a holding point from which the path jumps to 1, and then moves with uniform speed to the right. Show that this is a Hunt process for which Hypothesis (*B*) does not hold.

5. Let A be a thin compact set, G_n open, and $G_n \Downarrow A$. Show that under Hypothesis (*B*), we have under P^x, $x \notin A$:

$$\forall n: T_{G_n} < T_A, \qquad T_{G_n} \uparrow T_A, \quad \text{a.s. on } \{T_A < \infty\}.$$

Thus T_A is predictable under P^x, $x \notin A$.

6. Assume that the set A in Problem 5 has the stronger property that

$$\sup_{x \in \mathbf{E}_\partial} E^x\{e^{-T_A}\} < 1.$$

Then A is polar. [Hint: use Problem 2 of §3.4.]

NOTES ON CHAPTER 3

§3.1. Hunt process is named after G. A. Hunt, who was the author's classmate in Princeton. The basic assumptions stated here correspond roughly with Hypothesis (*A*), see Hunt [2]. The Borelian character of the semigroup is not always assumed. A more general process, called "standard process", is treated in Dynkin [1] and Blumenthal and Getoor [1]. Another technical variety called "Ray process" is treated in Getoor [1].

§3.2. The definition of an excessive function and its basic properties are due to Hunt [2]. Another definition via the resolvents given in (14) is used in Blumenthal and Getoor [1]. Though the former definition is more restrictive than the latter, it yields more transparent results. On the other hand, there are results valid only with the resolvents.

§3.3. We follow Hunt's exposition in Hunt [3]. Although the theory of analytical sets is essential here, we follow Hunt in citing Choquet's theorem without details. A proof of the theorem may be found in Helms [1]. But for the probabilistic applications (projection and section) we need an algebraic version of Choquet's theory without endowing the sample space Ω with a topology. This is given in Meyer [1] and Dellacherie and Meyer [1].

§3.4 and 3.5. These sections form the core of Hunt's theory. Theorem 6 is due to Doob [3], who gave the proof for a "subparabolic" function. Its extension to an excessive function is carried over by Hunt using his Theorem 5. The transfinite induction used there may be concealed by some kind of maximal argument, see Hunt [3] or Blumenthal and Getoor [1]. But the quickest proof of Theorem 6 is by means of a projection onto the optional field, see Meyer [2]. This is a good place to learn the new techniques alluded to in the Notes on §1.4.

Dellacherie's deep result (Theorem 8) is one of the cornerstones of what may be called "random set theory". Its proof is based on a random version of the Cantor-Bendixson theorem in classical set theory, see Meyer [2]. Oddly enough, the theorem is not included in Dellacherie [2], but must be read in Dellacherie [1], see also Dellacherie [3].

§3.6. Theorem 1 originated with H. Cartan for the Newtonian potential (Brownian motion in R^3); see e.g., Helms [1]. This and several other important results for Hunt processes have companions which are valid for left continuous moderate Markov processes, see Chung and Glover [1]. Since a general Hunt process reversed in time has these properties as mentioned in the Notes on §2.4, such developments may be interesting in the future.

"Last exit time" is hereby rechristened "quitting time" to rhyme with "hitting time". Although it is the obvious concept to be used in defining recurrence and transience, it made a belated entry in the current vocabulary of the theory. No doubt this is partly due to the former prejudice against a random time that is not optional, despite its demonstrated power in older theories such as random walks and Markov chains. See Chung [2] where a whole section is devoted to last exits. The name has now been generalized to "co-optional" and "co-terminal"; see Meyer-Smythe-Walsh [1]. Intuitively the quitting time of a set becomes its hitting time when the sense of time is reversed ("from infinity", unfortunately). But this facile interpretation is only a heuristic guide and not easy to make rigorous. See §5.1 for a vital application.

§3.7. For a somewhat more general discussion of the concepts of transience and recurrence, see Getoor [2]. The Hunt theory is mainly concerned with the transient case, having its origin in the Newtonian potential. Technically, transience can always be engineered by considering (P_t^α) with $\alpha > 0$, instead of (P_t). This amounts to killing the original process at an exponential rate $e^{-\alpha t}$, independently of its evolution. It can be shown that the resulting killed process is also a Hunt process.

§3.8. Hunt introduced his Hypothesis (B) in order to characterize the balayage (hitting) operator P_A in a way recognizable in modern potential theory, see Hunt [1;§3.6]. He stated that he had not found "simple and general conditions" to ensure its truth. Meyer [3] showed that it is implied by the duality assumptions and noted its importance in the dual theory. It may be regarded as a subtle generalization of the continuity of the paths. Indeed Azéma [1] and Smythe and Walsh [1] showed that it is equivalent to the quasi left continuity of a suitably reversed process. We need this hypothesis in §5.1 to extend a fundamental result in potential theory from the continuous case to the general case. A new condition for its truth will also be stated there.

Chapter 4

Brownian Motion

4.1. Spatial Homogeneity

In this chapter the state space $\mathbf{E}$ is the d-dimensional Euclidean space R^d, $d \geq 1$; $\mathscr{E}$ is the classical Borel field on R^d. For $A \in \mathscr{E}$, $B \in \mathscr{E}$, the sets $A \pm B$ are the vectorial sum and difference of A and B, namely the set of $x \pm y$ where $x \in A$, $y \in B$. When $B = \{x\}$ this is written as $A \pm x$; the set $o - A$ is written as $-A$, where o is the zero point (origin) of R^d.

Let $(P_t; t \geq 0)$ be a strictly stochastic transition semigroup such that for each $t \geq 0$, $x \in \mathbf{E}$ and $x_0 \in \mathbf{E}$ we have

$$P_t(x, A) = P_t(x_0 + x, x_0 + A). \tag{1}$$

Then we say (P_t) is spatially homogeneous. Temporally homogeneous Markov processes with such a semigroup constitute a large and important class. They are called "additive processes" by Paul Lévy, and are now also known as "Lévy processes"; see Lévy [1]. We cannot treat this class in depth here, but some of its formal structures will be presented here as a preparation for a more detailed study of Brownian motion in the following sections.

We define a family of probability measures as follows.

$$\pi_t(A) = P_t(o, A). \tag{2}$$

The semigroup property then becomes the convolution relation

$$\pi_{s+t}(A) = \int \pi_s(dy)\pi_t(A - y) = (\pi_s * \pi_t)(A) \tag{3}$$

where the integral is over $\mathbf{E}$ and $*$ denotes the convolution. This is valid for $s \geq 0$, $t \geq 0$, with π_0 the unit mass at 0. We add the condition:

$$\pi_t \to \pi_0 \quad \text{vaguely as } t \downarrow 0. \tag{4}$$

For $f \in \mathscr{E}_+$ and $t \geq 0$ we have

$$P_t f(x) = \int f(x + y) \pi_t(dy). \tag{5}$$

As a consequence of bounded convergence, we see that if $f \in b\mathbb{C}$, then $P_t f \in b\mathbb{C}$; if $f \in \mathbb{C}_0$, then $P_t f \in \mathbb{C}_0$. Moreover $\lim_{t \downarrow 0} P_t f = f$ for $f \in \mathbb{C}_0$, by (4). Thus (P_t) is a Feller semigroup. Hence a spatially homogeneous process may and will be supposed to be a Hunt process.

For $A \in \mathscr{E}$, we define

$$U^\alpha(A) = \int_0^\infty e^{-\alpha t} \pi_t(A) \, dt;$$

and $U^\alpha(x, A) = U^\alpha(A - x)$. Then U^α is the α-potential kernel and we have for $f \in \mathscr{E}_+$:

$$U^\alpha f = \int f(x + y) U^\alpha(dy). \tag{6}$$

We shall denote the Lebesgue measure in $\mathbf{E}$ by m. If $f \in L^1(m)$, then we have by (5), Fubini's theorem and the translation-invariance of m:

$$\int m(dx) P_t f(x) = \int \left[\int m(dx) f(x + y) \right] \pi_t(dy)$$

$$= \int \left[\int m(dx) f(x) \right] \pi_t(dy) = \int m(dx) f(x).$$

For a general measure μ on $\mathscr{E}$, we define the measure μP_t or μU^α by

$$\mu P_t(\cdot) = \int \mu(dx) P_t(x, \cdot); \qquad \mu U^\alpha(\cdot) = \int \mu(dx) U^\alpha(x, \cdot). \tag{7}$$

The preceding result may then be recorded as

$$m P_t = m, \qquad \forall t \geq 0. \tag{8}$$

We say m is an invariant measure for the semigroup (P_t). It follows similarly from (6) that

$$m U^\alpha = \frac{1}{\alpha} m, \qquad \forall \alpha > 0; \tag{9}$$

namely m is also an invariant measure for the resolvent operators (αU^α).

Proposition 1. *Let $\alpha > 0$. If $U^\alpha f$ is lower semi-continuous for every $f \in b\mathscr{E}_+$, then $U^\alpha \ll m$. If $U^\alpha \ll m$, then $U^\alpha f$ is continuous for every $f \in b\mathscr{E}$.*

Proof. To prove the first assertion, suppose $A \in \mathcal{E}$ and $m(A) = 0$. Then $U^{\alpha}(x, A) = 0$ for m-a.e. x by (9). Since $U^{\alpha}1_A$ is lower semi-continuous, this implies $U^{\alpha}(A) = U^{\alpha}(o, A) = 0$. Next, if $U^{\alpha} \ll m$ let u^{α} be a Radon-Nikodym derivative of U^{α} with respect to m. We may suppose $u^{\alpha} \geq 0$ and $u^{\alpha} \in \mathcal{E}$. Then (6) may be written as

$$U^{\alpha}f(x) = \int f(x + y)u^{\alpha}(y)m(dy) = \int f(z)u^{\alpha}(z - x)m(dz). \tag{10}$$

Since $u^{\alpha} \in L^1(m)$, a classical result in the Lebesgue theory asserts that

$$\lim_{x' \to x} \int |u^{\alpha}(z - x) - u^{\alpha}(z - x')|m(dz) = 0 \tag{11}$$

(see e.g. Titchmarsh [1], p. 377). Since f is bounded, the last term in (10) shows that $U^{\alpha}f$ is continuous by (11). □

Proposition 2. *The three conditions below are equivalent:*

(a) *For all $\alpha \geq 0$ all α-excessive functions are lower semi-continuous.*
(b) *For some $\alpha > 0$, all α-excessive functions are lower semi-continuous.*
(c) *A set A (in $\mathcal{E}$) is of zero potential if $m(A) = 0$.*

Proof. If (b) is true then $U^{\alpha} \ll m$ by Proposition 1. Hence if $m(A) = 0$, then for every x, $m(A - x) = 0$ and $U^{\alpha}(x, A) = U^{\alpha}(A - x) = 0$. Thus A is of zero potential.

If (c) is true, then $U^{\alpha} \ll m$. Hence by Proposition 1, $U^{\alpha}f$ is continuous for every $f \in b\mathcal{E}_+$. It follows that $U^{\alpha}f$ is lower semicontinuous for every $f \in \mathcal{E}_+$, by an obvious approximation. Hence (a) is true by Theorem 9 of §3.2, trivially extended to all α-excessive functions. □

We proceed to discuss a fundamental property of spatially homogeneous Markov processes (X_t): the existence of a *dual* process $(\hat{X}_t)$. The latter is simply defined as follows:

$$\hat{X}_t = -X_t. \tag{12}$$

It is clear that $(\hat{X}_t)$ is a spatially homogeneous process. Let $\hat{P}_t$, $\hat{U}^{\alpha}$, $\hat{\pi}_t$, $\hat{u}^{\alpha}$ be the quantities associated with it. Thus

$$\hat{P}_t(x, A) = P_t(-x, -A)$$

$$\hat{\pi}_t(A) = \pi_t(-A), \qquad \hat{u}^{\alpha}(x) = u^{\alpha}(-x).$$

It is convenient to introduce

$$u^{\alpha}(x, y) = u^{\alpha}(y - x), \qquad \hat{u}^{\alpha}(x, y) = \hat{u}^{\alpha}(y - x).$$

Thus we have

$$\hat{u}^{\alpha}(x, y) = u^{\alpha}(y, x) \tag{13}$$

and

$$\hat{U}^{\alpha}(x, A) = \int_A m(dy)u^{\alpha}(y, x). \tag{14}$$

(In the last relation it is tempting to write $\hat{U}^{\alpha}(A, x)$ for $\hat{U}^{\alpha}(x, A)$. This is indeed a good practice in many lengthy formulas.) Now let $f \in b\mathscr{E}_+$, then

$$\hat{U}^{\alpha}f(y) = \int m(dx)f(x)u^{\alpha}(x, y). \tag{15}$$

Recall the notation

$$U^{\alpha}\mu(x) = \int u^{\alpha}(x, y)\mu(dy) \tag{16}$$

for a measure μ. If follows from (15) and (16) that if μ is a σ-finite measure as well as m, we have by Fubini's theorem:

$$\int \hat{U}^{\alpha}f(y)\mu(dy) = \int m(dx)f(x)U^{\alpha}\mu(x);$$

or perhaps more legibly:

$$\int \hat{U}^{\alpha}f \, d\mu = \int (f \cdot U^{\alpha}\mu) \, dm. \tag{17}$$

This turns out to be a key formula worthy to be put in a general context.

We abstract the situation as follows. Let (X_t) and $(\hat{X}_t)$ be two Hunt processes on the same general $(\mathbf{E}, \mathscr{E})$; and let $(P_t), (\hat{P}_t); (U^{\alpha}), (\hat{U}^{\alpha})$ be the associated semigroups and resolvents. Assume that there is a (reference) measure m and a function $u^{\alpha} \geq 0$ such that for every $A \in \mathscr{E}$ we have

$$U^{\alpha}(x, A) = \int_A u^{\alpha}(x, y)m(dy), \qquad \hat{U}^{\alpha}(x, A) = \int_A m(dy)u^{\alpha}(y, x). \tag{18}$$

Under these conditions we say the processes are in duality. In this case the relation (17) holds for any $f \in \mathscr{E}_+$ and any σ-finite measure μ; clearly it also holds with U^{α} and $\hat{U}^{\alpha}$ interchanged. Further conditions may be put on the dual potential density function $u(\cdot, \cdot)$. We content ourselves with one important result in potential theory which flows from duality.

Theorem 3. *Assume duality. Let μ and v be two σ-finite measures such that for some $\alpha \geq 0$ we have*

$$U^{\alpha}\mu = U^{\alpha}v < \infty. \tag{19}$$

Then $\mu \equiv v$.

Proof. We have by the resolvent equation for $\beta \geq \alpha$:

$$U^\alpha \mu = U^\beta \mu + (\beta - \alpha)U^\beta U^\alpha \mu.$$

Thus if (19) is true it is also true when α is replaced by any greater value. To illustrate the methodology in the simplest situation let us first suppose that μ and ν are finite measures. Take $f \in b\mathbb{C}$, then

$$\lim_{\alpha \to \infty} \alpha \hat{U}^\alpha f = f$$

boundedly because $(\hat{X}_t)$ is a Hunt process. It follows from this, the duality relation (17), and the finiteness of μ that

$$\int f \, d\mu = \lim_{\alpha \to \infty} \alpha \int \hat{U}^\alpha f \, d\mu = \lim_{\alpha \to \infty} \alpha \int (f \cdot U^\alpha \mu) \, dm.$$

This is also true when μ is replaced by ν, hence by (19) for all large values of α we obtain

$$\int f \, d\mu = \int f \, d\nu.$$

This being true for all $f \in b\mathbb{C}$, we conclude that $\mu \equiv \nu$. $\qquad\square$

In the general case the idea is to find a function h on $\mathbf{E}$ such that $h > 0$ and $\int h \, d\mu < \infty$; then apply the argument above to the finite measure $h \cdot d\mu$. If μ is a Radon measure, a bounded continuous h satisfying the conditions above can be easily constructed, but for a general σ-finite μ we use a more sophisticated method involving an α-excessive function for the dual process (see Blumenthal and Getoor [1]). Since $U^\alpha \mu < \infty$, it is easy to see that there exists $g \in b\mathscr{E}$, $g > 0$, such that by (17):

$$\infty > \int (g \cdot U^\alpha \mu) \, dm = \int \hat{U}^\alpha g \, d\mu.$$

Put $h = \hat{U}^\alpha g$; then clearly $h \in b\mathscr{E}$, $h > 0$ and $\int h \, d\mu < \infty$. Moreover, h being an α-potential for $(\hat{X}_t)$, is α-excessive for $(\hat{P}_t)$ (called "α-co-excessive"). Thus it follows from (11) of §3.2 that $\lim_{\alpha \to \infty} \alpha \hat{U}^\alpha h = h$. But we need a bit more than this. We know from the Corollary to Theorem 1 of §3.5 that h is finely continuous with respect to $(\hat{X}_t)$ (called "co-finely continuous"); hence so is fh for any $f \in b\mathbb{C}$, because continuity implies fine continuity and continuity in any topology is preserved by multiplication. It follows (see Exercise 7 of §3.5) that

$$\lim_{\alpha \to \infty} \alpha \hat{U}^\alpha (fh) = fh.$$

Now we can apply (17) with f replaced by fh to conclude as before that $\int fh\,d\mu = \int fh\,dv$; hence $h\,d\mu = h\,dv$; hence $\mu \equiv v$.   □

Next, we shall illustrate the method of Fourier transforms by giving an alternative characterization of a spatially homogeneous and temporally homogeneous Markov process (without reference to sample function regularities). A stochastic process $\{X_t, t \geq 0\}$ is said *to have stationary independent increments* iff it has the following two properties:

(a) For any $0 \leq t_0 < t_1 < \cdots < t_n$, the random variables $\{X(t_0), X(t_k) - X(t_{k-1}), 1 \leq k \leq n\}$ are independent;
(b) For any $s \geq 0$, $t \geq 0$, $X(s + t) - X(t)$ has the same distribution as $X(s) - X(0)$.

Theorem 4. *A spatially homogeneous and temporally homogeneous Markov process is a process having stationary independent increments, and conversely.*

Proof. We leave the converse part to the reader (cf. Theorem 9.2.2 of *Course*). To prove the direct assertion we use Fourier transform (characteristic functions). We write for $x \in R^d$, $y \in R^d$:

$$\langle x, y \rangle = \sum_{j=1}^{d} x_j y_j \tag{20}$$

if $x = (x_1, \ldots, x_d)$, $y = (y_1, \ldots, y_d)$. Let $i = \sqrt{-1}$ and $u_k \in R^d$; and set $X(t_{-1}) = o$. We have

$$E\left\{\exp\left[i \sum_{k=0}^{n+1} \langle u_k, X(t_k) - X(t_{k-1}) \rangle\right] \middle| \mathscr{F}_{t_n}^0\right\}$$

$$= \exp\left[i \sum_{k=0}^{n} \langle u_k, X(t_k) - X(t_{k-1}) \rangle\right]$$

$$\cdot E\{\exp[i\langle u_{n+1}, X(t_{n+1}) - X(t_n)\rangle]\,|\,X(t_n)\}. \tag{21}$$

Next we have if $u \in R^d$, $s \geq 0$, $t \geq 0$:

$$E\{\exp[i\langle u, X(s + t) - X(t)\rangle]\,|\,X(t)\}$$

$$= \exp[-i\langle u, X(t)\rangle] \int P_s(X(t), dy)\exp[i\langle u, y\rangle]$$

$$= \exp[-i\langle u, X(t)\rangle] \int P_s(o, dz)\exp[i\langle u, z + X(t)\rangle]$$

$$= \int P_s(o, dz)\exp[i\langle u, z\rangle] = \tilde{\pi}_s(u) \tag{22}$$

where $\tilde{\pi}_s$ is the characteristic function of the probability measure π_s defined in (2). Taking expectations in (21) we obtain

$$E\left\{\exp\left[i\sum_{k=0}^{n+1}\langle u_k, X(t_k) - X(t_{k-1})\rangle\right]\right\}$$

$$= E\left\{\exp\left[i\sum_{k=0}^{n}\langle u_k, X(t_k) - X(t_{k-1})\rangle\right]\right\} \cdot \tilde{\pi}_{t_{n+1}-t_n}(u_{n+1}).$$

Hence by induction on n, this is equal to

$$\check{\mu}_{t_0}(u_0)\prod_{k=1}^{n+1}\check{\pi}_{t_k-t_{k-1}}(u_k),$$

where $\check{\mu}_t$ is the characteristic function of the distribution μ_t of $X(t)$. Assertion (a) follows from this by a standard result on independence (see Theorem 6.6.1 of *Course*), and assertion (b) from taking expectations in (22). Note that π_s is the distribution of $X(s) - X(0)$.   □

We close this section by mentioning an important property of a process having stationary independent increments: its remote field is trivial. The precise result together with some useful consequences are given below in Exercises 4, 5, and 6.

Exercises

1. Is m the unique σ-finite measure satisfying (8), apart from a multiplicative constant? [This is a hard question in general. For the Brownian motion semigroup the answer is "yes"; for the Poisson semigroup the answer is "no".]

2. For the uniform motion semigroup (P_t) (Example 2 of §1.2), show that $U^\alpha \ll m$ and find $u^\alpha(x, y)$. In this case $P_t(x, \cdot)$ is singular with respect to m for each t and x. For the Poisson semigroup (P_t) (Example 3 of §1.2), find an invariant measure m_0 (on $\mathbb{N}$). In this case $P_t(n, \cdot) \ll m_0$ for each t and n.

3. The property of independent increments in Theorem 4 may be extended as follows. Let $T_0 = 0$; for $n \geq 1$ suppose $T_{n-1} < T_n$ and $T_n - T_{n-1}$ is optional relative to the σ-fields $(\mathscr{A}_t)$, where $\mathscr{A}_t$ is generated by $\{X(T_{n-1} + s) - X(T_{n-1}), 0 < s \leq t\}$. Then the random variables $X(T_n) - X(T_{n-1})$, $n \geq 1$, are independent.

The next three exercises are valid for a process having stationary independent increments.

4. Put

$$\mathcal{G}_t = \sigma(X_s, s \geq t)^{\sim}, \qquad \mathcal{G} = \bigwedge_{0 \leq t < \infty} \mathcal{G}_t$$

where $\sim$ denotes augmentation as in §2.3. The σ-field $\mathcal{G}$ is called the "remote field of the process. Prove that if $\Lambda \in \mathcal{G}$, then $P^x\{\Lambda\}$ is zero or one for every x. Is the result true for a process having arbitrary independent increments? [Hint: for discrete time analogues see §8.1 of *Course*.]

5. Suppose there exists $t > 0$ such that π_t has a density. If $f \in b\mathcal{E}$ and $f = P_t f$, then f is a constant. [Hint: show $f \in \mathbb{C}$ and use the Choquet-Deny-Hunt theorem (*Course*, p. 355).]

6. Under the hypothesis on (P_t) in Problem 6, a set $\mathcal{E}$ is recurrent or transient (see §3.6) according as $P_A 1 \equiv 1$ or $P_A 1 \not\equiv 1$. [Hint: let $f = \lim_{n \to \infty} \downarrow P_n P_A 1$, then f is a constant by Exercises 5; now use Problem 4.]

7. Suppose that for each $t > 0$, there exists a density π'_t of π_t with respect to m such that $\pi'_t > 0$ in $\mathbf{E}$. Then if $A \in \mathcal{E}$ and $P_A 1(x_0) = 0$ for some x_0, A must be polar. [Hint: show that $P_A 1(x) = 0$ for π_t-a.e. x, hence $P_t P_A 1(x) = 0$ for all x.]

8. Suppose that for each y, $u^\alpha(\cdot, y)$ is α-excessive. (It is possible to polish up u^α to achieve this.) Prove that for any measure μ, $U^\alpha \mu$ is α-excessive. Under this condition show that the condition (19) in Theorem 3 may be required to hold only m-a.e.

4.2. Preliminary Properties of Brownian Motion

The Brownian motion process in R^d was introduced in Example 2 of §3.7. For $d = 1$ it was introduced in Example 4 of §1.2. It is a Markov process with transition function:

$$P_t(x, dy) = p_t(x, y) \, dy, \qquad p_t(x, y) = \prod_{j=1}^{d} \frac{1}{\sqrt{2\pi t}} \left(\exp\left[-\frac{(y_j - x_j)^2}{2t} \right] \right). \quad (1)$$

Here

$$dy = dy_1 \cdots dy_d$$

is the Lebesgue measure in R^d, also written as $m(dy)$. We shall write "a.e." for "m-a.e.", and "density" for "density with respect to m". The product form in (1) shows that the Brownian motion $X(t)$ in R^d may be defined as the vector $(X_1(t), \ldots, X_d(t))$ where each coordinate $X_j(t)$ is a Brownian motion in R^1, and the d coordinates are stochastically independent. It is spatially homogeneous, and so by Theorem 4 of §4.1 may be characterized as a process having stationary independent increments, such that $X(t) - X(0)$ has the d-dimensional Gaussian (normal) distribution with mean the zero vector and covariance matrix equal to t times the $d \times d$ identity matrix. For $d = 1$, another useful characterization is by way of a Gaussian process, which is defined to be a process having Gaussian distributions for all finite-dimensional (marginal) distributions. The Brownian motion is then characterized as a Gaussian process such that for every s and t:

$$E(X(t)) = 0; \qquad E(X(s)X(t)) = s \wedge t. \qquad (2)$$

The proof requires Exercise 2 of §6.6 of *Course*, which states that mutually orthogonal Gaussian random variables are actually independent.

A trivial extension of the definition consists of substituting $\sigma^2 t$ for t in (1), where $\sigma^2 > 0$. We deem this an unnecessary nuisance. In this section we review and adduce a few basic results for the Brownian motion. We begin with the most important one.

(I) There is a version of the Brownian motion with continuous sample paths.

For $d = 1$ the proof is given in the Example in §3.1. The general case is an immediate consequence of the coordinate representation (even without independence among the coordinates). This fundamental property is often taken as part of the definition of the process. For simplicity we may suppose *all* sample functions to be continuous.

(II) The Brownian motion semigroup has both the Feller property and the strong Feller property.

The latter is verified in §3.7, and the former in Exercise 2 of §2.2. As defined, the strong Feller property does not imply the Feller property! Recall also from Example 4 of §3.7 that Uf is bounded continuous if $f \in b\mathscr{E}$ and f has compact support, and $U^\alpha f$ for $\alpha > 0$ is bounded continuous for $f \in b\mathscr{E}$. These properties can often be used as substitutes for the Feller properties.

(III) The Brownian motion in R^d is recurrent for $d = 1$ and $d = 2$; it is transient for $d \geq 3$.

This is verified in Examples 2 and 4 of §3.7. Let us mention that for $d = 1$ and $d = 2$, there are discrete-time versions of recurrence which supplement the conclusions of Theorem 1 of §3.7. For each $\delta > 0$, let $\{X(n\delta), n \geq 1\}$ be a "skeleton" of the Brownian motion $\{X(t), t \geq 0\}$. Then $\{X(n\delta) - X((n-1)\delta), n \geq 1\}$ is a sequence of independent and identically distributed random variables, with mean zero and finite second moment (the latter being needed only in R^2). Hence the recurrence criteria for a random walk in R^1 or R^2 (§8.3 of *Course*) yield the following result. For any nonempty open set G:

$$P^x\{X(n\delta) \in G \text{ for infinitely many values of } n\} = 1.$$

This is just one illustration of the close tie between the theories of random walk and of Brownian motion. In fact, all the classical limit theorems have their analogues for Brownian motion, which are frequently easier to prove because there are ready sharp estimates. Since our emphasis here is on the *process* we shall not delve into those questions.

Let us also recall that for any $d \geq 1$, each set is either recurrent or transient by Exercise 6 of §4.1.

(IV) A singleton is recurrent for Brownian motion in R^1, and polar in R^d for $d \geq 2$.

The statement for R^1 is trivial by recurrence and continuity of paths. The statement for R^d, $d \geq 2$, is proved in the Example of §3.6. Another proof will be given in §4.4 below.

As a consequence, we can show that the fine topology in R^d, $d \geq 2$, is strictly finer than the Euclidean topology. For example, if Q is the set of points in R^d with rational coordinates then $R^d - Q$ is a finely open set because Q is a polar set. It is remarkable that all the paths "live" on this set full of holes.

(V) Let

$$F = \{x \in R^d \mid x_1 = 0\}, \qquad G_1 = \{x \in R^d \mid x_1 > 0\}, \qquad G_2 = \{x \in R^d \mid x_1 < 0\}.$$
(3)

Then $F \subset G_1^r \cap G_2^r$.

To prove this we need consider only the case $d = 1$, but we will argue in general. Let $x \in F$. By the symmetry of the distribution of $X_1(t)$, we have

$$P^x\{X(t) \in G_1\} = P^x\{X(t) \in G_2\}.$$

Since $P^x\{X(t) \in F\} = 0$ by the continuity of the distribution of $X_1(t)$, we deduce that $P^x\{X(t) \in \bar{G}_1\} = \frac{1}{2}$. Hence for each $u > 0$:

$$P^x\{X(t) \in \bar{G}_1 \text{ for all } t \in [0, u]\} \leq \frac{1}{2},$$

which implies $P^x\{T_{G_2} \leq u\} \geq \frac{1}{2}$. It follows that $P^x\{T_{G_2} = 0\} \geq \frac{1}{2}$, and consequently $x \in G_2^r$ by Blumenthal's zero-or-one law (Theorem 6 of §2.3). Interchanging G_1 and G_2 we obtain $x \in G_1^r$.

Corollary. *For the Brownian motion in R^1, each x is regular for $\{x\}$.*

By (V), the path starting at o must enter G_1 and G_2 in an arbitrarily short time interval $[0, \varepsilon]$, and therefore must cross F infinitely many times by continuity. Thus o is a point of accumulation of the set $\{o\}$ in the fine topology; the same is true for any x by spatial homogeneity.

(VI) For any Borel set B, we have

$$X(T_B) \in \partial B \quad \text{on } \{0 < T_B < \infty\}.$$
(4)

This is a consequence of the continuity of paths, but let us give the details of the argument. By the definition of T_B, on the set $\{T_B < \infty\}$ for any $\varepsilon > 0$ there exists $t \in [T_B, T_B + \varepsilon)$ such that $X(t) \in B$. Hence either $X(T_B) \in B$, or by the right continuity of the path,

$$X(T_B) = \lim_{t \downarrow\downarrow 0} X(T_B + t) \in \bar{B}.$$

On the other hand, on the set $\{0 < T_B\}$, $X(t) \in B^c$ for $0 < t < T_B$, hence by the left continuity of the path,

$$X(T_B) = \lim_{t \uparrow\uparrow 0} X(T_B - t) \in \overline{B^c}.$$

Since $\partial B = \bar{B} \cap \overline{B^c}$, (4) is proved.

If $x \in B^r$, then $P^x\{X(T_B) = X(0) = x\} = 1$, but of course x need not be in ∂B, for instance if B is open and $x \in B$. This trivial possibility should be remembered in quick arguments.

(VII) Let C_n be decreasing closed sets and $\bigcap_n C_n = C$. Then we have for each $x \in C^c \cup C^r$:

$$P^x\left\{\lim_n T_{C_n} = T_C \leq \infty\right\} = 1. \tag{5}$$

This is contained in the Corollary to Theorem 5 of §2.4. We begin by the alert that (5) may not be true for $x \in C - C^r$! Take for example a sequence of closed balls shrinking to a single point. If $x \in C^r$, (5) is trivial. Let $x \in C^c$. Then there exists k such that $x \in C_k^c$. Since C_k^c is open x is not regular for C_n for all $n \geq k$. The rest of the argument is true P^x-a.s. We have $T_{C_n} > 0$ for all $n \geq k$, and $T_{C_n}\uparrow$. Let $S = \lim_n \uparrow T_{C_n}$, then $0 < S \leq T_C$. On $\{0 < T_{C_n} < \infty\}$, we have $X(T_{C_n}) \in C_n$ by (VI). Hence on $\{0 < S < \infty\}$, we have by continuity of paths:

$$X(S) = \lim_n X(T_{C_n}) \in \bigcap_n C_n = C.$$

Thus $S \geq T_C$ and so $S = T_C$. On $\{S = \infty\}$, $T_C = \infty$.

(VIII) Let $\mathscr{F}_t$ be either $\mathscr{F}_t^0$ or $\mathscr{F}_t^\sim$ (see §2.3). Then for any x,

$$\{X(t), \mathscr{F}_t, P^x\} \quad \text{and} \quad \{\|X(t)\|^2 - td, \mathscr{F}_t, P^x\} \tag{6}$$

are martingales. (Here $\|x\|^2 = \sum_{j=1}^d x_j^2$.)

To verify the second, note that since

$$\|X(t)\|^2 - td = \sum_{j=1}^d (X_j(t)^2 - t)$$

we need only verify it for $d = 1$, which requires a simple computation. Observe that in the definition of a martingale $\{X_t, \mathscr{F}_t^0, P^x\}$, if the σ-field $\mathscr{F}_t^0$ is completed to $\mathscr{F}_t^x$ with respect to P^x, then $\{X_t, \mathscr{F}_t^x, P^x\}$ is also a martingale. Hence we may as well use the completed σ-field.

See Exercise 14 below for a useful addition.

(IX) For any constant $c > 0, (1/c)X(c^2t), t \geq 0\}$ is also a Brownian motion. If we define

$$\tilde{X}(t) = tX\left(\frac{1}{t}\right) \quad \text{for } t > 0;$$
$$= 0 \qquad \text{for } t = 0; \tag{7}$$

then $\{\tilde{X}(t), t \geq 0\}$ is also a Brownian motion.

The first case is referred to as "scaling". The second involves a kind of reversing the time since $1/t$ decreases as t increases, and is useful in transforming limit behavior of the path as $t \uparrow \infty$ to $t \downarrow 0$. The proof that $\tilde{X}$ is a Brownian motion is easiest if we first observe that each coordinate is a Gaussian process, and then check in R^1 for $s > 0$, $t > 0$:

$$E^0\{\tilde{X}(s)\tilde{X}(t)\} = stE\left\{X\left(\frac{1}{s}\right)X\left(\frac{1}{t}\right)\right\} = st\left(\frac{1}{s} \wedge \frac{1}{t}\right) = s \wedge t.$$

Thus (2) is true and the characterization mentioned there yields the conclusion.

(X) Let B be a Borel set such that $m(B) < \infty$. Then there exists $\varepsilon > 0$ such that

$$\sup_{x \in \mathbf{E}} E^x\{\exp(\varepsilon T_{B^c})\} < \infty. \tag{8}$$

Proof. If $x \in (\bar{B})^c$, then $P^x\{T_{B^c} = 0\} = 1$ so (8) is trivial. Hence it is sufficient to consider $x \in \bar{B}$. For any $t > 0$, we have

$$\sup_{x \in \bar{B}} P^x\{T_{B^c} > t\} \leq \sup_{x \in \bar{B}} P^x\{X(t) \in B\}$$
$$= \sup_{x \in \bar{B}} \int_B p_t(x, y)\,dy \leq \frac{m(B)}{(2\pi t)^{d/2}}. \tag{9}$$

This number may be made $\leq \frac{1}{2}$ if t is large enough. Fix such a value of t from here on. It follows from the Markov property that for $x \in \bar{B}$ and $n \geq 1$;

$$P^x\{T_{B^c} > (n + 1)t\} = P^x\{T_{B^c} > nt; P^{X(nt)}[T_{B^c} > t]\}$$
$$\leq \frac{1}{2}P^x\{T_{B^c} > nt\}$$

since $X(nt) \in B$ on $\{T_{B^c} > nt\}$. Hence by induction the probability above is $\leq 1/2^{n+1}$. We have therefore by elementary estimation:

$$E^x\{\exp(\varepsilon T_{B^c})\} \leq 1 + \sum_{n=0}^{\infty} e^{\varepsilon(n+1)t} P^x\{nt < T_{B^c} \leq (n+1)t\}$$

$$\leq 1 + \sum_{n=0}^{\infty} e^{\varepsilon(n+1)t} 2^{-n}.$$

This series converges for sufficient small ε and (8) is proved. $\qquad \square$

Corollary. *If* $m(B) < \infty$ *then* $E^x\{T_{B^c}^n\} < \infty$ *for all* $n \geq 1$; *in particular* $P^x\{T_{B^c} < \infty\} = 1$.

This simple method yields other useful results, see Exercise 11 below.

For each $r > 0$, let us put

$$T_r = \inf\{t > 0 \mid \|X(t) - X(0)\| \geq r\}. \tag{10}$$

Let $B(x, r)$ denote the open ball with center x and radius r, namely

$$B(x, r) = \{y \in \mathbf{E} \mid \|x - y\| < r\}; \tag{11}$$

and $S(x, r)$ the boundary sphere of $B(x, r)$:

$$S(x, r) = \partial B(x, r) = \{y \in \mathbf{E} \mid \|x - y\| = r\}. \tag{12}$$

Then under P^x, T_r is just T_{B^c} where $B = B(x, r)$; it is also equal to $T_{\partial B}$. The next result is a useful consequence of the rotational symmetry of the Brownian motion.

(XI) For each $r > 0$, the random variables T_r and $X(T_r)$ are independent under any P^x. Furthermore $X(T_r)$ is uniformly distributed on $S(x, r)$ under P^x.

A rigorous proof of this result takes longer than might be expected, but it will be given. To begin with, we identify each ω with the sample function $X(\cdot, \omega)$, the space Ω being the class of all continuous functions in $\mathbf{E} = R^d$. Let φ denote a rotation in $\mathbf{E}$, and $\varphi\omega$ the point in Ω which is the function $X(\cdot, \varphi\omega)$ obtained by rotating each t-coordinate of $X(\cdot, \omega)$, namely:

$$X(t, \varphi\omega) = \varphi X(t, \omega). \tag{13}$$

Since φ preserves distance it is clear that

$$T_r(\varphi\omega) = T_r(\omega). \tag{14}$$

It follows from (VI) that $X(T_r) \in S(X(0), r)$. Let $X(0) = x$, $t \geq 0$ and A be a Borel subset of $S(x, r)$; put

$$\Lambda = \{\omega \mid T_r(\omega) \leq t; \ X(T_r(\omega), \omega) \in A\}. \tag{15}$$

Then on general grounds we have

$$\varphi^{-1}\Lambda = \{\omega \mid T_r(\varphi\omega) \leq t; \ X(T_r(\varphi\omega), \varphi\omega) \in A\}.$$

Using (13) and (14), we see that

$$\varphi^{-1}\Lambda = \{\omega \mid T_r(\omega) \leq t; \ X(T_r(\omega), \omega) \in \varphi^{-1}A\}. \tag{16}$$

Observe the double usage of φ in $\varphi^{-1}\Lambda$ and $\varphi^{-1}A$ above. We now claim that if x is the center of the rotation φ, then

$$P^x\{\Lambda\} = P^x\{\varphi^{-1}\Lambda\}. \tag{17}$$

Granting this and substituting from (15) and (16), we obtain

$$P^x\{T_r \leq t; \ X(T_r) \in A\} = P^x\{T_r \leq t; \ X(T_r) \in \varphi^{-1}A\}. \tag{18}$$

For fixed x and t, if we regard the left member in (18) as a measure in A, then (18) asserts that it is invariant under each rotation φ. It is well known that the unique probability measure having this property is the *uniform distribution on $S(x, r)$* given by

$$\frac{\sigma(A)}{\sigma(r)}, \qquad A \in \mathcal{E}, \ A \subset S(x, r) \tag{19}$$

where σ is the Lebesgue (area) measure on $S(x, r)$; and

$$\sigma(r) = \sigma(S(x, r)) = r^{d-1}\sigma(1), \qquad \sigma(1) = \frac{2\pi^{d/2}}{\Gamma\left(\dfrac{d}{2}\right)}. \tag{20}$$

Since the total mass of the measure in (18) is equal to $P\{T_r \leq t\}$, it follows that the left member of (18) is equal to this number multiplied by the number in (19). This establishes the independence of T_r and $X(T_r)$ since t and A are arbitrary, as well as the asserted distribution of $X(T_r)$.

Let us introduce also the following notation for later use:

$$v(r) = m(B(x, r)) = \int_0^r \sigma(s)\,ds = \frac{r^d}{d}\sigma(1). \tag{21}$$

It remains to prove (17), which is better treated as a general proposition (suggested by R. Durrett), as follows.

Proposition. *Let $\{X_t\}$ be a Markov process with transition function (P_t). Let φ be a Borel mapping of $\mathbf{E}$ into $\mathbf{E}$ such that for all $t \geq 0$, $x \in \mathbf{E}$ and $A \in \mathscr{E}$, we have*

$$P_t(x, \varphi^{-1}A) = P_t(\varphi x, A).$$

Then the process $\{\varphi(X_t)\}$ under P^x has the same finite-dimensional distributions as the process $\{X_t\}$ under $P^{\varphi(x)}$.

Remark. Unless φ is one-to-one, $\{\varphi(X_t)\}$ is not necessarily Markovian.

Proof. For each $f \in b\mathscr{E}$, we have

$$\int P_t(x, dy) f(\varphi y) = \int P_t(\varphi x, dy) f(y)$$

because this true when $f = 1_A$ by hypothesis, hence in general by the usual approximation. We now prove by induction on l that for $0 \leq t_1 < \cdots < t_l$ and $f_j \in b\mathscr{E}$:

$$E^x\{f_1(\varphi X_{t_1}) \cdots f_l(\varphi X_{t_l})\} = E^{\varphi(x)}\{f_1(X_{t_1}) \cdots f_l(X_{t_l})\}. \tag{22}$$

The left member of (22) is, by the Markov property and the induction hypothesis, equal to

$$E^x\{f_1(\varphi X_{t_1}) P^{X_{t_1}}[f_2(\varphi X_{t_2-t_1}) \cdots f_l(\varphi X_{t_l-t_{l-1}})]\}$$
$$= E^x\{f_1(\varphi X_{t_1}) P^{\varphi X_{t_1}}[f_2(X_{t_2-t_1}) \cdots f_l(X_{t_l-t_{l-1}})]\}$$

which is equal to the right member of (22). $\qquad \square$

The proposition implies (17) if $\Lambda \in \mathscr{F}^0$ or more generally if $\Lambda \in \mathscr{F}^{\sim}$.

We conclude this section by discussing some questions of measurability. After the pains we took in §2.4 about augmentation it is only fair to apply the results to see why they are necessary, and worthwhile.

(XII) Let B be a (nearly) Borel set, f_1 and f_2 universally measurable, bounded numerical function on $[0, \infty)$ and $\mathbf{E}_\partial$ respectively; then the functions

$$x \to E^x\{f_1(T_B)\}, \qquad x \to E^x\{f_2(X(T_B))\} \tag{23}$$

are both universally measurable, namely in $\mathscr{E}^{\sim}/\mathscr{B}$ where $\mathscr{B}$ is the Borel field on R^1.

That $T_B \in \mathscr{F}^\sim/\mathscr{B}$ is a consequence of Theorem 7 of §3.4; next, $X(T_B) \in \mathscr{F}^\sim/\mathscr{E}$ follows from the general result in Theorem 10 of §1.3. The assertions of (XII) are then proved by Exercise 3 of §2.4. In particular the functions in (23) are Lebesgue measurable. This will be needed in the following sections.

It turns out that if f_1 and f_2 are Borel measurable, then the functions in (23) are Borel measurable; see Exercise 6 and 7 below.

To appreciate the problem of measurability let f be a bounded Lebesgue measurable function from R^d to R^1, and $\{X_t\}$ the Brownian motion in R^d. Can we make $f(X_t)$ measurable in some sense? There exist two bounded Borel functions f_1 and f_2 such that $f_1 \leq f \leq f_2$ and $m(\{f_1 \neq f_2\}) = 0$. It follows that for any finite measure μ on $\mathscr{E}$ and $t > 0$, we have

$$E^\mu\{f_2(X_t) - f_1(X_t)\} = \iint \mu(dx) p_t(x, y)[f_2(y) - f_1(y)] m(dy) = 0.$$

Thus by definition $f(X_t) \in \bigwedge_\mu \mathscr{F}^\mu = \mathscr{F}^\sim$. Now let T_r be as in (10), then under P^x, $X(T_r) \in S(x, r)$ by (VI). But the Lebesgue measurability of f does not guarantee its measurability with respect to the area measure σ on $S(x, r)$, when f is restricted to $S(x, r)$. In particular if $0 \leq f \leq 1$ we can alter the f_1 and f_2 above to make $f_1 = 0$ and $f_2 = 1$ on $S(x, r)$, so that $E^\sigma\{f_2(X_t) - f_1(X_t)\} = \sigma(S(x, r))$. It should now be clear how the universal measurability of f is needed to overcome the difficulty in dealing with $f(X(T_r))$. Let us observe also that for a general Borel set B the "surface" ∂B need not have an area. Yet $X(T_{\partial B})$ induces a measure on ∂B under each P^μ, and if f is universally measurable $E^\mu\{f(X(T_{\partial B}))\}$ may be defined.

Exercises

Unless otherwise stated, the process discussed in the problems is the Brownian motion in R^d, and (P_t) is its semigroup.

1. If $d \geq 2$, each point x has uncountably many fine neighborhoods none of which is contained in another. For $d = 1$, the fine topology coincides with the Euclidean topology.

2. For $d = 2$, each line segment is a recurrent set; and each point on it is regular for it. For $d = 3$, each line is a polar set, whereas each point of a nonempty open set on a plane is regular for the set. [Hint: we can change coordinates to make a given line a coordinate axis.]

3. Let D be an open set and $x \in D$. Then $P^x\{T_{D^c} = T_{\partial D}\} = 1$. Give an example where $P^x\{T_{D^c} < T_{(\bar{D})^c}\} > 0$. Prove that if at each point y on ∂D there exists a line segment $\overline{yy'} \in D^c$, then $P^x\{T_{D^c} = T_{(\bar{D})^c}\} = 1$ for every $x \in D$. [Hint: use (V) after a change of coordinates.]

4. Let B be a ball. Compute $E^x\{T_{B^c}\}$ for all $x \in \mathbf{E}$.

5. If $B \in \mathscr{E}$ and $t > 0$, then $P^x\{T_B = t\} = 0$. [Hint: show $\int P^x\{T_B = s\}\, dx = 0$ for all but a countable set of s.]

6. For any Hunt process with continuous paths, and any closed set C, we have $T_C \in \mathscr{F}^0$. Consequently $x \to P^x\{T_C \leq t\}$ and $x \to E^x\{f(X(T_C))\}$ are in $\mathscr{E}$ for each $t \geq 0$ and $f \in b\mathscr{E}$ or $\mathscr{E}_+$. Extend this to any $C \in \mathscr{E}$ by Proposition 10 of §3.5. [Hint: let C be closed and $G_n \downarrow\downarrow C$ where G_n is open; then $\bigcap_n \{\exists t \in [a,b] : X(t) \in G_n\} = \{\exists t \in [a,b] : X(t) \in C\}$.]

7. For any Hunt process satisfying Hypothesis (L), $B \in \mathscr{E}$, $f \in \mathscr{B}$ and $f \in \mathscr{E}_+$ respectively, the functions $x \to E^x\{f(T_B)\}$ and $x \to E^x\{f(X(T_B))\}$ are both in $\mathscr{E}$. [Hint: by Exercise 3 of §3.5, $x \to E^x\{e^{-\alpha T_B}\}$ is in $\mathscr{E}$ for each $\alpha > 0$; use the Stone-Weierstrass theorem to approximate any function in $\mathbb{C}_0([0, \infty))$ by polynomials of e^{-x}. For the second function (say φ) we have $\alpha U^\alpha \varphi \to \varphi$ if $f \in b\mathbb{C}_+$. This is due to Getoor.]

8. If $f \in L^1(R^d)$ and $t > 0$, then $P_t f$ is bounded continuous.

9. If $f \in \mathscr{E}$ and $P_t|f| < \infty$ for every $t > 0$, then $P_t f$ is continuous for every $t > 0$. [Hint: for $\|x\| \leq 1$ we have $|P_t f(x) - P_t f(o)| \leq A[P_t|f|(o) + P_{4t}|f|(o)]$. This is due to T. Liggett.]

10. If $f \in b\mathscr{E}$, then $\lim_{t \to \infty} [P_t f(x) - P_t f(y)] = 0$ for every x and y. [Hint: put $x - z = \sqrt{2t}\zeta$ in $\int|p_t(x, z) - p_t(y, z)| \, dz$.]

11. In (X) if ε is fixed and $m(B) \to 0$, then the quantity in (8) converges to one.

12. Let X be the Brownian motion in R^1, $a > 0$ and B be a Borel set contained in $(-\infty, a]$. Prove that for each $t > 0$:
$$P^0\{T_{\{a\}} \leq t; X(t) \in B\} = P^0\{X(t) \in 2a - B\}$$
and deduce that
$$P^0\{T_{\{a\}} \leq t\} = 2P^0\{X(t) > a\}.$$

This is André's reflection principle. A rigorous proof may be based on Theorem 3 of §2.3.

13. For Brownian motion in R^1, we define the "last exit from zero before time t" as follows:
$$\gamma(t) = \sup\{s \leq t : X(s) = 0\}.$$
Prove that for $s \in (0, t)$ and $x \in R^1$, we have
$$P^0\{\gamma(t) \in ds\} = \frac{ds}{\pi\sqrt{s(t - s)}},$$
$$P^0\{\gamma(t) \in ds; X(t) \in dx\} = \frac{x}{2\pi\sqrt{s(t - s)^3}} e^{-x^2/2(t-s)} \, ds \, dx.$$

[This gives a glimpse of the theory of *excursions* initiated by P. Lévy. See Chung [8] for many such explicit formulas. The literature is growing in this area.]

14. For each $\lambda \in R^d$, $\{\exp(\langle \lambda, X_t \rangle) - \|\lambda\|^2 t/2, \mathscr{F}_t, P^x\}$ is a martingale, where $\langle \lambda, x \rangle = \sum_{j=1}^d \lambda_j x_j$.

4.3. Harmonic Function

A *domain* in $R^d (= \mathbf{E})$ is an open and connected (nonempty!) set. Any open set in R^d is the union of at most a countable number of disjoint domains, each of which called a component. It will soon be apparent that there is some advantage in considering a domain instead of an open set. Let D be a domain. We denote the hitting time of its complement by τ_D, namely:

$$\tau_D = T_{D^c} = \inf\{t > 0 \,|\, X(t) \in D^c\}. \tag{1}$$

This is called the "(first) exit time" from D. The usual convention that $\tau_D = +\infty$ when $X(t) \in D$ for all $t > 0$ is meaningful; but note that $X(\infty)$ is generally not defined. By (X) of §4.2, we have $P^x\{\tau_D < \infty\} = 1$ for all $x \in D$ if $m(D) < \infty$, in particular if D is bounded.

The class of Borel measurable functions from R^d to $[-\infty, +\infty]$ or $[0, \infty]$ will be denoted by $\mathscr{E}$ or $\mathscr{E}_+$; with the prefix "b" when it is bounded. We say f is locally integrable in D and write $f \in L^1_{\mathrm{loc}}(D)$ iff $\int_K |f| \, dm < \infty$ for each compact $K \subset D$, where m is the Lebesgue measure. We say A is "strictly contained" in B and write $A \Subset B$ iff $\bar{A} \subset D$.

Let $f \in \mathscr{E}_+$, and define a function h by

$$h(x) = E^x\{f(X(\tau_D)); \tau_D < \infty\}. \tag{2}$$

This has been denoted before by $P_{\tau_D} f$ or $P_{D^c} f$. Although h is defined for all $x \in \mathbf{E}$, and is clearly equal to $f(x)$ if $x \in (D^c)^\circ = (\bar{D})^c$, we are mainly concerned with it in $\bar{D}$. It is important to note that h is universally measurable, hence Lebesgue measurable by (XII) of §4.2. More specifically, it is Borel measurable by Exercise 6 of §4.2 because D^c is closed and the paths are continuous. Finally if f is defined (and Borel or universally measurable) only on ∂D and we replace τ_D by $T_{\partial D}$ in (2), the resulting function agrees with h in D (Exercise 1).

Recall the notation (11), (12) and (20) from §4.2.

Theorem 1. *If $h \not\equiv \infty$ in a domain D, then $h < \infty$ in D. For any ball $B(x, r) \Subset D$, we have*

$$h(x) = \frac{1}{\sigma(r)} \int_{S(x,r)} h(y)\sigma(dy). \tag{3}$$

Furthermore, h is continuous in D.

Proof. Write B for $B(x, r)$, then almost surely $\tau_B < \infty$; under P^x, $\tau_B < \tau_D$ so that $\tau_D = \tau_B + \tau_D \circ \theta_{\tau_B}$. Hence it follows from the fundamental balayage formula (Proposition 1 of §3.4) that

$$P_{\tau_D} f(x) = P_{\tau_B} P_{\tau_D} f(x) \tag{4}$$

which is just

$$h(x) = P_{\tau_B} h(x) \tag{5}$$

This in turn becomes (3) by the second assertion in (XI) of §4.2. Let us observe that (4) and (5) are valid even when both members of the equations are equal to $+\infty$. This is seen by first replacing f by $f \wedge n$ and then letting $n \to \infty$ there. A similar remark applies below. Replacing r by s in (3), then multiplying by $\sigma(s)$ and integrating we obtain

$$h(x) \int_0^r \sigma(s)\, ds = \int_0^r \int_{S(x,s)} h(y)\sigma(dy)\, ds = \int_{B(x,r)} h(y)m(dy). \tag{6}$$

where in the last step a trivial use of polar coordinates is involved. Recalling (21) of §4.2 we see that (6) may be written as

$$h(x) = \frac{1}{v(r)} \int_{B(x,r)} h(y)m(dy), \tag{7}$$

whether $h(x)$ is finite or not.

Now suppose $h \not\equiv \infty$ in D. Let $h(x_0) < \infty$, $x_0 \in D$; and $B(x,\rho) \subset B(x_0, r) \Subset D$. It follows from (7) used twice for x and for x_0 that

$$h(x) = \frac{1}{v(\rho)} \int_{B(x,\rho)} h(y)m(dy) \le \frac{1}{v(\rho)} \int_{B(x_0,r)} h(y)m(dy)$$

$$\le \frac{v(r)}{v(\rho)} h(x_0) < \infty. \tag{8}$$

Thus $h < \infty$ in $B(x_0, r)$. This shows that the set $F = D \cap \{x \mid h(x) < \infty\}$ is open. But it also shows that F is closed in D. For if $x_n \in F$ and $x_n \to x_\infty \in D$, then $x_\infty \in B(x_n, \delta) \Subset D$ for some $\delta > 0$ and some large n, so that $x_\infty \in F$ by the argument above. Since D is connected and our hypothesis is that F is not empty, we conclude that $F = D$. The relation (7) then holds for every x in D with the left member finite, showing that $h \in L^1_{loc}(D)$.

To prove that h is continuous in D, let r be so small that both $B(x,r)$ and $B(x',r)$ are strictly contained in D. Applying (7) for x and x' we obtain

$$|h(x) - h(x')| \le \frac{1}{v(r)} \int_C h(y)m(dy) \tag{9}$$

where $C = B(x,r) \triangle B(x',r)$. If x is fixed and $x' \to x$ then it is obvious that $m(C) \to 0$, and the integral in (8) converges to zero by the proven integrability of h. Thus h is continuous and Theorem 1 is proved. $\qquad\square$

We have shown above that the "sphere-averaging" property given in (3) entails the "ball-averaging" property given in (7). The converse is also true.

For we may suppose $h \not\equiv \infty$ in D; then h is finite continuous in D by the proof above. Now write (7) as follows:

$$h(x)v(r) = \int_0^r \int_{S(x,s)} h(y)\sigma(dy)\,ds.$$

Since h is continuous its surface integral over $S(x, s)$ is also continuous in s. Differentiating the equation above with respect to r we obtain (3) since $v'(r) = \sigma(r)$. We now proceed to unravel the deeper properties of the function h.

Definition 1. Let D be an open set. A function h is called *harmonic in D* iff

(a)   it is finite continuous in D;
(b)   the sphere-averaging property holds for any $B(x, r) \Subset D$.

By the proof of Theorem 1, if D is a domain and $h \in \mathscr{E}_+$, $h \not\equiv \infty$ in D, then (b) entails (a). Another easy consequence is as follows.

Corollary. *Let D be a domain and f be a Borel measurable function on ∂D. If $P_{D^c}|f| \not\equiv \infty$ in D, then $P_{D^c}f$ is harmonic in D. In particular this is the case if f is bounded on ∂D.*

It is remarkable that there is another quite different characterization of the class of harmonic functions. Recall that $C^{(k)}(D)$ is the class of k-times continuously differentiable functions in D;

$$\Delta = \sum_{j=1}^d \left(\frac{\partial}{\partial x_j}\right)^2$$

is the *Laplacian* in R^d.

Definition 2. A function h is harmonic in the open set D iff it belongs to $C^{(2)}(D)$, and satisfies in D the Laplace equation:

$$\Delta h = 0. \tag{10}$$

Theorem 2. (Gauss-Koebe) *Definitions 1 and 2 are equivalent. Moreover a harmonic function in D belongs to $C^{(\infty)}(D)$.*

Proof. Let h be harmonic in D according to Definition 1. We show first that $h \in C^{(\infty)}(D)$. For any $\delta > 0$ there exists a function φ in $C^{(\infty)}(E)$ with the following properties: $\varphi(x) = \varphi(\|x\|): \varphi(x) > 0$ for $\|x\| < \delta$; $\varphi(x) = 0$ for $\|x\| \geq \delta$; and $\int_E \varphi(x)m(dx) = 1$. See Exercise 6 for an example. We have then

$$\int_0^\infty \varphi(r)\sigma(r)\,dr = 1. \tag{11}$$

For each x in D, there exists $\delta > 0$ such that h is bounded in $\overline{B(x, \delta)}$ and (3) holds for $0 < r < \delta$. It follows that

$$
\begin{aligned}
h(x) &= \int_0^\infty \left[\frac{1}{\sigma(r)} \int_{S(x,r)} h(y)\sigma(dy) \right] \varphi(r)\sigma(r)\, dr \\
&= \int_0^\infty \int_{S(x,r)} h(y)\varphi(\|x - y\|)\sigma(dy)\, dr \\
&= \int_{\mathbf{E}} h(y)\varphi(\|x - y\|)m(dy).
\end{aligned}
\tag{12}
$$

Since φ has support in $B(o, \delta)$, and all partial derivatives of φ are bounded continuous in $B(o, \delta)$, we may differentiate the last-written integral with respect to x under the integral. Since $\varphi(\|x - y\|)$ is infinitely differentiable in x for each y, we see that h is infinitely differentiable.

To prove (10) let us recall Gauss's "divergence formula" from calculus, in a simple situation. Let B be a ball and h be twice continuously differentiable in a neighborhood of $\bar{B}$, then the formula may be written as follows:

$$
\int_B \Delta h(y)m(dy) = \int_{\partial B} \frac{\partial h}{\partial n}(y)\sigma(dy)
\tag{13}
$$

where $\partial h / \partial n$ is the outward normal derivative. Now let us denote the right member of (3) by $A(x, r)$. Making the change of variables $y = x + rz$ we have

$$
A(x, r) = \frac{1}{\sigma(1)} \int_{S(o,1)} h(x + rz)\sigma(dz).
\tag{14}
$$

Straightforward differentiation gives for fixed x:

$$
h'(x + rz) = \sum_{j=1}^d z_j \frac{\partial h}{\partial z_j}(x + rz) = \frac{\partial h}{\partial n}(x + rz);
\tag{15}
$$

$$
\begin{aligned}
A'(x, r) &= \frac{1}{\sigma(1)} \int_{S(o,1)} \frac{\partial h}{\partial n}(x + rz)\sigma(dz) \\
&= \frac{1}{\sigma(r)} \int_{S(x,r)} \frac{\partial h}{\partial n}(y)\sigma(dy) \\
&= \frac{1}{\sigma(r)} \int_{B(x,r)} \Delta h(y)m(dy).
\end{aligned}
\tag{16}
$$

The differentiation of (13) under the integral is correct because h has bounded continuous partial derivatives in a neighborhood of $S(o, 1)$. The first member of (16) vanishes for all sufficiently small values of r by (3), hence so does the integral in the last member (rid of the factor $1/\sigma(r)$). This being true for all sufficiently small values of r, we conclude $\Delta h(x) = 0$ by continuity. Thus h is harmonic according to Definition 2. Conversely if that is the hypothesis, then for any $B(x,r) \Subset D$, the formula (16) is valid because $h \in C^{(2)}(D)$. Since $\Delta h = 0$ in D the last member of (16) vanishes, hence so does the first member. Hence $A(x,r) = \lim_{r \downarrow 0} A(x,r)$, which is equal to $h(x)$ by the continuity of h. Therefore (3) is true and Theorem 2 is proved. $\square$

Definition 2 allows a "local" characterization of harmonicity. The function h is harmonic at x iff it belongs to $C^{(2)}$ in a neighborhood of x and $\Delta h(x) = 0$. It also yields an improvement of Definition 1, by retaining condition (a) there but weakening condition (b) as follows:

(b') the sphere-averaging property holds for each x in D, and all sufficiently small balls (strictly) contained in D with center at x.

Actually, even weaker conditions than (b') suffice, but these are technical problems; see Rao [1], p. 49. We add two basic propositions about harmonic functions below, both deducible from Definition 1 alone. The first is sometimes referred to as a "principle of maximum." However, there are some other results in potential theory called by that name.

Proposition 3. *Let h be harmonic in the domain D, and put*

$$M = \sup_{x \in D} h(x), \qquad m = \inf_{x \in D} h(x).$$

If there exists an x in D such that $h(x) = M$ or $h(x) = m$, then h is a constant in D.

Proof. Suppose $h(x) = M$. Then it follows at once from the ball-averaging property (7) that $h(y) = M$ first for a.e. y in a neighborhood of x, then for all y there since h is continuous. Thus the set $D \cap \{x \mid h(x) = M\}$ is open in D; it is also relatively closed in D by continuity. Hence $h = M$ in D. The proof for the infimum is exactly the same. $\square$

Corollary. *If h is harmonic in a bounded domain D, and continuous in $\bar{D}$, then the maximum and minimum values of h are taken on ∂D. In particular if $h \equiv 0$ on ∂D, then $h \equiv 0$ in $\bar{D}$.*

The next proposition is one of Harnack's theorems which hark back to Theorem 1. We relegate two other Harnack's theorems to Exercises 8 and 9.

Proposition 4. *Let D be a domain and $\{h_n, n \geq 1\}$ a sequence of harmonic functions in D. Suppose h_n increases to h_∞ in D. Then either $h_\infty \equiv +\infty$ in D, or h is harmonic in D.*

Proof. Even without the use of Definition 2, it is easy to see that we need only prove that h_∞ is harmonic in each bounded subdomain $D_1 \Subset D$. On D_1,

$h_1 \geq m > -\infty$ by continuity. Let $\tilde{h}_n = h_n - m$; then $\tilde{h}_n$ increases to $h_\infty = h_\infty - m \geq 0$. Each $\tilde{h}_n$ has the sphere-averaging property in D_1, hence so does $\tilde{h}_\infty$ by monotone convergence. Therefore by the proof of Theorem 1 from (6) on, either $\tilde{h}_\infty \equiv +\infty$ in D_1, or $\tilde{h}_\infty$ is finite continuous, hence harmonic in D_1.   □

Thanks to Definition 2, it is trivial to find explicit harmonic functions. It is well known that the real part of a complex analytic function is harmonic (in R^2), see e.g. Ahlfors [1]. Indeed the two characterizations have their analogues in the Cauchy integral formula and holomorphy, respectively. It is obvious from Definition 2 that in R^1, harmonicity is tantamount to linearity.

EXAMPLE. In R^2, $\log\|x\|$ is harmonic in any open set not containing o. In R^d, $d \geq 3$, $1/\|x\|^{d-2}$ is harmonic in any open set not containing o. Let K be a compact set, μ a measure such that $\mu(K) < \infty$; then

$$\int_K \log\|x - y\|\mu(dy), \qquad \int_K \|x - y\|^{2-d}\mu(dy) \tag{17}$$

are harmonic functions of x in $R^2 - K$ and $R^d - K$ respectively.

The first two assertions are standard exercises in calculus upon using Definition 2. The last assertion then follows either by differentiation under the integrals, or perhaps better by integration using Definition 1 and Fubini's theorem. Now consider the particular case of (17) when μ is the uniform distribution on the sphere $S(o, r)$. If $\|x\| > r$, then $z \to \log\|z\|$ is harmonic in a neighborhood of $B(x, r)$; hence the sphere-averaging property yields

$$\log\|x\| = \frac{1}{\sigma(r)} \int_{S(x,r)} \log\|z\|\sigma(dz) = \frac{1}{\sigma(r)} \int_{S(o,r)} \log\|x - y\|\sigma(dy);$$

similarly for the second integral in (17). However, to evaluate the integrals when $\|x\| < r$ we need the following proposition.

Proposition 5. *If h is harmonic and is a function of $\|x\|$ alone, then it must be of the form*

$$c_1 \log\|x\| + c_2, \qquad c_1\|x\|^{2-d} + c_2 \tag{18}$$

respectively in R^2 and R^d, $d \geq 3$: where c_1 and c_2 are constants.

Proof. Writing ρ for $\|x\|$, we see that the Laplace equation for h reduces to

$$\frac{d^2}{d\rho^2} h + \frac{d - 1}{\rho} \frac{d}{d\rho} h = 0. \tag{19}$$

Solving this simple differential equation we obtain the general solutions given in (18). These are called *fundamental radial solutions* of the Laplace equation.   □

Proposition 6. *We have for any r > 0:*

$$\frac{1}{\sigma(r)} \int_{S(o,r)} \log\|x - y\| \sigma(dy) = \log(\|x\| \vee r), \qquad d = 2; \qquad (20)$$

$$\frac{1}{\sigma(r)} \int_{S(o,r)} \|x - y\|^{2-d} \sigma(dy) = (\|x\| \vee r)^{2-d}, \qquad d \geq 3. \qquad (21)$$

Proof. Denote the left member of (20) by $f(x)$. It is harmonic in $B(o, r)$ and is a function of $\|x\|$ alone upon a geometric inspection. Hence by Proposition 5, $f(x) = a \log\|x\| + b$. For $x = o$ we have $f(o) = \log r$ by inspection; hence $a = 0$ and $b = \log r$. It follows that $f(x) = \log r$ for $\|x\| < r$. Similarly, $f(x) = a' \log\|x\| + b'$ for $\|x\| > r$. It is easily seen that $\lim_{\|x\| \to \infty} [f(x) - \log\|x\|] = 0$; hence $a' = 1$ and $b' = 0$. Thus $f(x) = \log\|x\|$ for $\|x\| > r$. We have already observed this above. It remains to show that f is finite continuous at $\|x\| = r$. This is a good exercise in calculus, not quite trivial and left to the reader. The evaluation of (21) is similar.   □

The exact calculation of formulas like (20) and (21) forms a vital part of classical potential theory, and is the source of many interesting results.

Exercises

1. For any open set D show that $P^x\{X(\tau_D) \in \partial D\} = 1$ for all $x \in \bar{D}$. However, if $x \in \partial D$ it is possible that $P^x\{\tau_D < T_{\partial D}\} = 1$. An example is known as *Lebesgue's thorn*, see Port and Stone [1], p. 68.

2. Show that if h is harmonic then so are all its partial derivatives of all orders. Verify the harmonicity of the following functions in R^3:

$$\frac{x_1}{\|x\|}, \qquad \frac{1}{\|x\|} \tan^{-1} \frac{x_2}{x_1}, \qquad x_3 \tan^{-1} \frac{x_2}{x_1}, \qquad 2x_3^2 - x_1^2 - x_2^2.$$

3. Let h be harmonic in R^d, $d \geq 1$ such that $P_t|h| < \infty$ for all $t > 0$. Then $h = P_t h$ for all $t \geq 0$, where (P_t) is the semigroup of the Brownian motion. The process $\{h(X_t), \mathscr{F}_t, t \geq 0\}$ is a continuous martingale. [Hint: use polar coordinates and the sphere-averaging property; a by-product is the value of $\sigma(1)$.]

4. Let $h \in \mathscr{E}, 0 \leq h < \infty$ and $h = P_1 h$. Then h is a constant. [Hint: if $h \not\equiv \infty$ then h is locally integrable; now compare the values of $P_n h$ at two points as $n \to \infty$, by elementary analysis. This solution is due to Hsu Pei.]

5. (Picard's Theorem.) If h is harmonic and ≥ 0 in R^d, then it is a constant. [Hint: there is a short proof by ball-averaging. A longer proof uses martingale and Exercise 4 of §4.1 to show that $\lim_{t \to \infty} h(X_t)$ is a constant a.s.]

6. Let $\varphi_1(x) = c_1 \exp((\|x\|^2 - 1)^{-1})$ for $\|x\| < 1$, $\varphi_1(x) = 0$ for $\|x\| \geq 1$, and c_1 is so chosen that $\int_{R_d} \varphi_1(x)\,dx = 1$. Put $\varphi_\delta(x) = \varphi(x/\delta)(1/\delta^d)$. Show that φ_δ has all the properties required of φ in the proof of Theorem 2.

7. Let D be a domain and h harmonic in D. If D is unbounded we adjoin a point at infinity ∞ to ∂D, and say that $x \to \infty$ when $\|x\| \to \infty$. Suppose now that

$$\varlimsup_{D \ni x \to \partial D \cup \{\infty\}} h(x) \leq M. \tag{22}$$

Then $h(x) \leq M$ for all $x \in D$. Give an example in which the conclusion becomes false if $\{\infty\}$ is omitted in (22).

8. Let $\{h_\alpha\}$ be a family of harmonic functions in an open set D. Suppose that the family is uniformly bounded on each compact subset of D. Then from any sequence from the family we can extract a subsequence which converges to a harmonic function in D, and the convergence is uniform on each compact subset of D. [Hint: prove first equi-continuity of the family on each compact subset by using (9), then apply the Ascoli-Arzelà theorem.]

9. Let D be a domain. For each compact $K \subset D$ there exists a constant $c(K)$ such that for any harmonic function $h > 0$ in D we have for any $x_1 \in K$, $x_2 \in K$:

$$\frac{h(x_1)}{h(x_2)} \geq c(K).$$

[Hint: the classical proof uses inequalities which are derived from the Poisson representation for a harmonic function in a ball (see §4.4). But a proof follows from (8), followed by the usual "chain argument."].

10. Let $a > 0$, $b > 0$. Evaluate the following integrals by means of Proposition 6:

$$\int_0^{2\pi} \log(a^2 + b^2 - 2ab \cos \theta)\,d\theta;$$

$$\int_0^{2\pi} (a^2 + b^2 - 2ab \cos \theta)^{-1/2}\,d\theta.$$

Compare the method with contour integration in complex variables.

11. Prove the continuity of the left members of (20) and (21) as functions of x, as asserted in Proposition 6.

12. Compute

$$\int_{B(o,r)} \log\|x - y\| m(dy), \qquad \int_{B(o,r)} \|x - y\|^{2-d} m(dy)$$

in R^2 and R^d, $d \geq 3$, respectively. [Answer for the second integral:

$$\frac{\sigma(1)}{d} \frac{r^d}{\|x\|^{d-2}} \qquad \text{if } \|x\| \geq r;$$

$$\frac{\sigma(1)r^2}{2} + \sigma(1)\|x\|^2\left(\frac{1}{d} - \frac{1}{2}\right), \qquad \text{if } \|x\| \leq r.]$$

13. (Gauss) In R^d, $d \geq 3$, let $\lfloor \mu \subset K \subset B(o,r)$, then

$$\mu(K) = \frac{r^{d-2}}{\sigma_d(r)} \int_{S(o,r)} U\mu(y)\sigma(dy).$$

Here $\sigma_d(r) = \sigma(r)$ in (20) of §4.2.

14. Extend the result in Exercise 13 to R^2, using U^* (given in (14) of §4.6 below) for U.

4.4. Dirichlet Problem

Let D be an open set and $f \in b\mathscr{E}$ on ∂D, and consider the function h defined in $\bar{D}$ by (2) of §4.3. Changing the notation, we shall denote h by $H_D f$, thus

$$H_D f(x) = E^x\{f(X(\tau_D)); \tau_D < \infty\}, \qquad x \in \bar{D}. \tag{1}$$

This formula defines a measure $H_D(x, \cdot)$ on the boundary ∂D, called the *harmonic measure* (of x with respect to D). For each Borel subset A of ∂D, the function $H(\cdot, A)$ is harmomic in D by the Corollary to Theorem 1 of §4.3. We may regard the measure $H(x, \cdot)$ as defined for all Borel sets, though it is supported by ∂D. The measure $H(x, \cdot)$ is also defined for $x \in D^c$ and is harmonic in $(\bar{D})^c = (D^c)^o$. For example if ∂D is the horizontal axis in R^2, $H_D f$ is a harmonic function both in the upper and lower open half-planes, but not necessarily in the whole plane.

We now study the behavior of $h(x)$ as x approaches the boundary. Since h is defined only in $\bar{D}$, we shall not repeat the restriction that $x \in \bar{D}$ below. Recall that a point z on ∂D is *regular* or *irregular for D^c* according as $P^z\{\tau_D = 0\} = 1$ or 0 by Blumenthal's zero-or-one law. It turns out that this is the same as its being a "regular boundary point of D" in the language of the non-probablistic treatment of the Dirichlet problem. Since z may be regular for D^c but irregular for D, the difference of terminology should be borne in mind. For instance, it is easy to construct a domain with a boundary point z which is irregular for D^c, but not so easy for z to be irregular for D.

Proposition 1. *For each* $t > 0$, *the function*

$$x \to P^x\{\tau_D \leq t\}$$

is lower semi-continuous in **E**.

Proof. Since D is open and X is continuous, we have $\{\tau_D = t\} \subset \{X(t) \in D^c\}$. It follows that

$$\{\tau_D > t\} = \{\forall s \in (0, t] : X(s) \in D\}$$

$$= \bigcap_{n=1}^{\infty} \left\{\forall s \in \left(\frac{t}{n}, t\right] : X(s) \in D\right\}.$$

The set above is in $\mathcal{F}_t^0$, and

$$P^x\{\tau_D > t\} = \lim_{n \to \infty} \downarrow P^x\left\{\forall s \in \left(\frac{t}{n}, t\right] : X(s) \in D\right\}. \tag{2}$$

The probability on the right side of (2) is of the form $P_{t/n}\varphi(x)$, where $\varphi(x) \in b\mathcal{E}$. Hence by the strong Feller property ((II) of §4.2), $x \to P_{t/n}\varphi(x)$ is bounded continuous. Therefore the left member of (2) is upper semi-continuous in x, which is equivalent to the proposition. $\qquad\square$

Note that on account of Exercise 5 of §4.2, $P^x\{\tau_D = t\} = 0$, but we do not need this.

Theorem 2. *If* z *on* ∂D *is regular for* D^c, *and if* f *is continuous at* z, *then we have*

$$\lim_{\bar{D} \ni x \to z} h(x) = f(z). \tag{3}$$

Proof. We have by Proposition 1, for each $\varepsilon > 0$:

$$\lim_{x \to z} P^x\{\tau_D \leq \varepsilon\} \geq P^z\{\tau_D \leq \varepsilon\} = 1. \tag{4}$$

Hence the corresponding limit exists and is equal to one. Note that here x need not be restricted to $\bar{D}$. Let T_r be as in (10) of §4.2. For $r > 0$ and $\varepsilon > 0$, $P^x\{T_r > \varepsilon\}$ does not depend on x, and for any fixed r we have

$$\lim_{\varepsilon \downarrow 0} P^o\{T_r > \varepsilon\} = 1, \tag{5}$$

because $T_r > 0$ almost surely. Now for any x, an elementary inequality gives

$$P^x\{\tau_D \leq \varepsilon < T_r\} \geq P^x\{\tau_D \leq \varepsilon\} + P^x\{\varepsilon < T_r\} - 1. \tag{6}$$

Consequently we have by (4)

$$\lim_{x \to z} P^x\{\tau_D \le \varepsilon < T_r\} \ge P^o\{\varepsilon < T_r\} \tag{7}$$

and so by (5):

$$\lim_{x \to z} P^x\{\tau_D < T_r\} = 1. \tag{8}$$

On the set $\{\tau_D < T_r\}$, we have $\|X(\tau_D) - X(0)\| < r$ and so $\|X(\tau_D) - z\| < \|X(0) - z\| + r$. Since f is continuous at z, given $\delta > 0$, there exists η such that if $y \in \partial D$ and $\|y - z\| < 2\eta$, then $|f(y) - f(z)| < \delta$. Now let $x \in \bar{D}$ and $\|x - z\| < \eta$; put $r = \eta$. Then under P^x, $X(\tau_D) \in \partial D$ (Exercise 1 of §4.3) and $\|X(\tau_D) - z\| < 2\eta$ on $\{\tau_D < T_\eta\}$; hence $|f(X(\tau_D)) - f(z)| < \delta$. Thus we have

$$E^x\{|f(X(\tau_D)) - f(z)|; \tau_D < \infty\} < \delta P^x\{\tau_D < T_\eta\} + 2\|f\|P^x\{T_\eta \le \tau_D < \infty\}. \tag{9}$$

When $x \to z$, the last term above converges to zero by (8) with η for r. Since δ is arbitrary, we have proved that the left member of (9) converges to zero. As a consequence,

$$\lim_{x \to z} h(x) = f(z) \lim_{x \to z} P^x\{\tau_D < \infty\} = f(z)$$

by (4). □

Let us supplement this result at once by showing that the condition of regularity of z cannot be omitted for the validity of (3) in R^d, $d \ge 2$. Note: in R^1 every $z \in \partial D$ is regular for $\{z\} \subset D^c$.

Theorem 3. *Suppose $d \ge 2$ and z is irregular for D^c. Then there exists $f \in b\mathbb{C}$ on ∂D such that*

$$\lim_{D \ni x \to z} h(x) < f(z). \tag{10}$$

Proof. Indeed (10) is true for any $f \in b\mathbb{C}(\partial D)$ such that $f(z) = 1$ and $f < 1$ on $\partial D - \{z\}$. An example of such a function is given by $(1 - \|x - z\|) \vee 0$. Since $\{z\}$ is polar by (IV) of §4.2, and z is irregular for D^c, we have $P^z\{X(\tau_D) = z\} = 0$. This implies the first inequality below:

$$1 > E\{f(X(\tau_D)); \tau_D < \infty\}$$
$$= \lim_{r \downarrow 0} E^z\{T_r < \tau_D; h(X(T_r))\}$$
$$\ge \lim_{r \downarrow 0} P^z\{T_r < \tau_D\} \inf_{x \in B(z,r) \cap D} h(x).$$

where the equation follows from $\lim_{r \downarrow 0} P^z\{T_r < \tau_D\} = 1$, and the strong Markov property applied at T_r on $\{T_r < \tau_D\}$. The last member above is equal to the left member of (10). □

We proceed to give a sufficient condition for z to be regular for D^c. This is a sharper form of a criterion known as Zaremba's "cone condition". We shall replace his solid cone by a flat one. Let us first observe that if $z \in \partial D$ and B is any ball with z as center, then z is regular for D^c if and only if it is regular for $(D \cap B)^c$. In this sense regularity is a local property. A "flat cone" in R^d, $d \geq 2$, is a cone in a hyperplane of dimension $d - 1$. It is "truncated" if all but a portion near its vertex has been cut off by a hyperplane perpendicular to its axis. Thus in R^2, a truncated cone reduces to a (short) line segment; in R^3, to a (narrow) fan.

Theorem 4. *The boundary point z is regular for D^c if there is a truncated flat cone with vertex at z and lying entirely in D^c.*

Proof. We may suppose that z is the origin and the flat cone lies in the hyperplane of the first $d - 1$ coordinate variables. In the following argument all paths issue from o. Let

$$T_n = \inf\left\{t > \frac{1}{n} \,\middle|\, X_d(t) = 0\right\}.$$

Since o is regular for $\{o\}$ in R^1 by (V) of §4.2, we have $\lim_{n \to \infty} T_n = 0$, P^o-a.s. Since $X_d(\cdot)$ is independent of $Y(\cdot) = \{X_1(\cdot), \ldots, X_{d-1}(\cdot)\}$, T_n is independent of $Y(\cdot)$. It follows (proof ?) that $Y(T_n)$ has a $(d - 1)$-dimensional distribution which is seen to be rotationally symmetric. Let C be a flat cone with vertex o, and C_0 its truncation which lies in D^c. Then rotational symmetry implies that $P^o\{Y(T_n) \in C\} = \theta$, where θ is the ratio of the "angle" of the flat cone to a "full angle". Since $\lim_{n \to \infty} Y(T_n) = 0$ by continuity, it follows that $\lim_{n \to \infty} P^o\{Y(T_n) \in C_0\} = \theta$. Since $C_0 \in D^c$, this implies

$$P^o\{\tau_D = 0\} = \lim_{n \to \infty} P^o\{\tau_D \leq T_n\} \geq \theta > 0.$$

Therefore, o is regular for D^c by the zero-or-one law. $\qquad\square$

Let us call a domain D *regular* when every point on ∂D is regular for D^c. As examples of Theorem 4, in R^2 an open disk with a radius deleted is regular; similarly in R^3 an open ball minus a sector of its intersection with a plane is regular, see Figure (p. 165). Of course, a ball or a cube is regular; for a simpler proof, see Exercise 1.

The most classical form of the *Dirichlet boundary value problem* may be stated as follows. Given an open set D and a bounded continuous function f on ∂D, to find a function which is harmonic in D, continuous in $\bar{D}$, and equal to f on ∂D. We shall refer to this problem as (D, f). The Corollary to Theorem 1 of §4.3, and Theorem 2 above taken together establish that the function $H_D f$ is a solution to the problem, provided that D is regular as just defined. When B is a ball, for instance, this is called the "interior Dirichlet problem" if $D = B$, and the "exterior Dirichlet problem" if $D = (\bar{B})^c$. The necessity of regularity was actually discovered long after the problem was posed. If there are irregular boundary points, the problem may have no solution. The simplest example is as follows. Let $D = B(o, 1) - \{o\}$ in R^2, namely the punctured unit disk; and define $f = 0$ on $\partial B(o, 1)$, $f(o) = 1$. The function $H_D f$ reduces in this case to the constant zero (why ?), which does not satisfy the boundary condition at o. But is there a true solution to the Dirichlet problem? Call this h_1; then h_1 is harmonic in D, continuous in $\bar{D}$ and equal to 1 and 0 respectively on $\partial B(o, 1)$ and at o. Let φ be an arbitrary rotation about the origin, and put $h_2(x) = h_1(\varphi x)$ for $x \in \bar{D}$. Whether by Definition 1 or 2 of §4.3, it is easy to see that h_2 is harmonic in D and satisfies the same boundary condition as h_1. Hence h_1 is identical with h_2 (see Proposition 5 below). In other words, we have shown that h_1 is rotationally invariant. Therefore by Proposition 5 of §4.3, it must be of the form $c_1 \log\|x\| + c_2$ in D. The boundary condition implies that $c_1 = 0$, $c_2 = 1$. Thus h_1 is identically equal to one in D. It cannot converge to zero at the origin. [The preceding argument is due to Ruth Williams.] It is obvious that o is an irregular boundary point of D, but one may quibble about its being isolated so that the continuity of f on ∂D is a fiction. However, we shall soon discuss a more general kind of example by the methods of probability to show where the trouble lies. Let us state first two basic results in the simplest case, the first of which was already used in the above.

Proposition 5. *If D is bounded open and regular, then $H_D f$ is the unique solution to the original Dirichlet problem (D, f).*

Proof. We know that $h = H_D f$ is harmonic in D. Since all points of ∂D are regular, and f is continuous in ∂D, h is continuous in $\bar{D}$ by Theorem 2. Let h_1 be harmonic in D, continuous in $\bar{D}$, and equal to f on ∂D. Then $h - h_1$ is harmonic in D, continuous in $\bar{D}$, and equal to zero on ∂D. Applying the Corollary to Proposition 3 of §4.3, we obtain $h - h_1 = 0$ in $\bar{D}$.   □

The following result is to be carefully distinguished from Proposition 5.

Proposition 6. *Let D be a bounded open set. Suppose that h is harmonic in D and continuous in $\bar{D}$. Then*

$$h(x) = H_D h(x), \qquad \forall x \in \bar{D}.$$

Proof. There exist regular open sets D_n such that $D_n \uparrow\uparrow D$. In fact each D_n may be taken to be the union of cells of a grid on R^d, seen to be regular by the cone condition. We have $h = H_{D_n}h$ in $\bar{D}_n$ by Proposition 5. Let first $x \in D$, then there exists m such that $x \in D_n$ for $n \geq m$; hence

$$h(x) = E^x\{h(X(\tau_{D_n}))\}.$$

Letting $n \to \infty$, we have $\tau_{D_n} \uparrow \tau_D$ P^x-a.s. by (VII) of §4.2. Hence the right member above converges to $H_D h(x)$ by bounded convergence. Next if $x \in \partial D$ and x is regular for D^c, then $H_D h(x) = h(x)$ trivially. Finally if $x \in \partial D$ and x is not regular for D^c, then $P^x\{\tau_D > 0\} = 1$. We have for each $t > 0$:

$$\begin{aligned}
E^x\{t < \tau_D; h(X(\tau_D))\} &= E^x\{t < \tau_D; E^{X(t)}[h(X(\tau_D))]\} \\
&= E^x\{t < \tau_D; h(X(t))\}
\end{aligned}$$

because $X(t) \in D$ on $\{t < \tau_D\}$, so the second equation follows from what we have already proved. As $t \downarrow 0$, the first term above converges to $H_D h(x)$, the third term to $E^x\{0 < \tau_D; h(X(0))\} = h(x)$ by bounded convergence. $\square$

We are now ready to discuss a general kind of unsolvable Dirichlet problem. Let D_0 be a bounded open domain in R^d, $d \geq 2$, and S a compact polar set contained in D_0. For instance, if $d = 2$, S may be any countable compact set; if $d = 3$, S may be a countable collection of linear segments. Let $D = D_0 - S$, and define $f = 1$ on ∂D_0, $f = 0$ on S. Note that $\partial D = (\partial D_0) \cup S$ since a polar set cannot have interior points. There exists a sequence of open neighborhoods $B_n \downarrow\downarrow S$ such that $\bar{B}_n \subset D_0$. Let $D_n = D_0 - B_n$. Suppose that h is harmonic in D, continuous in $\bar{D}$, and equal to one on ∂D_0. Nothing is said about its value on S. Since h is harmonic in D_n and continuous in $\bar{D}_n$, we have by Proposition 6 for $x \in D_n$:

$$\begin{aligned}
h(x) &= E^x\{h(X(\tau_{D_n}))\} \\
&= P^x\{T_{\partial D_0} < T_{\partial B_n}\} + E^x\{T_{\partial B_n} < T_{\partial D_0}; h(X(T_{\partial B_n}))\}.
\end{aligned} \tag{11}$$

Since S is polar, we have

$$\lim_n P^x\{T_{\partial B_n} < T_{\partial D_0}\} = P^x\{T_S < T_{\partial D_0}\} = 0. \tag{12}$$

Since h is bounded in $\bar{D}$, it follows that the last term in (11) converges to zero as $n \to \infty$. Consequently we obtain

$$h(x) = \lim_n P^x\{T_{\partial D_0} < T_{\partial B_n}\} = P^x\{T_{\partial D_0} < \infty\} = 1. \tag{13}$$

This being true for x in D_n for every n, we have proved that $h \equiv 1$ in D. Thus the Dirichlet problem (D, f) has no solution.

More generally, let f be continuous on ∂D_0, and arbitrary on S. Then $H_D f$ is harmonic in D, and the argument above shows that $H_{D_0} f$ is the unique continuous extension of $H_D f$ to D_0 (which is harmonic in D_0 to boot). In particular, the Dirichlet problem (D, f) is solvable only if $f = H_{D_0} f$ on S.

In the example above we constructed an irregular part of ∂D by using the polar set S. We shall see in §4.5 below that for any open D, the set of irregular points of ∂D is always a polar set. This deep result suggests a reformulation of the Dirichlet problem. But let us consider some simple examples first.

EXAMPLE 1. In R^d, $d \geq 3$, let $B = \overline{B(o, b)}$. Consider the function $P_B 1(x)$ for $x \in B^c$. This function is harmonic in B^c and is obviously a function of $\|x\|$ alone. Hence by Proposition 5 of §4.3, it is of the form $c_1 \|x\|^{2-d} + c_2$. The following equation illustrates a useful technique and should be argued out directly:

$$\forall x \in B^c: U(x, B) = P^x\{T_B < \infty\} U(y, B) \tag{14}$$

where y is any fixed point on ∂B. But actually this is just the identity $U 1_B = P_B U 1_B$ contained in Theorem 3(b) of §3.4, followed by the observation that $U(y, B)$ is a constant for $y \in \partial B$ by rotational symmetry. Now the left member of (14), apart from a constant, is equal to

$$\int_B \frac{dy}{\|x - y\|^{d-2}}$$

which converges to zero as $\|x\| \to \infty$. Hence $P_B 1(x) \to 0$ as $\|x\| \to \infty$, and so $c_2 = 0$. Since $P_B 1 = 1$ on ∂B, $c_1 = b^{d-2}$. Thus

$$P^x\{T_B < \infty\} = \left(\frac{b}{\|x\|}\right)^{d-2}, \qquad \|x\| \geq b. \tag{15}$$

Next, let $0 < a < b < \infty$ and

$$D = \{x \mid a < \|x\| < b\}. \tag{16}$$

Consider the Dirichlet problem (D, f), where f is equal to one on $S(o, a)$, zero on $S(o, b)$. The probabilistic solution given by $P_{D^c} f$ signifies the probability that the path hits $S(o, a)$ before $S(o, b)$. It is the unique solution by Proposition 5. Since it is a function of $\|x\|$ alone it is of the form $c_\perp \|x\|^{2-d} + c_2$. The constants c_1 and c_2 are easily determined by its value for $\|x\| = a$ and $\|x\| = b$. The result is recorded below:

$$P^x\{T_{S(o,a)} < T_{S(o,b)}\} = \frac{\|x\|^{2-d} - b^{2-d}}{a^{2-d} - b^{2-d}}, \qquad a \leq \|x\| \leq b \tag{17}$$

If $b \uparrow +\infty$ then $T_{S(o,b)} \uparrow \infty$ (why?), and we get (15) again with b replaced by a.

Exactly the same method yields for $d = 2$;

$$P^x\{T_{S(o,a)} < T_{S(o,b)}\} = \frac{\log\|x\| - \log b}{\log a - \log b}, \qquad a \leq \|x\| \leq b. \tag{18}$$

If we let $b \uparrow +\infty$, this time the limit of the above probability is 1, as it should be on account of recurrence. On the other hand, if we fix b and let $a \downarrow 0$ in (18), the result is $P^x\{T_{\{o\}} < T_{S(o,b)}\} = 0$ (why?). Now let $b \uparrow \infty$ to conclude that $P_{\{o\}}1(x) = 0$ for $x \neq o$. Since $P_t(x, \{o\}) = 0$ for any x in $\mathbf{E}$ and $t > 0$, it follows that $P_t P_{\{o\}}1(x) = 0$; hence $P_{\{o\}}1(x) = \lim_{t \downarrow 0} P_t P_{\{o\}}1(x) = 0$. Thus $\{o\}$ is a polar set. This is the second proof of a fundamental result reviewed in (IV) of §4.2.

EXAMPLE 2. The harmonic measure $H_B(x, \cdot)$ for a ball $B = B(o, r)$ can be explicitly computed. It is known as *Poisson's formula* and is one of the key formulas in the analytical developments of potential theory. In R^2 it is an analogue of, and can be derived from, the Cauchy integral formula for an analytic function in B:

$$f(x) = \frac{1}{2\pi i} \int_{S(o,r)} \frac{f(z)}{x - z} \, dz \tag{19}$$

where the integral is taken counterclockwise in the sense of complex integration. We leave this as an exercise and proceed to derive the analogue in R^d, $d \geq 3$. The method below works also in R^2.

For fixed z we know that $\|x - z\|^{2-d}$ is harmonic at all x except z. Hence so are its partial derivatives, and therefore also the following linear combination:

$$\sum_{j=1}^{d} \frac{2z_j}{d-2} \frac{\partial}{\partial x_j} \frac{1}{\|x - z\|^{d-2}} - \frac{1}{\|x - z\|^{d-2}} = \frac{\|z\|^2 - \|x\|^2}{\|x - z\|^d}.$$

Now consider the integral:

$$\int_{S(o,r)} \frac{r^2 - \|x\|^2}{\|x - z\|^d} \, \sigma(dz).$$

Then this is also harmonic for $x \in B(o, r)$ (why?). Inspection shows that it is a function of $\|x\|$ alone, hence it is of the form $c_1\|x\|^{2-d} + c_2$. Since its value at $x = 0$ is equal to $r^{2-d}\sigma(r)$ we must have $c_1 = 0$ and $c_2 = r^{2-d}\sigma(r) = r\sigma(1)$. Recall the value of $\sigma(1)$ from (20) of §4.2 which depends on the dimension d and will be denoted by $\sigma_d(1)$ in this example. Now let f be any continuous function on $S(o, r)$ and put

$$I(x, f) = \frac{1}{r\sigma_d(1)} \int_{S(o,r)} \frac{r^2 - \|x\|^2}{\|x - z\|^d} f(z)\sigma(dz). \tag{20}$$

This is the *Poisson integral* of f for $B(o, r)$. We have just shown that $I(x, 1) = 1$ for $x \in B(o, r)$. We will now prove that $I(x, f) = H_B(x, f)$ for all $f \in \mathbb{C}(\partial B)$, namely that the measure $H_B(x, \cdot)$ is given explicitly by $I(x, A)$ for all $A \in \mathscr{E}$, $A \subset \partial B$.

The argument below is general and will be so presented. Consider

$$h(x, z) = \frac{1}{r \sigma_d(1)} \frac{r^2 - \|x\|^2}{\|x - z\|^d}, \qquad \|x\| < r, \|z\| = r,$$

then $h > 0$ where it is defined, and $\int_{S(o, r)} h(x, z) \sigma(dz) = 1$. Furthermore if $\|y\| = r, y \neq z$, then $\lim_{x \to z} h(x, y) = 0$ boundedly for y outside any neighborhood of z. It follows easily that the probability measures $I(x, dz) = h(x, z) \sigma(dz)$ converge vaguely to the unit mass at z as $x \to z$. Namely for any $f \in \mathbb{C}(\partial B)$, $I(x, f)$ converges to $f(z)$ as $x \to z$. Since $I(\cdot, f)$ is harmonic in $B(o, r)$, $I(\cdot, f)$ is a solution to the Dirichlet problem (D, f). Hence it must coincide with $H_B f = H_B(\cdot, f)$ by Proposition 5.

Let us remark that $H_B(x, \cdot)$ is the distribution of the "exit place" of the Brownian motion path from the ball $B(o, r)$, if it starts at x in the ball. This is a simple-sounding problem in so-called "geometric probability". An explict formula for the joint distribution of τ_B and $X(\tau_B)$ is given in Wendel [1].

In view of the possible unsolvability of the original Dirichlet problem, we will generalize it as follows. Given D and f as before, we say that h is a solution to the *generalized Dirichlet problem* (D, f) iff h is harmonic in D and converges to f at every point of ∂D which is regular for D^c. Thus we have proved above that this generalized problem always has a solution given by $H_D f$. If D is bounded, this is the unique bounded solution. The proof will be given in Proposition 11 of §4.5. We shall turn our attention to the general question of non-uniqueness for an unbounded domain D.

Consider the function g defined by

$$g(x) = P^x\{\tau_D = \infty\} = 1 - P_{D^c} 1(x). \tag{21}$$

If D is bounded, then $\tau_D < \infty$ almost surely so that $g \equiv 0$. In general, g is a solution of the generalized Dirichlet problem $(D, 0)$. The question is whether it is a trivial solution, i.e., $g \equiv 0$ *in D*. Recall from §3.6 that (for a Hunt process) a Borel set A is recurrent if and only if $P_A 1 \equiv 1$. It turns out that this identity holds if it holds in A^c. For then we have for any x,

$$P^x\{T_A < \infty\} \geq P^x\{X(t) \in A\} + E^x\{X(t) \in A^c; P^{X(t)}[T_A < \infty]\}$$
$$= P^x\{X(t) \in A\} + P^x\{X(t) \in A^c\} = 1.$$

Applying this remark to $A = D^c$, we obtain the following criterion.

Proposition 7. *We have $g \equiv 0$ in D if and only if D^c is recurrent.*

As examples: in R^d, $d \geq 3$, if D is the complement of a compact set, then $g \not\equiv 0$ in D. In R^2, $g \not\equiv 0$ in D if and only if D^c is a polar set, in which case $g \equiv 1$.

When $g \not\equiv 0$ in D, the generalized Dirichlet problem (D, f) has the solutions $H_D f + cg$ for any constant c. The next result implies that there are no other bounded solutions. We state it in a form dictated by the proof.

Theorem 8. *Let D be an open set and f be a bounded Borel measurable function on ∂D. If h is bounded and harmonic in D, and $\lim_{D \ni x \to z} h(x) = f(z)$ for all $z \in \partial D$, then h must be of the form below:*

$$h(x) = H_D f(x) + cg(x), \qquad x \in D; \tag{22}$$

where g is defined in (21) and c is any constant. This h in fact converges to f at each point z of ∂D which is regular for D^c and at which f is continuous. In particular if f is continuous on ∂D, h is a generalized solution to the Dirichlet problem (D, f).

Proof. There exists a sequence of bounded regular open sets D_n such that $D_n \Subset D_{n+1}$ for all n, and $\bigcup_n D_n = D$. Such a sequence exists by a previous remark in the proof of Proposition 6. Suppose that h is bounded and harmonic in D. Consider the Dirichlet problem (D_n, h). It is plain that h is a solution to the problem, hence we have by Proposition 5:

$$h(x) = H_{D_n} h(x), \qquad x \in \bar{D}_n. \tag{23}$$

This implies the first part of the lemma below, in which we write T_n for τ_{D_n}, T for τ_D to lighten the typography; also $T_0 = 0$.

Lemma 9. *For each $x \in D$, the sequence $\{h(X(T_n)), \mathscr{F}(T_n), P^x; n \geq 0\}$ is a bounded martingale. We have P^x-a.s.:*

$$\lim_{n \to \infty} h(X(T_n)) = \begin{cases} f(X(T)) & \text{on } \{T < \infty\}; \\ c & \text{on } \{T = \infty\}; \end{cases} \tag{24}$$

where c is a constant not depending on x.

Proof. Since h is bounded in D by hypothesis, we have by the strong Markov property, for each $x \in D$ and $n \geq 1$:

$$E^x\{h(X(T_n)) \mid \mathscr{F}(T_{n-1})\} = E^{X(T_{n-1})}\{h(X(T_n))\}$$
$$= H_{D_n} h(X(T_{n-1})) = h(X(T_{n-1}))$$

where the last equation comes from (23). This proves the first assertion, and consequently by martingale convergence theorem the limit in (24) exists.

We now prove that there exists a random variable Z belonging to the remote field $\mathcal{G}$ of the Brownian motion process, such that the limit in (24) is equal to Z almost surely on the set $\{T = \infty\}$. [Readers who regard this as "intuitively obvious" should do Exercise 10 first.] Recall that $\mathcal{G}$ may be defined as follows. From each integer $k \geq 1$ let $\mathcal{G}_k$ be the σ-field generated by $X(n)$ for $n \geq k$, and augmented as done in §2.3, namely $\mathcal{G}_k = \sigma(X(n), n \geq k)^{\sim}$. Then $\mathcal{G} = \bigwedge_{k=1}^{\infty} \mathcal{G}_k$. By the Corollary to Theorem 8.1.4 of *Course* (see also the discussion on p. 258 there), $\mathcal{G}$ is trivial. [This is the gist of Exercise 4 of §4.1.]

Define $\tilde{h}$ in R^d to be h in D, and zero in D^c. We claim that there is a random variable Z (defined in Ω) such that

$$\lim_{n \to \infty} \tilde{h}(X(T_n)) = Z, \qquad P^x\text{-a.s.} \tag{25}$$

for every $x \in R^d$. For $x \in D$, under P^x we have $\tilde{h}(X(T_n)) = h(X(T_n))$, so the limit in (25) exists and is the same as that in (24). For $x \in D^c$, under P^x we have $\tilde{h}(X(T_n)) = \tilde{h}(x) = 0$, hence the limit in (25) exists trivially and is equal to zero. Since (25) holds P^x-a.s. for every $x \in R^d$, we can apply the shift to obtain for each $k \geq 1$:

$$\lim_{n \to \infty} \tilde{h}(X(T_n)) \circ \theta_k = Z \circ \theta_k, \qquad P^x\text{-a.s.} \tag{26}$$

for every $x \in R^d$. Let $x \in D$; then under P^x and on the set $\{T = \infty\}$, we have $T_n > k$ and consequently $\tilde{h}(X(T_n)) \circ \theta_k = h(X(T_n))$ for all sufficiently large values of n. Hence in this case the limit in (26) coincides with that in (24). Therefore the latter is equal to $Z \circ \theta_k$ which belongs to $\mathcal{G}_k$. This being true for each $k \geq 1$, the limit in (24) is also equal to $\overline{\lim}_{k \to \infty} Z \circ \theta_k$ which belongs to $\mathcal{G}$. Since $\mathcal{G}$ is trivial, the upper limit is a constant P^x-a.s. for each $x \in R^d$. This means: for each x there is a number $c(x)$ such that

$$P^x\{T = \infty; Z = c(x)\} = 1. \tag{27}$$

By Exercise 4 below, $g > 0$ in at most one component domain D_0 of D, and $g \equiv 0$ in $D - D_0$. Choose any x_0 in D_0 and put for all x in D:

$$\varphi(x) = P^x\{T = \infty; Z \neq c(x_0)\}.$$

It is clear from the definition of Z that for any ball $B = B(x, r) \Subset D$, $\varphi(x) = P_{B^c}\varphi(x)$ as in Theorem 1 of §4.3. Since φ is bounded, it is harmonic in D. Since $\varphi \geq 0$ and $\varphi(x_0) = 0$, as shown above, it follows by Proposition 3 of §4.3 (minimum principle) that $\varphi \equiv 0$ in D_0. As $g \equiv 0$ in $D - D_0$, we have $\varphi \equiv 0$ in D. Thus we may replace $c(x)$ by $c(x_0)$ in (27), proving the second line in (24). For $x \in D$, P^x-a.s. on the set $\{T < \infty\}$, we have $T_n \uparrow T$, $X(T_n) \in D$ and $X(T_n) \to X(T)$; since $h(y) \to f(z)$ as $y \in D$, $y \to z \in \partial D$ by hypothesis, we have $h(X(T_n)) \to f(X(T))$. This proves the first line of (24). The lemma is proved.

Now it follows by (23), (25) and bounded convergence that for each $x \in D$:

$$h(x) = \lim_{n \to \infty} E^x\{h(X(T_n))\}$$

$$= E^x\left\{\lim_{n \to \infty} h(X(T_n))\right\} = E^x\{Z\} = E^x\{Z; T < \infty\} + E^x\{Z; T = \infty\}$$

which reduces to (22) by (24). The rest of Theorem 8 is contained in Theorem 2 and is stated only for a recapitulation.

Exercises

1. Without using Theorem 4, give a simple proof that a ball or a cube in R^d is regular. Generalize to any solid with a "regular" surface so that there is a normal pointing outward at each point. [Hint: (V) of §4.2 is sufficient for most occasions.]

2. Let D be the domain obtained by deleting a radius from the ball $B(o, 1)$ in R^3; and let $f = \|x\|$ on ∂D. The original Dirichlet problem (D, f) has no solution.

3. In classical potential theory, a bounded domain D is said to be regular iff the Dirichlet problem (D, f) is solvable for every continuous f on D. Show that this definition is equivalent to the definition given here.

4. For any open set D, show that the function g defined in (21) is either identically zero or strictly positive in each connected component (domain) of D. Moreover there is at most one component in which $g > 0$. Give examples to illustrate the various possibilities.

5. What is the analogue of (17) and (18) for R^1?

6. Derive the Poisson integral in R^2 from Cauchy's formula (19), or the Taylor series of an analytic function (see Titchmarsh [1]).

7. Derive the Poisson integral corresponding to the exterior Dirichlet problem for the ball $B = B(o, r)$; namely find the explicit distribution of $X(T_{\partial B})1_{\{T_{\partial B} < \infty\}}$ under P^x, where $x \in B^c$.

8. Let $d \geq 2$; A be the hyperplane $\{x \in R^d \mid x_d = 0\}$; and let $D = \{x \in R^d \mid x_d > 0\}$. Compute $H_D(x, \cdot)$, namely the distribution of $X(T_A)$ under P^x, for $x \in D$. [Hint: we need the formula for Brownian motion in $R^1 : P^x\{T_{\{o\}} \in dt\} = |x|/(2\pi t)^{3/2} \exp[-x^2/2t] dt$; the rest is an easy computation. Answer: $H_D(x, dy) = \Gamma(d/2)x_d \pi^{-d/2}\|x - y\|^{-d}\sigma(dy)$ where σ is the area on A.]

9. Let D be open in $R^d, d \geq 2$. Then D^c is polar if and only if there does not exist any nonconstant bounded harmonic function in D. [Hint: if D^c is not polar, there exists a bounded continuous function f on ∂D which is not constant.]

10. Let $\{x_n, n \geq 1\}$ be independent, identically distributed random variables taking the values ± 1 with probability $1/2$ each; and let $X_n = \sum_{j=1}^{n} x_j$. Define for $k \geq 1: T_k = T_{\{k\}}$ on $\{X_1 = +1\}$; $T_k = T_{\{-k\}}$ on $\{X_1 = -1\}$; and $\Lambda = \{\lim_{k \to \infty} X(T_k) = +\infty\}$. Show that each T_k is optional with respect to $\{\mathscr{F}_n\}$, where $\mathscr{F}_n = \sigma(X_j, 1 \leq j \leq n)$; $T_k \uparrow \infty$ almost surely; but Λ does not belong to the remote field $\bigwedge_{m=1}^{\infty} \sigma(X_n, n \geq m)$. Now find a similar example for the Brownian motion. (This is due to Durrett.)

4.5. Superharmonic Function and Supermartingale

We introduce a class of functions which stand in the same relation to harmonic functions as supermartingales to martingales.

Definition. Let D be an open set. A function u is called *superharmonic in D* iff

(a) $-\infty < u \leq +\infty$; $u \not\equiv +\infty$ in each component domain of D; and u is lower semi-continuous in D;

(b)   for any ball $B(x, r) \Subset D$ we have

$$u(x) \geq \frac{1}{\sigma(r)} \int_{S(x,r)} u(y)\sigma(dy). \tag{1}$$

[A function u is *subharmonic in D* iff $-u$ is superharmonic in D; but we can dispense with this term here.] Thus the sphere-averaging property in (3) of §4.3 is replaced here by an inequality. The assumption of lower semi-continuity in (a) implies that u is bounded below on each compact subset of D. In particular the integral in the right member of (1) is well defined and not equal to $-\infty$. We shall see later that it is in fact finite even if $u(x) = +\infty$.

In consequence of (1), we have also for any $B(x, r) \Subset D$:

$$u(x) \geq \frac{1}{v(r)} \int_{B(x,r)} u(y)m(dy). \tag{2}$$

Since u is bounded below in $B(x, r)$ we may suppose $u \geq 0$ there, and $u(x) < \infty$. We then obtain (2) from (1) as we did in (6) of §4.3. It now follows that if u is superharmonic in a *domain D*, then it is locally integrable there. To see this we may again suppose $u \geq 0$ because u is bounded below on each compact subset of D. Let A be the set of points x in D such that u has a finite integral over some ball with center at x. It is trivial that A is open. Observe that if $x \in A$, then there is x_n arbitrarily near x for which $u(x_n) < \infty$, and consequently u has a finite integral over any $B(x_n, r) \Subset D$. Therefore this is also true for any $B(x, r) \Subset D$ by an argument of inclusion. Now let $x_n \in A$, $x_n \to x \in D$. Then for sufficiently large n and small r we have $x \in B(x_n, r) \Subset D$. Thus $x \in A$ and so A is closed in D. Since D is a domain and A is not empty we have proved $A = D$ as claimed.

We now follow an unpublished method of Doob's to analyse superharmonic functions by means of supermartingales. Suppose first that D is bounded and u is superharmonic in an open set containing $\bar{D}$. Let ρ denote the distance function in R^d. For a fixed $\varepsilon > 0$ and each x in D we put

$$r(x) = \tfrac{1}{2}\rho(x, \partial D) \wedge \varepsilon, \qquad S(x) = S(x, r(x));$$

thus $S(x)$ is the sphere with center x and radius $r(x)$ varying with x as prescribed. Define a sequence of space-dependent hitting times $\{T_n, n \geq 0\}$ as follows: $T_0 = 0$ and for $n \geq 1$:

$$T_n = \inf\{t > T_{n-1} \mid X(t) \in S(X(T_{n-1}))\} \tag{3}$$

where $\inf \varnothing = +\infty$ as usual. Each T_n is optional with respect to $\{\mathscr{F}_t\}$ (exercise!). The strong Markov property of X and the inequality (1) imply the key relation:

$$u(X(T_{n-1})) \geq E^x\{u(X(T_n)) \mid \mathscr{F}(T_{n-1})\}, \tag{4}$$

provided that $u(x) < \infty$. The integrability of $u(X(T_n))$ is proved by induction on n under this proviso. Thus $\{u(X(T_n)), \mathscr{F}(T_n), n \geq 1\}$ is a supermartingale under P^x. Since D is bounded all the terms of the supermartingale are bounded below by a fixed constant. Hence it may be treated like a positive supermartingale; in particular, it may be extended to the index set "$1 \leq n \leq \infty$" in a trivial way.

Let us first prove the crucial fact that for each $x \in D$, we have

$$P^x\left\{\lim_{n \to \infty} T_n = \tau_D\right\} = 1. \tag{5}$$

For it is clear that P^x-a.s., $T_n \uparrow$ and $T_n < \tau_D < \infty$ because D is bounded. Let $T_\infty = \lim_n \uparrow T_n$. Then $X(T_\infty) = \lim_{n \to \infty} X(T_n)$ by continuity and so $X(T_\infty) \in \bar{D}$. Suppose if possible that $X(T_\infty) \in D$. Then for all sufficiently large values of n, we have $\rho(X(T_{n-1}), \partial D) > 2^{-1}\rho(X(T), \partial D) > 0$ and $\rho(X(T_{n-1}), X(T_n)) < 4^{-1}\rho(X(T), \partial D) \wedge \varepsilon$. But by (3), $\rho(X(T_{n-1}), X(T_n)) = 2^{-1}\rho(X(T_{n-1}), \partial D) \wedge \varepsilon$. These inequalities are incompatible. Hence $X(T_\infty) \in \partial D$ and consequently $T_\infty = \tau_D$.

Now fix $t > 0$ and define a stopping random variable as follows:

$$N = \inf\{n \geq 1 \mid T_n \geq t\}. \tag{6}$$

where $\inf \varnothing = +\infty$. For each integer $k \geq 1$, $\{N = k\}$ belongs to the σ-field generated by $T_1, \ldots, T_k$; hence to $\mathscr{F}(T_k)$. Thus N is optional with respect to $\{\mathscr{F}(T_n), n \geq 1\}$. Therefore by Doob's stopping theorem for a positive (hence closable) supermartingale (Theorem 9.4.5 of *Course*):

$$\{u(X(T_{n \wedge N})), \mathscr{F}(T_{n \wedge N}), 1 \leq n \leq \infty\} \tag{7}$$

is a supermartingale under P^x provided $u(x) < \infty$. Here $X(T_\infty) = X(\tau_D)$, $\mathscr{F}(T_\infty) = \mathscr{F}(\tau_D)$ by (5). It follows that

$$u(x) \geq \lim_n E^x\left\{u(X(T_{n \wedge N}))\right\} \geq E^x\left\{\lim_n u(X(T_{n \wedge N}))\right\} \tag{8}$$

by Fatou's lemma since all the terms are bounded below. Now $\lim_n X(T_{n \wedge N}) = X(T_N)$ whether N is finite or infinite, and by the lower semi-continuity of u in $\bar{D}$ (because u is superharmonic in an open set containing $\bar{D}$) we conclude that

$$u(x) \geq E^x\{u(X(T_{N(\varepsilon)}))\}, \tag{9}$$

where we have indicated the dependence of N on ε. On the set $\{t < \tau_D\}$, there exists n such that $t < T_n$; hence $N(\varepsilon) < \infty$ and $T_{N(\varepsilon)-1} < t \leq T_{N(\varepsilon)}$ by (6). By the definition of $T_{N(\varepsilon)}$, this implies $|X(t) - X(T_{N(\varepsilon)})| \leq \varepsilon$. Thus we have proved

$$\lim_{\varepsilon \downarrow 0} X(T_{N(\varepsilon)}) = X(t \wedge \tau_D) \tag{10}$$

P^x-a.s. on $\{t < \tau_D\}$. On the set $\{t \geq \tau_D\}$, $N(\varepsilon) = +\infty$ for all $\varepsilon > 0$ and $X(T_{N(\varepsilon)}) = X(\tau_D)$ by (5). Hence (10) is also true. Using (10) in (9) and the lower semi-continuity of u again, we obtain

$$u(x) \geq E^x\{u(X(t \wedge \tau_D))\}. \tag{11}$$

The next step is standard: if $0 < s < t$, we have P^x-a.s.

$$E^x\{u(X(t \wedge \tau_D)) \mid \mathscr{F}_s\} = E^{X(s)}\{u(X((t - s) \wedge \tau_D))\} \leq u(X(s))$$
$$= u(X(s \wedge \tau_D)) \tag{12}$$

on $\{s < \tau_D\}$, since $X(s) \in D$; whereas on $\{s \geq \tau_D\}$ the two extreme members of (12) both reduce to $u(X(\tau_D))$ and there is equality. We record this important result below.

Theorem 1. *If D is bounded and u is superharmonic in an open set containing $\bar{D}$, then*

$$\{u(X(t \wedge \tau_D)), \mathscr{F}(t), 0 \leq t \leq \infty\} \tag{13}$$

is a supermartingale under P^x for each $x \in D$ for which $u(x) < \infty$. When u is harmonic in an open set containing $\bar{D}$, then (13) is a martingale under P^x for each $x \in D$.

We proceed to extend this result to an unbounded D and a function u which is superharmonic in D only. At the same time we will prove that the associated supermartingale has right continuous paths with left limits. To do this we use the basic machinery of a Hunt process given in §3.5. First, we must show that the relevant process is a Hunt process. For the Brownian motion in R^d this is trivial; indeed we verified that it is a Feller process. But what we need here is a Brownian motion somehow living on the open set D, as follows (see Exercise 9 for another way).

Let D be any open subset of R^d, and $D_\partial = D \cup \{\partial\}$ the one-point compactification of D. Thus ∂ plays the role of "the point at infinity": $x \to \partial$ iff x leaves all compact subsets of D, namely $x \to \partial D$; and all points of ∂D are identified with ∂. Define a process $\{\tilde{X}(t); t \geq 0\}$ living on the state space D_∂ as follows:

$$\tilde{X}(t) = \begin{cases} X(t), & \text{if } t < \tau_D; \\ \partial, & \text{if } \tau_D \leq t \leq \infty. \end{cases} \tag{14}$$

By this definition ∂ is made an absorbing state, and τ_D the lifetime of $\tilde{X}$; see (13) and (14) of §1.2. Define a transition semigroup $(Q_t, t \geq 0)$ as follows; for any bounded Borel function φ on D_∂ and $x \in D_\partial$:

$$Q_t\varphi(x) = E^x\{t < \tau_D; \varphi(X_t)\} + P^x\{t \geq \tau_D\}\varphi(\partial). \tag{15}$$

In particular, $Q_0\varphi = \varphi$. We call $\tilde{X}$ the "Brownian motion killed outside D" or "killed at the boundary ∂D". It is easy to check that (Q_t) is the transition semigroup of $\tilde{X}$, but the next theorem says much more. Recall $\{\mathscr{F}_t\}$ is the family of augmented σ-fields associated with the unrestricted Brownian motion $\{X_t, t \geq 0\}$.

Theorem 2. $\{\tilde{X}_t, \mathscr{F}_t, t \geq 0\}$ *is a Hunt process.*

Proof. First of all, almost all sample paths are continuous in $[0, \infty)$, in the topology of D_∂. To see this we need only check the continuity at τ_D when $\tau_D < \infty$. This is true because as $t \uparrow \tau_D < \infty$, $X(t) \to \partial = X(\tau_D)$. Since continuity clearly implies quasi left continuity, it remains only to check the strong Markov property. Let f be a continuous function on D_∂, then it is bounded because D_∂ is compact. Let T be optional with respect to $\{\mathscr{F}_t\}$, and $\Lambda \in \mathscr{F}_T$.

We have for $x \in D_\partial$ and $t \geq 0$:

$$E^x\{A; Q_t f(\tilde{X}_T); T < \infty\}$$
$$= E^x\{A; T < \tau_D; Q_t f(X_T)\} + P^x\{A; \tau_D \leq T < \infty\} f(\partial). \qquad (16)$$

since $Q_t f(\partial) = f(\partial)$ for $t \geq 0$. By (15), the first term on the right side above is equal to

$$E^x\{A; T < \tau_D; E^{X(T)}[t < \tau_D; f(X_t)]\} + E^x\{A; T < \tau_D; E^{X(T)}[t \geq \tau_D]\} f(\partial),$$

which is equal to

$$E^x\{A; T + t < \tau_D; f(X_{T+t})\} + P^x\{A; T < \tau_D \leq T + t\} f(\partial)$$

by the strong Markov property of the unrestricted Brownian motion X applied at T. Substituting into (16) we obtain

$$E^x\{A; T + t < \tau_D; f(X_{T+t})\} + P^x\{A; \tau_D \leq T + t < \infty\} f(\partial).$$

This is none other than $E^x\{A; f(\tilde{X}_{T+t}); T + t < \infty\}$ upon inspection. Since A is an arbitrary set from $\mathscr{F}_T$, we have verified that

$$E^x\{f(\tilde{X}_{T+t}) | \mathscr{F}_T\} = Q_t f(\tilde{X}_T)$$

for every continuous f on D_∂. Therefore, $\tilde{X}$ has the strong Markov property (cf. Theorem 1 of §2.3), indeed with respect to a larger σ-field than that generated by $\tilde{X}$ itself. $\qquad \square$

We are ready for the *dénouement*.

Theorem 3. *Let D be an open set and u be superharmonic and ≥ 0 in D; put $u(\partial) = 0$. Then u is excessive with respect to (Q_t). Conversely if u is excessive with respect to (Q_t), and $\not\equiv \infty$ in each component domain of D, then u restricted to D is superharmonic and ≥ 0.*

Proof. Let each D_n be bounded open and $D_n \uparrow\uparrow D$. Write $\tau_n = \tau_{D_n}$, $\tau = \tau_D$ below. Theorem 1 above is applicable to each D_n. Hence if $x \in D$ and $u(x) < \infty$ we have

$$u(x) \geq E^x\{u(X(t \wedge \tau_n))\} \geq E^x\{u(X(t))1_{\{t < \tau_n\}}\} \qquad (17)$$

where the second inequality is trivial because $u \geq 0$. As $n \to \infty$ we have P^x-a.s. $\tau_n \uparrow \tau$ and $\{t < \tau_n\} \uparrow \{t < \tau\}$. Hence it follows by monotone convergence that

$$u(x) \geq E^x\{u(X(t))1_{\{t < \tau\}}\} = Q_t u(x) \qquad (18)$$

where the equation is due to $u(\partial) = 0$. Since u is lower semi-continuous at x we have P^x-a.s. $u(x) \leq \underline{\lim}_{t \downarrow 0} u(X(t))$; while $\lim_{t \downarrow 0} 1_{\{t < \tau\}} = 1$. Hence by Fatou:

$$u(x) \geq E^x \left\{ \underline{\lim}_{t \downarrow 0} u(X(t)) 1_{\{t < \tau\}} \right\} \geq \underline{\lim}_{t \downarrow 0} Q_t u(x). \tag{19}$$

The two relations (18) and (19), together with the banality $u(\partial) = Q_t u(\partial)$, show that u is excessive with respect to (Q_t).

Conversely, if u is excessive with respect to (Q_t), then $u \geq 0$; and the inequality (1) is a very special case of $u \geq P_A u$ in Theorem 4 of §3.4. Observe that $Q_{A^c}(x, \cdot) = P_{A^c}(x, \cdot)$ if $x \in D$ and $\bar{A} \subset D$, where Q_A is defined for $\tilde{X}$ as P_A is for X. To show that u is superharmonic in D it remains to show that u is lower semi-continuous at each x in D. Let $B(x, 2r) \Subset D$. Then for all $y \in B(x, r) = B$, we have $P^y\{\tau \leq t\} \leq P^y\{T_r \leq t\}$ and the latter does not depend on y (T_r is defined in (10) of §4.2). It follows that

$$\lim_{t \downarrow 0} P^y\{\tau \leq t\} = 0 \quad \text{uniformly for } y \in B. \tag{20}$$

Now write for $y \in B$:

$$Q_t u(y) = E^y\{t < \tau; u(X_t)\} + P^y\{\tau \leq t\} u(\partial)$$
$$= P_t u(y) - E^y\{\tau \leq t; u(X_t) - u(\partial)\}. \tag{21}$$

Suppose first that u is bounded, then $P_t u$ is continuous by the strong Feller property; while the last term in (21) is less than ε if $0 < t < \delta(\varepsilon)$, for all $y \in B$. We have then

$$u(y) \geq Q_t u(y) \geq P_t u(y) - \varepsilon;$$
$$\underline{\lim}_{y \to x} u(y) \geq \lim_{y \to x} P_t u(y) - \varepsilon = P_t u(x) - \varepsilon \geq Q_t u(x) - \varepsilon.$$

Letting $t \downarrow 0$ we see that u is lower semi-continuous at x. For a general excessive u, we have $u = \lim_n \uparrow (u \wedge n)$, hence its lower semi-continuity follows at once from that of $u \wedge n$. $\quad\square$

Here is an application to a well known result in potential theory known as the *barrier theorem*. It gives a necessary and sufficient condition for the regularity at a boundary point of an open set. We may suppose the set to be bounded since regularity is a local property. A function u defined in D is called a *barrier* at z iff it is superharmonic and > 0 in D, and $u(x)$ converges to zero as x in D tends to z.

Proposition 4. *Let $z \in \partial D$; then z is regular for D^c if and only if there exists a barrier at z.*

Proof. Let u be a barrier at z and put $u(\partial) = 0$. Then u is excessive with respect to (Q_t). We may suppose u bounded by replacing it with $u \wedge 1$. Suppose z is not regular for D^c. Let B_1 and B_2 be two balls with center at z, $B_1 \subset B_2$. We have then

$$E^z\{\tau_{B_2} < \tau_D; u(X(\tau_{B_2}))\} = E^z\{\tau_{B_1} < \tau_D; E^{X(\tau_{B_1})}[\tau_{B_2} < \tau_D; u(X(\tau_{B_2}))]\}. \quad (22)$$

On $\{\tau_{B_1} < \tau_D\}$, we have $X(\tau_{B_1}) \in D$ and

$$E^{X(\tau_{B_1})}[\tau_{B_2} < \tau_D; u(X(\tau_{B_2}))] \leq E^{X(\tau_{B_1})}[u(\tilde{X}(\tau_{B_2} \wedge \tau_D)]$$
$$= Q_C u(\tilde{X}(\tau_{B_1})) \leq u(\tilde{X}(\tau_{B_1})) = u(X(\tau_{B_1})),$$

where $C = (B_2 \cap D)^c \cup \{\partial\}$, explicitly. Substituting into (22) we obtain

$$E^z\{\tau_{B_2} < \tau_D; u(X(\tau_{B_2}))\} \leq E^z\{\tau_{B_1} < \tau_D; u(X(\tau_{B_1}))\}. \quad (23)$$

Since P^z-a.s., $\tau_D > 0$ while $\tau_{B_2} \downarrow 0$ if B_2 shrinks to z. Hence we may choose B_2 so that the left member of (23) has a value > 0 because $u > 0$ in D. Now fix B_2 and let B_1 shrink to z. Then $X(\tau_{B_1}) \to z$ and so $u(X(\tau_{B_1})) \to 0$ by hypothesis; hence the right member converges to zero by bounded convergence. This contradiction proves that z must be regular for D^c.

Conversely if z is regular for D^c, put $f(x) = \|x - z\|$ on ∂D. Then f is bounded on ∂D since D is bounded; and $f(x) = 0$ if and only if $x = z$. By the solution of the Dirichlet problem for (D, f), $u = P_{D^c} f$ is harmonic in D and converges to $f(z) = 0$ as $x \to z$. Since $f > 0$ on ∂D except at one point (which forms a polar set), it is clear that $u > 0$ in D. Thus u is the desired barrier at z. $\square$

Proposition 5. *Let u be superharmonic and bounded below in a domain D. Let B be open and $\bar{B} \subset D$. Then $P_{B^c} u$ is harmonic in B, and superharmonic in D.*

Proof. We may suppose $u \geq 0$ by adding a constant and put $u(\partial) = 0$. Then u is excessive with respect to (Q_t) and so by Theorem 4 of §3.4, for every $x \in D$:

$$u(x) \geq Q_{B^c} u(x) = P_{B^c} u(x). \quad (24)$$

Since u is locally integrable, there exists x in each component of B for which $u(x) < \infty$. Theorem 1 of §4.3 applied with $f = u$ then establishes the harmonicity of $P_{B^c} u$ in B. $P_{B^c} u$ is excessive with respect to (Q_t) by Theorem 4 of §3.4, hence superharmonic in D by Theorem 3 above. $\square$

When B is a ball with $\bar{B} \subset D$, the function P_{B^c} is known as the *Poisson integral of u for B* and plays a major role in classical potential theory. Since B is regular we have $P_{B^c} u = u$ in $D - B$. Thus the superharmonic function u is transformed into another one which has the nicer property of being harmonic in B. This is the simplest illustration of the method of balayage: "sweeping all the charge in B onto its boundary". (There is no charge where $\Delta u = 0$, by a law of electromagnetism.) Poisson's formula in Example 2 of §4.4 gives the exact form of the Poisson integral. A closely related notion will now be discussed.

EXAMPLE. Recall from Example 4 of §3.7, for $d \geq 3$:

$$u(x, y) = \frac{A_d}{\|x - y\|^{d-2}}, \qquad A_d = \frac{\Gamma\left(\dfrac{d}{2} - 1\right)}{2\pi^{d/2}}.$$

Then $u(x, y)$ denotes the potential density at y for the Brownian motion starting from x. For a fixed y, $u(\cdot, y)$ is harmonic in $R^d - \{y\}$ by Proposition 5 of §4.3. Since it equals $+\infty$ at y it is trivial that the inequality (1) holds for $x = y$; since it is continuous in the extended sense, $u(\cdot, y)$ is superharmonic in R^d. Now let D be an open set and for a fixed $y \in D$ put

$$g_D(x, y) = u(x, y) - P_{D^c}u(x, y) \tag{25}$$

where P_{D^c} operates on the function $x \to u(x, y)$. It follows from the preceding discussion that $g_D(\cdot, y)$ is superharmonic in D, harmonic in $D - \{y\}$, and vanishes in $(\bar{D})^c$ and at points of ∂D which are regular for D^c; if D is regular then it vanishes in D^c. Since $u(\cdot, y)$ is superharmonic and ≥ 0 in R^d, it is excessive with respect to (P_t) by an application of Theorem 3 with $D = R^d$. Of course this fact can also be verified directly without the intervention of superharmonicity; cf. the Example in §3.6. Now $g_D \geq 0$ by Theorem 4 of §3.4. Hence it is excessive with respect to (Q_t) by another application of Theorem 3. For any Borel measurable function f such that $U|f|(x) < \infty$ we have

$$\int_{R^d} g_D(x, y)f(y)\,dy = Uf(x) - P_{D^c}Uf(x) = E^x\left\{\int_0^{\tau_D} f(X_t)\,dt\right\} \tag{26}$$

where X_t may be replaced by $\tilde{X}_t$ in the last member above. Thus $g_D(x, y)$ plays the same role for $\tilde{X}$ as $u(x, y)$ for X; it is the potential density at y for the Brownian motion starting from x and killed outside D.

The function g_D is known as the *Green's function* for D and the quantity in (26), which may be denoted by $G_D f$, is the associated *Green's potential* of f. It is an important result that $g_D(x, y) = g_D(y, x)$ for all x and y; indeed Hunt [1] proved that $Q_t(x, dy)$ has a density $q(t; x, y)$ which is symmetric in (x, y) and continuous in $(0, \infty) \times D \times D$. This is a basic result in the deeper study of the killed process, which cannot be undertaken here. In R^2, Green's function can also be defined but the situation is complicated by the fact that the Brownian motion is recurrent so that the corresponding

u in (25) is identically $+\infty$. [This is excessive but not superharmonic!] Finally, for a ball in any dimension Green's function can be obtained by a device known as Kelvin's transformation; see e.g. Kellogg [1]. The case $d = 1$ is of course quite elementary.

As a corollary of Proposition 5, we see now that the sphere-average in the right member of (1) is finite for every x in D. For it is none other than $P_{B^c}u(x)$ when $B = B(x, r)$. The next proposition is somewhat deeper.

Proposition 6. *If u is excessive with respect to (Q_t), and $\not\equiv \infty$ in each component of D, with $u(\partial) < \infty$, then $Q_t u < \infty$ for each $t > 0$.*

Proof. Let $x \in D$, $B = B(x, r) \Subset D$, and $\tau = \tau_D$. Then we have

$$E^x\{t < \tau; u(X_t)\} = E^x\{t \leq \tau_B; u(X_t)\} + E^x\{\tau_B < t < \tau; u(X_{\tau_B \vee t})\}. \quad (27)$$

The first term on the right side of (27) does not exceed

$$E^x\{X_t \in \bar{B}; u(X_t)\} = \int_B p(t; x, y)u(y)\,dy \leq \frac{1}{(2\pi t)^{d/2}} \int_B u(y)\,dy < \infty$$

because u is locally integrable. The second term does not exceed

$$E^x\{u(X(\tau_B))\} = P_{B^c}u(x) < \infty$$

by the supermartingale stopping theorem applied to $\{u(\tilde{X}_t)\}$. Since

$$E^x\{t \geq \tau; \; u(X_t)\} = P^x\{t \geq \tau\}u(\partial) < \infty$$

$Q_t u(x) < \infty$ by (15).   □

We can now extend Theorem 1 as follows.

Theorem 7. *Let D be an open set and u be superharmonic and ≥ 0 in D. Then for any $x \in D$,*

$$w = \lim_{t \uparrow\uparrow \tau_D} u(X(t)) \quad \text{exists } P^x\text{-a.s.} \quad (28)$$

Put for $0 \leq t \leq \infty$:

$$w(t) = \begin{cases} u(X(t)), & t < t_D; \\ w, & t \geq \tau_D; \end{cases} \quad (29)$$

thus $w(\infty) = w$. Then for any $x \in D$, we have P^x-a.s.: $t \to w(t)$ is right continuous in $[0, \infty)$ and has left limits in $(0, \infty]$; and $\{w(t), \mathscr{F}(t), 0 < t \leq \infty\}$ is a supermartingale under P^x. In case $u(x) < \infty$, the parameter value $t = 0$ may be added to the supermartingale. If u is harmonic and bounded in D, then $\{w(t), \mathscr{F}(t), 0 \leq t \leq \infty\}$ is a continuous martingale under P^x for any $x \in D$.

Proof. By Theorem 3, u is excessive with respect to (Q_t). Hence by Theorem 6 of §3.4 applied to the Hunt process $\tilde{X}$, $t \to u(\tilde{X}_t)$ is right continuous in $[0, \infty)$ and has left limits in $(0, \infty]$, P^x-a.s. for any $x \in D$. It follows that the limit in (28) exists. The asserted continuity properties of $t \to w(t)$ then follow trivially. Next, let D_n be bounded open, $\bar{D}_n \subset D_{n+1}$ and $\bigcup_n D_n = D$. Given x in D there exists n for which $x \in D_n$. It follows from Proposition 5 that $E^x\{u(X(\tau_{D_m}))\} < \infty$ for all $m \geq n$, and is decreasing since u is excessive. Hence, $E^x\{w\} < \infty$ by Fatou. We have either by Theorem 1 or by the stopping theorem for excessive functions, under P^x:

$$w(0) = u(x) \geq \varliminf_n E^x\{u(X(t \wedge \tau_{D_n}))\} \geq E^x\left\{\varliminf_n u(X(t \wedge \tau_{D_n}))\right\}$$

$$= E^x\{u(X(t)); t < \tau_D\} + E^x\{w; t \geq \tau_D\}$$

$$= E^x\{w(t)\}.$$

For any $x \in D$ and $t > 0$;

$$E^x\{w(t)\} = Q_t u(x) + E^x\{t \geq \tau_D; w\} < \infty.$$

These relations imply the assertions of the theorem, the details of which are left to the reader.   □

There is a major improvement on Theorem 7 to the effect that almost all sample functions of $w(t)$ defined in (29) are actually *continuous*. This is Doob's theorem and the proof given below makes use of a new idea, the reversal of time. We introduce it for a spatially homogeneous Markov process discussed in §4.1, as follows. The transition probability function is said to have a *density* iff for every $t > 0$ there exists $p_t \geq 0$, $p_t \in \mathscr{E} \times \mathscr{E}$ such that

$$P_t(x, dy) = p_t(x, y)\,dy. \tag{30}$$

Recall that the Lebesgue measure m is invariant in the sense of (8) of §4.1. In the case of (30) this amounts to $\int m(dx)p_t(x, y) = 1$ for m-a.e. y. Although m is not a finite measure, it is trivial to define P^m and E^m in the usual way; thus $P^m(\Lambda) = \int m(dx)P^x(\Lambda)$ for each $\Lambda \in \mathscr{F}^\sim$. Now for a fixed $c > 0$ define a *reverse process* $\tilde{X}_c$ in $[0, c]$ as follows:

$$\tilde{X}_c(t) = X(c - t), \qquad t \in [0, c]. \tag{31}$$

There is no difficulty in the concept of a stochastic process on a parameter set such as $[0, c]$. The sample functions of $\tilde{X}_c$ are obtained from those of X by reversing the time from c. Since X as a Hunt process is right continuous and has left limits in $(0, c)$, $\tilde{X}_c$ is left continuous and has right limits there. This kind of sample functions has been considered earlier in Chapter 2, for instance in Theorem 6 of §2.2.

Proposition 8. *Let X be a spatially homogeneous Markov process satisfying* (30), *where for each $t > 0$, p_t is a symmetric function of (x, y). Then under* P^m, $\tilde{X}_c$ *has the same finite dimensional distributions as X in* $[0, c]$.

Proof. Let $0 \le t_1 < \cdots < t_n \le c$, and $f_j \in \mathscr{E}_+$ for $1 \le j \le n$. To show that

$$E^m \left\{ \prod_{j=1}^{n} f_j(X(t_j)) \right\} = E^m \left\{ \prod_{j=1}^{n} f_j(\tilde{X}_c(t_j)) \right\} \tag{32}$$

we rewrite the right member above as $E^m \{ \prod_{j=n}^{1} f_j(X(c - t_j)) \}$. Since $mP_{c-t_j} = m$, the latter can be evaluated as follows:

$$\int \cdots \int dx_n f_n(x_n) \prod_{j=n-1}^{1} p(t_{j+1} - t_j; x_{j+1}, x_j) f_j(x_j) \, dx_j$$

$$= \int \cdots \int f_1(x_1) \, dx_1 \prod_{j=1}^{n-1} p(t_{j+1} - t_j; x_j, x_{j+1}) f_{j+1}(x_{j+1}) \, dx_{j+1}.$$

The last expression is equal to the left member of (32) by an analogous evaluation. □

We return to Brownian motion X below.

Theorem 9. *Let u be a positive superharmonic function in R^d. Then $t \to u(X(t))$ is a.s. continuous in* $[0, \infty]$ *(i.e., continuous in* $(0, \infty)$ *and has limits at 0 and at ∞).*

Proof. Consider the reverse process $\tilde{X}_c$ in (31). Although under P^m it is a copy of Brownian motion in $[0, c]$ under P^m by Proposition 8, it is not absolutely clear that we can apply Theorem 6 of §3.4 to $u(\tilde{X}_c(t))$ since that theorem is stated for a Hunt process in $[0, \infty)$. However, it is easy to extend $\tilde{X}_t$ to $[0, \infty)$ as follows. Define

$$Y_c(t) = \begin{cases} \tilde{X}_c(t), & \text{for } t \in [0, c], \\ \tilde{X}_c(c) + X(t) - X(c), & \text{for } t \in [c, \infty). \end{cases} \tag{33}$$

It follows from the independent increments property of X and Proposition 8 that $\{Y_c(t), t \ge 0\}$ under P^m is a copy of Brownian motion under P^m. Therefore $t \to u(Y_c(t))$ is right continuous in $[0, \infty)$ and has left limits in $(0, \infty]$. In particular, $t \to u(\tilde{X}_c(t)) = u(X(c - t))$ is right continuous in $(0, c)$, which means that $t \to u(X(t))$ is left continuous in $(0, c)$. But $t \to u(X(t))$ is also right continuous in $(0, c)$, hence it is in fact continuous there. This being true for each c, we conclude that under P^m, $t \to u(X(t))$ is continuous in $[0, \infty]$. By the definition of P^m, this means that for m-a.e. x we have the same result under P^x. For each $s > 0$ and x, $P_s(x, \cdot)$ is absolutely continuous with respect to m. Hence for $P_s(x, \cdot)$-a.e. y, the result holds under P^y. For $s \geq 0$ let

$$\Lambda_s = \{\omega \,|\, u(X(\cdot, \omega)) \text{ is continuous in } [s, \infty]\}.$$

Then $\Lambda_s \in \mathscr{F}^\sim$ (proof?). We have for each x,

$$P^x\{\Lambda_s\} = E^x\{P^{X(s)}[\Lambda_0]\} = 1$$

because $X(s)$ has $P_s(x, \cdot)$ as distribution. Letting $s \downarrow 0$, we obtain $P^x\{\Lambda_0\} = 1$.
$\square$

Theorem 9 is stated for u defined in the whole space. It will now be extended as follows. Let D be an open set, and u be positive and superharmonic in D. The function $t \to u(X(t, \omega))$ is defined in the set $I(\omega) = \{t > 0 \,|\, X(t, \omega) \in D\}$, which is a.s. an open set in $(0, \infty)$. We shall say that $u(X)$ is (right, left) continuous *wherever defined* iff $t \to u(X(t, \omega))$ is so on $I(\omega)$ for a.e. ω. Let r and r' be rational numbers, $r < r'$. It follows from Theorem 7 that for each x

$$P^x\{X(t) \in D \text{ for all } t \in [r, r'], \text{ and } t \to u(X(t)) \text{ is right continuous}$$
$$\text{in } (r, r')\} = P^x\{X(t) \in D \text{ for all } t \in [r, r']\}. \tag{34}$$

For we can apply Theorem 7 to the Brownian motion starting at time r in D, on the set $\{r' < \tau_D\}$. Since every generic t in $I(\omega)$ is caught between a pair $[r, r']$ in the manner shown in (34), we have proved that for a Brownian motion X with any initial distribution, $u(X)$ is right continuous wherever defined. In particular, the last assertion is true for the Brownian motion Y_c defined in (33) under P^m. Thus $t \to u(Y_c(t))$ is right continuous wherever defined, hence $t \to u(X(c - t))$ is right continuous for $t \in (0, c)$ and $c - t \in I(\omega)$. Since c is arbitrary, this implies that $t \to u(X(t))$ is left continuous for $t \in I(\omega)$. Thus $u(X)$ is in fact continuous wherever defined under P^m. As in Theorem 9, we can replace P^m by P^x for every x. In the context of Theorem 7, we have proved that $t \to u(X(t))$ is continuous in $(0, \tau_D)$, hence $t \to w(t)$ is continuous in $[0, \infty]$. We state this as a corollary, although it is actually an extension of Theorem 9.

Corollary. *In Theorem 7, $t \to w(t)$ is continuous in $[0, \infty]$, a.s.*

Using the left continuity of an excessive function along Brownian paths, we can prove a fundamental result known as the Kellogg-Evans theorem in classical potential theory (see §5.1).

Theorem 10. *For the Brownian motion process, a semi-polar set is polar.*

Proof. By the results of §3.5, it is sufficient to prove that a thin compact set K is polar. Put $\varphi(x) = E^x\{e^{-T_K}\}$. Since K is thin, $\varphi(x) < 1$ for all x. Let first $x \notin K$, and G_n be open sets such that $G_n \Downarrow K$. Then $\lim_n \uparrow T_{G_n} = T_K \leq \infty$ and $T_{G_n} < T_K$ for all n, P^x-a.s. The strict inequality above is due to the continuity of the Brownian paths and implies $T_K = T_{G_n} + T_K \circ \theta(T_{G_n})$ on $\{T_{G_n} < \infty\}$. It follows by the strong Markov property that

$$E^x\{e^{-T_K} \,|\, \mathscr{F}(T_{G_n})\} = e^{-T_{G_n}}\varphi(X(T_{G_n})). \tag{35}$$

Since $T_{G_n} \in \mathscr{F}(T_{G_n})$, we have $T_K \in \bigvee_{n=1}^{\infty} \mathscr{F}(T_{G_n})$. Letting $n \to \infty$ in (35), we obtain a.s.

$$e^{-T_K} = e^{-T_K}\varphi(X(T_K)) \tag{36}$$

where Theorem 9 is crucially used to ensure that $\varphi(X(\cdot))$ is left continuous at T_K if $T_K < \infty$, and has a limit at T_K if $T_K = \infty$. In the latter case of course $\varphi(X(T_K))$ is meant to be $\lim_{t\uparrow\infty} \varphi(X(t))$. However, we need (36) only on the set $\{T_K < \infty\}$, on which the relation (36) entails $\varphi(X(T_K)) = 1$. Since $\varphi < 1$ everywhere this forces $P^x\{T_K < \infty\} = 0$. Now let x be arbitrary and $t > 0$. We have

$$P^x\{t < T_K < \infty\} = E^x\{P^{X(t)}[T_K < \infty]\} = 0$$

because $X(t) \in K^c$ on $\{t < T_K\}$. Letting $t \downarrow 0$ we obtain $P^x\{T_K < \infty\} = 0$ since x is not regular for K. Thus K is polar.   □

Corollary. *For any Borel set A, $A \backslash A^r$ is a polar set.*

This follows because $A \backslash A^r$ is semi-polar by Theorem 6 of §3.5. We can now prove the uniqueness of the solution of the generalized Dirichlet problem in §4.4.

Proposition 11. *Let D be a bounded open set; f be a bounded measurable function defined on ∂D and continuous on $\partial^* D = \partial D \cap (D^c)^r$. Then the function $H_D f$ in (1) of §4.4 is the unique bounded solution to the generalized Dirichlet problem.*

Proof. We know from §4.4 that $H_D f$ is a solution. Let h be another solution and put $u = h - H_D f$. Then u is harmonic in D and converges to zero on $\partial^* D$, the set of "regular boundary points". By the Corollary to Theorem 10, $D^c \backslash (D^c)^r$ is a polar set. Observe that it contains $\partial D - \partial^* D$. For each x in D, the following assertions are true P^x-a.s. As $t \uparrow\uparrow \tau_D$, $X(t) \to X(\tau_D) \in \partial^* D$ because $\partial D - \partial^* D$ being a subset of $D^c \backslash (D^c)^r$ cannot be hit. Hence $u(X(t))$ as $t \uparrow\uparrow \tau_D$ converges to zero, and consequently the random variable w defined in (28) is equal to zero. Applying the last assertion of Theorem 7 to u, we conclude that $u(x) = E^x\{w\} = 0$. In other words, $h \equiv H_D f$ in D.

                                                                                                                                    □

We close this section by discussing the analogue of Definition 2 for a harmonic function in §4.3. It is trivial that the limit of an increasing sequence of superharmonic functions in a domain D is superharmonic provided it is not identically infinite there. On the other hand, any superharmonic function in D is the limit of an increasing sequence of infinitely differentiable superharmonic functions in the sense made precise in Exercise 12 below. Finally, we have the following characterization for a twice continuously differentiable superharmonic function.

Theorem 12. *If* $u \in \mathbb{C}^{(2)}(D)$ *and* $\Delta u \leq 0$ *in* D, *then* u *is superharmonic in* D. *If* u *is superharmonic and belong to* $\mathbb{C}^{(2)}$ *in a neighborhood of* x, *then* $\Delta u(x) \leq 0$.

Proof. To prove the first assertion let $B(x, r) \Subset D$. Then u belongs to $\mathbb{C}^{(2)}$ in an open neighborhood of $B(x, r)$. Hence formula (16) of §4.3 holds with h replaced by u, and yields $(d/d\delta)A(x, \delta) \leq 0$, for $0 < \delta \leq r$. It follows that $A(x, r) \leq \lim_{\delta \downarrow 0} A(x, \delta) = u(x)$, which is the inequality (1). The other conditions are trivially satisfied and so u is superharmonic in D. Next, suppose that u is superharmonic and belongs to $\mathbb{C}^{(2)}$ in $B(x, r)$. We may suppose $u \geq 0$ there, and then by Theorem 3 u is excessive with respect to the Brownian motion killed outside $B(x, r)$. It follows by Theorem 4 of §3.4 that $P_{B(x,\delta)^c}u(x)$ for $0 < \delta < r$ decreases as δ increases. Hence in the notation above $(d/d\delta)A(x, \delta) \leq 0$ for m-a.e. δ in $(0, r)$, and so by (16) of §4.3 we have $\int_{S(x,\delta)} \Delta u(y)\sigma(dy) \leq 0$ for a sequence of $\delta \downarrow 0$. Therefore $\Delta u(x) \leq 0$ by continuity as $\delta \downarrow 0$. $\square$

Exercises

1. If u is superharmonic in D, prove that the right members of (1) and of (2) both increase to $u(x)$ as $r \downarrow 0$.

2. If u is superharmonic in D, then for each $x \in D$ we have $u(x) = \varliminf_{x \neq y \to x} u(y)$.

3. Let u be superharmonic in D, and $D_r = \{x \in D \mid \rho(x, \partial D) > r\}$, $r > 0$. Define

$$u_r(x) = \int_{B(x,r)} u(y)m(dy).$$

Prove that u_r is a continuous superharmonic function *in* D_r.

4. (Minimum principle). Let u be superharmonic in a domain D. Suppose that there exists $x_0 \in D$ such that $u(x_0) = \inf_{x \in D} u(x)$; then $u \equiv u(x_0)$ in D. Next, suppose that for every $z \in \partial D \cup \{\infty\}$ we have $\varliminf_{x \to z} u(x) \geq 0$, where $x \to \infty$ means $\|x\| \to \infty$. Then $u \geq 0$ in D.

5. Let D be a domain and u have the properties in (a) of the definition of a superharmonic function. Then a necessary and sufficient condition for u to be superharmonic in D is the following. For any ball $B \Subset D$, and any function h which is harmonic in B and continuous in $\bar{B}$, $u \geq h$ on ∂B implies $u \geq h$ in $\bar{B}$. [Hint: for the sufficiency part take f_n continuous

on ∂B and $\uparrow u$ there, then $H_B f_n \leq u$ on ∂B; for the necessity part use Proposition 5 of §4.4.]

6. Extend the necessity part of Exercise 5 to any open set B with compact $\bar{B} \subset D$. [Hint: if B is not regular, Proposition 5 of §4.4 is no longer sufficient, but Proposition 6 is. Alternatively, we may apply Exercise 4 to $u - h$.]

7. If u is superharmonic in R^d and $t > 0$, then $P_t u$ is continuous [Hint: split the exponential and use $P_{t/2} u(0) < \infty$.]

8. Let D be bounded, u be harmonic in D and continuous in $\bar{D}$. Then the process in (13) is a martingale. Without using the martingale verify directly that $u(x) = E^x\{u(X(t \wedge \tau_D))\}$ for $0 \leq t \leq \infty$. Note that D need not be regular.

9. Let D be an open set and define for $t \geq 0$:

$$\check{X}(t) = X(t \wedge \tau_D).$$

The process $\check{X}$ is called the *Brownian motion stopped at the boundary of D*. Its appropriate state space is $\bar{D}$. Find the transition semigroup of X. Are all the points of ∂D absorbing states? Prove that X is a Hunt process. [Hint: we need Theorem 10; but try first to see the difficulty!]

10. If D is open bounded and z is regular for D^c. Prove that the function $u(x) = E^x\{\tau_D\}$ is a barrier at z for D.

11. Show that the first function in (17) of §4.3 is subharmonic in R^2; the second superharmonic in $R^d, d \geq 3$.

12. Let u be superharmonic in D. For $\delta > 0$ put

$$u_\delta(x) = \int_{R^d} \varphi_\delta(\|x - y\|) u(y) \, dy$$

where φ_δ is given in Exercise 6 of §4.3. Let D_δ be as in Exercise 3 above. Prove that u_δ is superharmonic and infinitely differentiable in D_δ, and increases to u as $\delta \downarrow 0$.

13. For each $t_0 > 0$, the function $x \rightarrow P^x\{\tau_D > t_0\}$ is excessive with respect to (Q_t). Hence it is continuous in D. [Hint: use Proposition 1 of §4.4.]

14. For each bounded Borel measurable f, and $t > 0$, $Q_t f$ is continuous in D. Hence if D is regular, the Brownian motion killed outside D is a Feller process. [Hint: for $0 < s < t$ we have $Q_t f = P_s Q_{t-s} f - E^x\{s \geq \tau_D; Q_{t-s} f(X_s)\}$; use (20).]

15. Under the conditions of Theorem 1, prove that the function $P_{D^c} u$ is superharmonic in an open set D_0 containing $\bar{D}$. [Hint: u being superharmonic in D_0, $P_{D^c} u$ is excessive with respect to the Brownian motion killed outside D_0.]

16. In R^2, a positive superharmonic function is a constant.

In the next two exercises, proofs of some of the main results above are sketched which do not use the general Hunt theory as we did.

17. Let u be a positive superharmonic function in R^d. We may suppose it bounded by first truncating it. Prove that almost surely:
    (a) $t \to u(X(t))$ is lower semi-continuous;
    (b) if Q is the set of positive rationals, then

$$\lim_{Q \ni t \downarrow 0} u(X(t)) = u(X(0));$$

   (c) the preceding relation is true without restricting t to Q, on account of (a);
    (d) $t \to u(X(t))$ is right continuous by the transfinite induction used in Theorem 6 of §3.4.

18. Let $u(x) = E^x\{e^{-T_A}\}$ where $A \in \mathscr{E}$. Prove that u is lower semi-continuous (for the Brownian motion). Show that $\{e^{-t}u(X_t), \mathscr{F}_t, t \geq 0\}$ is a super-martingale, and that it is right continuous, hence also has left limits. [Hint: Use the *proof* in Problem 17; we do not need the superharmonicity.]

Theorem 9 follows from Problem 17 as before, and Theorem 10 from Problem 18.

4.6. The Role of the Laplacian

The Dirichlet problem deals with the solution of Laplace's equation $\Delta\varphi = 0$ in D with the boundary condition $\varphi = f$ on ∂D. The inhomogeneous case of this equation is called Poisson's equation: $\Delta\varphi = g$ in D where g is a given function. Formally this equation can be solved if there is an operator Δ^{-1} inverse to Δ, so that $\Delta(\Delta^{-1}g) = g$. We have indeed known such an inverse under a guise, and now proceed to unveil it.

Suppose that g is bounded continuous and $U|g| < \infty$. Then we have

$$\lim_{t \downarrow 0} \frac{1}{t}(P_t - I)Ug = \lim_{t \downarrow 0} \frac{-1}{t} \int_0^t P_s g \, ds = -g \tag{1}$$

where $I \equiv P_0$ is the identity operator. Thus the operator

$$\mathscr{A} = \lim_{t \downarrow 0} \frac{1}{t}(P_t - I) \tag{2}$$

acts as an inverse to the potential operator U. It turns out $\mathscr{A} = \frac{1}{2}\Delta$ when acting on a suitable class of functions specified below.

For $k \geq 0$ we denote by $\mathbb{C}^{(k)}$ the class of functions which have continuous partial derivatives of order $\leq k$. Note that it does not imply boundedness of the function or its partial derivatives. The subclass of $\mathbb{C}^{(k)}$ with bounded kth partial derivatives is denoted by $\mathbb{C}_b^{(k)}$; the subclass of $\mathbb{C}^{(k)}$ having compact supports is denoted by $\mathbb{C}_c^{(k)}$. Clearly $\mathbb{C}_c^{(k)} \subset \mathbb{C}_b^{(k)}$. These conditions may be restricted to an open subset D of R^d, in which case the notation $\mathbb{C}^{(k)}(D)$, etc. is used.

Theorem 1. *Let f be bounded Lebesgue measurable and belong to $\mathbb{C}_b^{(2)}$ in an open neighborhood of x. Then we have*

$$\lim_{t \downarrow 0} \left(\frac{P_t - I}{t} \right) f(x) = \frac{\Delta}{2} f(x), \tag{3}$$

boundedly in the neighborhood.

Proof. Let x be fixed and f belong to $\mathbb{C}_b^{(2)}$ in the ball $B(x, r)$. By Taylor's theorem with an integral remainder term, we have for $\|y\| < r$:

$$f(x + y) - f(x) = \sum_{i=1}^{d} y_i f_i(x) + \sum_{i,j=1}^{d} y_i y_j \int_0^1 f_{ij}(x + sy)(1 - s)\,ds \tag{4}$$

where f_i and f_{ij} denote the first and second partial derivatives of f. Note that they are bounded in $B(x, r)$. Now write

$$\frac{1}{t}\{P_t f(x) - f(x)\} = \frac{1}{t}\int_{R^d} p(t; o, y)[f(x + y) - f(x)]\,dy; \tag{5}$$

put $y = z\sqrt{t}$, and split the range of the integral into two parts: $\|y\| = \|z\|\sqrt{t} < r$ and $\|y\| = \|z\|\sqrt{t} \geq r$ respectively. Observe that $p(t; o, z\sqrt{t})\,d(z\sqrt{t}) = p(1; o, z)\,dz$. The second part of the integral is bounded by

$$\frac{2\|f\|}{t} \int_{\|z\| \geq r/\sqrt{t}} p(1; o, z)\,dz$$

which converges to zero boundedly as $t \downarrow 0$. Substituting (4) into the first part of the integral, we obtain

$$\frac{1}{t}\int_{\|z\| < r/\sqrt{t}} p(1; o, z)\left\{ \sqrt{t} \sum_{i=1}^{d} z_i f_i(x) + t \sum_{i,j=1}^{d} z_i z_j \int_0^1 f_{ij}(x + sz\sqrt{t})(1 - s)\,ds \right\}dz$$

$$= \int_{\|z\| < r/\sqrt{t}} p(1; o, z) \sum_{i,j=1}^{d} z_i z_j \int_0^1 f_{ij}(x + sz\sqrt{t})(1 - s)\,ds\,dz, \tag{6}$$

because by spherical symmetry we have for each i:

$$\int_{\|z\| < r/\sqrt{t}} p(1; o, z) z_i\,dz = 0.$$

The right member of (6) may be written as

$$\sum_{i,j=1}^{d} \int_{R^d} p(1;o,z)z_i z_j \varphi_{ij}(t,z)\,dz$$

$$\text{where } \varphi_{ij}(t,z) = \begin{cases} \int_0^1 f_{ij}(x + sz\sqrt{t})(1-s)\,ds, & \text{if } \sqrt{t}\|z\| < r; \\ 0, & \text{if } \sqrt{t}\|z\| \geq r. \end{cases} \tag{7}$$

We have by bounded convergence,

$$\lim_{t \downarrow 0} \varphi_{ij}(t,z) = \int_0^1 f_{ij}(x)(1-s)\,ds = \frac{1}{2} f_{ij}(x);$$

and by the properties of the normal distribution,

$$\int_{R^d} p(1;o,z)z_i z_j\,dz = \delta_{ij}.$$

It follows that as $t \downarrow 0$, the sum in (7) converges boundedly to

$$\frac{1}{2}\sum_{i,j=1}^{d} \delta_{ij} f_{ij}(x) = \frac{1}{2}\Delta f(x). \qquad \square$$

We wish to apply Theorem 1 to $f = Ug$. When does this function belong to $\mathbb{C}^{(2)}$? First of all, we need $U|g| < \infty$. When $U1 = +\infty$, it is not sufficient that g be bounded. Let $\mathbb{B}_c$ denote the class of functions on R^d which are Lebesgue measurable and bounded with compact supports. If $g \in \mathbb{B}_c$ then we have shown in §3.7, for $d = 3$, that

$$Ug(x) = \frac{1}{2\pi} \int \frac{g(y)}{\|x - y\|}\,dy$$

is bounded continuous in x. In fact $Ug \in \mathbb{C}^{(1)}$. The same argument there applies to any $d \geq 3$. Thus in this case Ug is smoother than g. In order that Ug belong to $\mathbb{C}^{(2)}$, we need a condition which is stronger than $\mathbb{C}^{(0)}$ (continuity) but weaker than $\mathbb{C}^{(1)}$. There are several forms of such a condition but we content ourselves with the following. The functions g is said to satisfy a *Hölder condition* iff for each compact K there exists $\alpha > 0$ and $M < \infty$ (α and M may depend on K) such that

$$|g(x) - g(y)| \leq M\|x - y\|^{\alpha} \tag{8}$$

for all x and y in K. If $\alpha = 1$ this is known as *Lipschitz condition*. If g also has compact support then it is clearly bounded in R^d. We shall denote this class of functions by $\mathbb{H}_c$. For a proof of the following result under slightly

more general assumptions, see Port and Stone [1], pp. 115–118. An easy
case of it is given in Exercise 1 below.

Theorem 2. In R^d, $d \geq 3$, if $g \in \mathbb{B}_c$ then $Ug \in \mathbb{C}^{(1)}$; if $g \in \mathbb{H}_c$ then $Ug \in \mathbb{C}^{(2)}$.

We are now ready to show that the two operators $-\Delta/2$ and U act as
inverses to each other on certain classes of functions.

Theorem 3 (for R^d, $d \geq 3$). If $g \in \mathbb{H}_c$, then $-(\Delta/2)(Ug) = g$. If $g \in \mathbb{C}_c^{(2)}$, then
$U(-(\Delta/2)g) = g$.

Proof. If $g \in \mathbb{H}_c$ or $g \in \mathbb{C}_c^{(2)}$ then $Ug \in \mathbb{C}^{(2)}$ by Theorem 2. In either case g and
$U|g|$ are both bounded continuous in R^d. Applying Theorem 1 with $f = Ug$,
together with (1), we obtain $(\Delta/2)(Ug) = -g$. Next if $g \in \mathbb{C}_c^{(2)}$, we have by
Theorem 1 applied to g and bounded convergence:

$$P_t g - g = \lim_{h \downarrow 0} \int_0^t P_s \left(\frac{P_h g - g}{h} \right) ds = \int_0^t P_s \left(\lim_{h \downarrow 0} \frac{P_h g - g}{h} \right) ds$$

$$= \int_0^t P_s \left(\frac{\Delta}{2} g \right) ds. \tag{9}$$

If K is the compact support of g, then $|P_t g| \leq \|g\| \|P_t 1_K\| \to 0$ as $t \to \infty$ because
K is transient in R^d, $d \geq 3$. Since $|\Delta g|$ is bounded and has compact support,
the last member of (9) converges to $U((\Delta/2)g)$ as $t \to \infty$. We have thus proved
that $-g = U((\Delta/2)g)$.   □

Since $\mathbb{C}_c^{(2)} \subset \mathbb{H}_c$, Theorem 3 implies that $-\Delta/2$ and U are inverses to each
other on $\mathbb{C}_c^{(2)}$.

Next, recalling (26) of §4.5 but writing g for the f there, we know that if
$U|g| < \infty$, then

$$Ug - P_{D^c}Ug = G_D g \tag{10}$$

where

$$G_D g(x) = E^x \left\{ \int_0^{\tau_D} g(X_t) \, dt \right\}, \tag{11}$$

and D is any bounded open set in R^d, $d \geq 3$. This is then true if $g \in \mathbb{H}_c$. Moreover if we apply the Laplacian to all the terms in (10), we obtain by Theorem 3

$$\Delta(G_D g) = -2g \quad \text{in } D \tag{12}$$

because $P_{D^c}Ug$ is harmonic in D, being the solution to the generalized
Dirichlet problem for (D, Ug). If D is regular, then $P_{D^c}Ug$ is continuous in
$\bar{D}$ as the classical solution to the Dirichlet problem. Hence $G_D g$ is continuous
in $\bar{D}$ and vanishes on ∂D, by (10). We summarize the result as follows.

Theorem 4 (for R^d, $d \geq 3$). *If $g \in \mathbb{H}_c$ and D is a bounded regular open set, then the function given in (11) is the unique solution of Poisson's equation $-(\Delta/2)\varphi = g$ in D which is continuous in $\bar{D}$ and vanishes on ∂D.*

To see the uniqueness, suppose φ is such a solution. Then $\varphi - G_D g$ is harmonic in D, continuous in $\bar{D}$, and vanishes on ∂D. Hence it vanishes in $\bar{D}$ by the maximum (minimum) principle for harmonic functions.

We proceed to obtain similar results for R^2. Since the Brownian motion in R^2 is recurrent, the potential kernel U is of no use. We use a substitute culled from Port and Stone [1]. Let $\|x_0\| = 1$ and put for all $x \in R^2$:

$$p_t^*(x) = p_t(x) - p_t(x_0), \qquad u^*(x) = \int_0^\infty p_t^*(x)\,dt.$$

We have

$$u^*(x) = \frac{1}{2\pi} \int_0^\infty (e^{-\|x\|^2/2t} - e^{-1/2t})t^{-1}\,dt$$

$$= \frac{1}{2\pi} \int_0^\infty (e^{-\|x\|^2 s} - e^{-s})s^{-1}\,ds$$

$$= \frac{1}{2\pi} \int_0^\infty \int_{\|x\|^2 s}^s e^{-r}\,dr\,s^{-1}\,ds$$

$$= \frac{1}{2\pi} \int_0^\infty \int_r^{r/\|x\|^2} s^{-1}\,ds\,e^{-r}\,dr = \frac{1}{\pi} \log \frac{1}{\|x\|}. \tag{13}$$

Now put

$$p_t^*(x, y) = p_t^*(y - x), \qquad u^*(x, y) = u^*(y - x),$$
$$P_t^*(x, dy) = p_t^*(x, y)\,dy;$$

$$U^*(x, dy) = u^*(x, y)\,dy = \frac{1}{\pi} \log \frac{1}{\|x - y\|}\,dy. \tag{14}$$

The kernel U^* is the *logarithmic potential*. Note that not only $u^*(x)$ is positive or negative according as $\|x\| \leq 1$ or ≥ 1; but the same is true for $p_t^*(x)$ for each $t \geq 0$. If g is bounded with compact support, then U^*g is bounded continuous in R^2. In this respect the analogy of U^* with U in R^d for $d \geq 3$ is perfect; Theorem 2 also goes over as follows.

Theorem 2′. *In R^2, if $g \in \mathbb{B}_c$, then $U^*g \in \mathbb{C}^{(1)}$; if $g \in \mathbb{H}_c$ then $U^*g \in \mathbb{C}^{(2)}$.*

The following computations should be handled with care because the quantities may be positive or negative. If g is Lebesgue integrable, we have

$$P_s^* g = P_s g - p_s(x_0) \int g(y)\,dy, \qquad P_t P_s^* g = P_{t+s} g - p_s(x_0) \int g(y)\,dy.$$

It follows that if $g \in \mathbb{H}_c$ so that $\int |u^*(x, y)g(y)| \, dy < \infty$, we have

$$U^*g - P_t U^*g = \int_0^t P_s g \, ds - \lim_{A \to \infty} \int_A^{A+t} P_s g \, ds = \int_0^t P_s g \, ds \qquad (15)$$

because $|P_s g| \leq 1/2\pi s \int |g(y)| \, dy \to 0$ as $s \to \infty$.

Theorem 5 (for R^2). *If* $g \in \mathbb{H}_c$, *then* $-(\varDelta/2)(U^*g) = g$. *If* $g \in \mathbb{C}_c^{(2)}$, *then* $U^*(-(\varDelta/2)g) = g$.

Proof. The first assertion follows from Theorems 1 and 2′ and (15), in exactly the same way as the corresponding case in Theorem 3. To prove the second assertion we can use an analogue of (9), see Exercise 3 below. But we will vary the technique by first proving the assertion when $\varDelta g \in \mathbb{H}_c$. Under this stronger assumption we may apply the first assertion to $\varDelta g$ to obtain

$$-\frac{\varDelta}{2} U^*(\varDelta g) = \varDelta g.$$

Hence $\varDelta(g + \frac{1}{2}U^*(\varDelta g)) = 0$. Since both g and $\frac{1}{2}U^*(\varDelta g)$ are bounded it follows by Picard's theorem (Exercise 5 of §4.3) that $g + \frac{1}{2}U^*(\varDelta g) = 0$ as asserted. For a general $g \in \mathbb{C}_c^{(2)}$, we rely on the following lemma which is generally useful in analytical questions of this sort. Its proof is left as an exercise.

Lemma 6. *Let* g *be bounded Lebesgue measurable and put for* $\delta > 0$:

$$g_\delta(x) = \int g(x - y)\varphi_\delta(y) \, dy$$

where φ_δ *is defined in Exercise 6 of §4.3. Then* $g_\delta \in \mathbb{C}^{(\infty)}$. *If in addition* g *belongs to* $\mathbb{C}_b^{(2)}$ *in an open set* D, *then* $g_\delta \to g$ *and* $\varDelta g_\delta \to \varDelta g$ *both boundedly in* D *as* $\delta \downarrow 0$.

We now return to the proof of the second assertion of Theorem 5. If $g \in \mathbb{C}_c^{(2)}$, then $\varDelta g$ and $\varDelta g_\delta$ for all $\delta > 0$ have a compact support K. Since $|U^*1_K| < \infty$ we have $U^*(\varDelta g_\delta) \to U^*(\varDelta g)$ by dominated convergence because of the bounded convergence in Lemma 6. Since $g_\delta \in \mathbb{C}^{(\infty)}$, we have already proved that $g_\delta + \frac{1}{2}U^*(\varDelta g_\delta) = 0$ for all $\delta > 0$; as $\delta \downarrow 0$ this yields

$$g + \frac{1}{2}U^*(\varDelta g) = 0. \qquad \square$$

Next we consider Poisson's equation in R^2. Since U^* is not a true potential, the analogue of (10) is in doubt. We make a detour as follows. Let $g \in \mathbb{C}_c^{(2)}$, then (9) is true. We now rewrite (9) in terms of the process

$$E^x\{g(X_t)\} - g(x) = E^x\left\{\int_0^t \frac{\varDelta}{2} g(X_s) \, ds\right\}. \qquad (16)$$

If we put

$$M_t = g(X_t) - \int_0^t \frac{\Delta}{2} g(X_s)\, ds \tag{17}$$

then a general argument (Exercise 10) shows that $\{M_t, \mathscr{F}_t, t \geq 0\}$ is a martingale under P^x for each x. Let D be a bounded open set, then $\{M_{t \wedge \tau_D}, \mathscr{F}_t, t \geq 0\}$ is a martingale by Doob's stopping theorem. Since g and Δg are bounded and $E^x\{\tau_D\} < \infty$ for $x \in D$, it follows by dominated convergence that if $x \in D$:

$$g(x) = E^x\{M_0\} = \lim_{t \to \infty} E^x\{M_{t \wedge \tau_D}\} = E^x\{M_{\tau_D}\}$$

$$= E^x\{g(X_{\tau_D})\} - E^x\left\{\int_0^{\tau_D} \frac{\Delta}{2} g(X_s)\, ds\right\}. \tag{18}$$

In particular if $g = U^*f$ where $f \in \mathbb{H}_c$, then g is bounded, $g \in \mathbb{C}^{(2)}$ by Theorem 2′, and $(\Delta/2)g = -f$ by Theorem 5. Hence (18) becomes

$$U^*f(x) = P_{D^c}U^*f + G_D f. \tag{19}$$

This is the exact analogue of (10). It follows that Theorem 4 holds intact in R^2.

We have made the assumption of compact support in several places above such as Theorems 2 and 2′, mainly to lighten the exposition. This assumption can be replaced by an integrability condition, but the following lemma (valid in R^d for $d \geq 1$) shows that in certain situations there is no loss of generality in assuming compact support.

Lemma 7. *Let $f \in \mathbb{C}^{(k)}$ ($k \geq 0$) or satisfy a Hölder condition, and D be a bounded open set. Then there exists g having compact support which satisfies the same condition and coincides with f in D.*

Proof. Let D_0 be a bounded open set such that $D_0 \supset \bar{D}$ and $\rho(\partial D, \partial D_0) = \delta_0$. Let $0 < \delta < \delta_0$ and put

$$\psi(x) = \int_{D_0} \varphi_\delta(x - y)\, dy$$

where φ_δ is as in Lemma 6. Then $\psi \in \mathbb{C}^{(\infty)}$. If $\rho(x, D_0) > \delta$, then $\psi(x) = 0$. If $x \in D$, then $\psi(x) = 1$. It is clear that the function $g = f \cdot \psi$ has the required properties. $\qquad \square$

EXAMPLE 1. Let D be a bounded regular open set in R^d, $d \geq 1$. Solve the equation

$$\Delta \varphi = 1 \quad \text{in } D; \qquad \varphi = 0 \quad \text{in } D^c.$$

By Lemma 7, there exists $g \in \mathbb{C}_c^{(\infty)}$ such that $g = 1$ in $\bar{D}$. Put

$$\varphi(x) = -\tfrac{1}{2}E^x\left\{\int_0^{\tau_D} g(X_t)\,dt\right\} = -\tfrac{1}{2}G_D g(x).$$

We do not need Theorem 2 or 2' here, only the easier Exercise 1 to conclude that Ug or U^*g belongs to $\mathbb{C}^{(\infty)}$; hence $G_D g \in \mathbb{C}^{(\infty)}$ by (10) or (19). We have by Theorem 4,

$$\Delta\varphi = g = 1 \quad \text{in } D.$$

It is clear that $\varphi(x) = 0$ if $x \in D^c$ by the regularity of D.

Since $g = 1$ in D, we have $\varphi(x) = -\tfrac{1}{2}E^x\{\tau_D\}$. Using (VIII) of §4.2, we have

$$E^x\{\tau_D\} = \frac{1}{d}E^x\{\|X(\tau_D)\|^2\} - \frac{1}{d}\|x\|^2. \tag{20}$$

Since the first term on the right side of (20) is harmonic in D, we have

$$-\frac{\Delta}{2}E^x\{\tau_D\} = \frac{1}{2d}\Delta\|x\|^2 = 1,$$

a neat verification.

The preceding discussion does not extend to R^1. In general, Brownian motion in R^1 is a rather special case owing to the simple topological structure of the line. On the analytical side, since Δ reduces to the second derivative many results simplify. For instance it is trivial that given any continuous g there exists f such that $\Delta f = g$; think of the high dimensional analogue! For these reasons it is possible to obtain far more specific results in R^1 by methods which do not easily generalize. We leave a couple of exercises to suggest the possibilities without developing them because a creditable treatment of the Brownian motion in R^1 on its own right would take a volume or two. However, we will describe the solution of Poisson's equation by a different method attuned to one dimension, which is useful for other harder problems. Let g be bounded continuous in (a, b), $a < b$; we put $\tau = \tau_{(a,b)}$ and

$$\varphi(x) = E^x\left\{\int_0^\tau g(X_t)\,dt\right\}.$$

Recall the notation $T_h = \inf\{t > 0 : |X(t) - X(0)| \geq h\}$. For each $x \in (a, b)$, let $[x - h, x + h] \subset (a, b)$. Then $P^x\{T_h < \tau < \infty\} = 1$. We have by the strong Markov property:

$$\varphi(x) = E^x\left\{\left(\int_0^{T_h} + \int_{T_h}^\tau\right)g(X_t)\,dt\right\}$$

$$= E^x\left\{\int_0^{T_h} g(X_t)\,dt\right\} + P_{T_h}\varphi(x). \tag{21}$$

It is obvious by symmetry that $P^x\{X(T_h) = x + h\} = P^x\{X(T_h) = x - h\} = \frac{1}{2}$; hence

$$P_{T_h}\varphi(x) = \frac{1}{2}\{\varphi(x + h) + \varphi(x - h)\}.$$

Substituting into (21) we obtain

$$\frac{1}{h^2}\{\varphi(x + h) - 2\varphi(x) + \varphi(x - h)\} = \frac{-2}{h^2} E^x\left\{\int_0^{T_h} g(X_t)\,dt\right\}. \qquad (22)$$

Without loss of generality we may suppose that $g \geq 0$. Then it follows from (21) that

$$\varphi(x) \geq \frac{1}{2}\{\varphi(x + h) + \varphi(x - h)\}.$$

Since φ is bounded this implies that φ is continuous, indeed concave (see e.g. Courant [1], Vol. 2, p. 326). Next as $h \downarrow 0$, it is easy to see by dominated convergence that the right member of (22) converges to

$$\lim_{h \downarrow 0} \frac{1}{h^2} E^x\{T_h\}g(x) = g(x)$$

because $E^x\{T_h\} = h^2$ by a well-known particular case of (20). Therefore, the left member of (22) converges as $h \downarrow 0$ to $g(x)$. Now the limit is known as the generalized second derivative. A classical result by Schwarz (a basic lemma in Fourier series, see Titchmarsh [1], p. 431) states that if φ is continuous and the limit exists as a continuous function of x in (a, b), then φ'' exists in (a, b) and is equal to the limit. Thus $\varphi'' = -2g$. $\square$

It turns out that in higher dimensions there are also certain substitutes for the Laplacian similar to the generalized second derivative above. Some of the analytical difficulties disappear if these substitutes are used instead; on the other hand under more stringent conditions they can be shown to be the Laplacian after all, as in the case above.

EXAMPLE 2. Let $0 < a < b < \infty$. Compute the expected occupation times of $B(o, a)$ before the first exit time from $B(o, b)$.

We will treat the problem in R^2 which is the most interesting case since the expected occupation time of $B(o, a)$ is infinite owing to the recurrence of the Brownian motion. In the notation of (19), the problem is to compute $G_D f$ when $D = B(o, b)$, $f = 1$ in $B(o, a)$ and $f = 0$ in $B(o, b) - B(o, a)$. This f is only "piecewise continuous" but we can apply (19) to two domains separately. We seek φ such that

$$\begin{aligned}
\Delta\varphi(x) &= -2, \quad \text{for } \|x\| < a, \\
\Delta\varphi(x) &= 0, \qquad \text{for } a < \|x\| < b, \\
\varphi(x) &= 0, \qquad \text{for } \|x\| \geq b.
\end{aligned}$$

Clearly, φ depends on $\|x\|$ only, nence we can solve the equations above by the polar form of the Laplacian given in (19) of §4.3. Let $\|x\| = r$; we obtain

$$\varphi(r) = -\frac{r^2}{2} + c_1 + c_2 \log r, \qquad 0 < r < a;$$

$$= c_3 + c_4 \log r, \qquad a < r < b.$$

How do we determine these four constants? Obviously $\varphi(0) < \infty$ gives $c_2 = 0$, and $\varphi(b) = 0$ gives $c_3 = -c_4 \log b$. It is now necessary to have re-course to the first assertion in Theorem 2′ to know that both φ and $\partial\varphi/\partial r$ are continuous. Hence $\varphi(a-) = \varphi(a+)$, $(\partial\varphi/\partial r)(a-) = (\partial\varphi/\partial r)(a+)$. This yields the solution:

$$\varphi(r) = \frac{a^2 - r^2}{2} + a^2 \log \frac{b}{a}, \qquad 0 \le r \le a;$$

$$= a^2 \log \frac{b}{r}, \qquad a \le r \le b.$$

Exercises

1. In R^d, $d \ge 3$, if $g \in \mathbb{C}_c^{(k)}$, $k \ge 1$, then $Ug \in \mathbb{C}^{(k)}$. In R^2, if $g \in \mathbb{C}_c^{(k)}$, $k \ge 1$, then $U*g \in \mathbb{C}^{(k)}$.

2. Prove Lemma 6. [Hint: use Green's formula for $B(x, \delta_0)$, $\delta_0 > \delta$ to show $\Delta f_\delta(x) = \int \Delta f(x - y)\varphi_\delta(y)\,dy$.]

3. Prove the analogue of (9) for R^2:

$$P_t g - g = \int_0^t P_s^* \left(\frac{\Delta}{2} g \right) ds$$

and deduce the analogue of (10) from it.

4. Is the following dual of (15) true?

$$U*g - U*P_t g = \int_0^t P_s g \, ds.$$

5. Define $u^*(x)$ in R^1 in the same way as in R^2. Compute $u^*(x)$ and extend Theorem 5 to R^1. [Hint: $u^*(x) = 1 - |x|$.]

6. Compute $u^\alpha(x)$ in R^1. Prove that if g is bounded continuous (in R^1), then for each $\alpha > 0$, $U^\alpha g$ has bounded first and second derivatives, and $(\Delta/2)(U^\alpha g) = \alpha U^\alpha g - g$. Conversely, if g and its first and second deriva-tives are bounded continuous, then $U^\alpha((\Delta/2)g) = \alpha U^\alpha g - g$. [Hint: $u^\alpha(x) = e^{-|x|\sqrt{2\alpha}}/\sqrt{2\alpha}$. These results pertain to the definition of an "infini-tesimal generator" which is of some use in R^1; see, e.g., Ito [1].

7. Let $D = (a, b)$ in R^1. Derive Green's function $g_D(x, y)$ by identifying the solution of Poisson's equation for an arbitrary bounded continuous f:

$$\int g_D(x, y)f(y)\,dy = E^x\left\{\int_0^{\tau_D} f(X_t)\,dt\right\}.$$

[Hint: $g_D(x, y) = 2(x - a)(b - y)/(b - a)$ if $a < x \le y \le b$; $2(b - x)(y - a)/(b - a)$ if $a < y \le x < b$.]

8. Solve the problem in Example 2 for R^3.

9. Let μ be a σ-finite measure in R^d, $d \ge 2$. Suppose $U\mu$ is harmonic in an open set D. Prove that $\mu(D) = 0$. [Hint: let $f \in C^{(2)}$ in $B \Subset D$ and $f = 0$ outside B where B is a ball; $\int_B (U\mu)\,\Delta f\,dm = \int_{R_d} U(\Delta f)\,d\mu$; use Green's formula.]

10. The martingale in (17) is a case of a useful general proposition. Let $\{M_t, t \ge 0\}$ be associated with the Markov process $\{X_t, \mathscr{F}_t, t \ge 0\}$ as follows: (i) $M_0 = 0$; (ii) $M_t \in \mathscr{F}_t$; (iii) $M_{s+t} = M_s + M_t \circ \theta_s$ where $\{\theta_s, s \ge 0\}$ is the shift; (iv) for each x, $E^x\{X_t\} = 0$. Then $\{M_t, \mathscr{F}_t, t \ge 0\}$ is a martingale under each P^x. Examples of M_t are $g(X_t) - g(X_0)$ and $\int_0^t \varphi(X_s)\,ds$, where $g \in b\mathscr{E}$, $\varphi \in b\mathscr{E}$; and their sum. Condition (iii) is the additivity in *additive functionals*.

4.7. The Feynman-Kac Functional and the Schrödinger Equation

In this section we discuss the boundary value problem for the Schrödinger equation. This includes Dirichlet's problem in §4.4 as a particular case. The probabilistic method is based on the following functional of the Brownian motion process. Let $q \in b\mathscr{E}$, and put for $t \ge 0$:

$$e_q(t) = \exp\left[\int_0^t q(X_s)\,ds\right], \tag{1}$$

where $\{X_t\}$ is the Brownian motion in R^d, $d \ge 1$. Let D be a bounded domain, τ_D the first exit time from D defined in (1) of §4.4, and $f \in \mathscr{E}_+(\partial D)$. We put for all x in R^d:

$$u(x) = E^x\{e_q(\tau_D)f(X(\tau_D))\}. \tag{2}$$

Since the integrand above belongs to $\mathscr{F}^\sim$, u is universally measurable, (hence Lebesgue measurable) by Exercise 3 of §2.4. In fact $u \in \mathscr{E}$ because $\tau_D \in \mathscr{F}^0$ by Exercise 6 of §4.2, and $e_q(t)f(X(t))$ as a function of (t, ω) belongs to $\mathscr{B} \times \mathscr{F}^0$. The details are left as Exercise 1. Of course $u \ge 0$ everywhere, but u may be $+\infty$. Our principal result below is the dichotomy that if $f \in b\mathscr{E}_+$ then either

$u \equiv +\infty$ in D, or u is bounded in $\bar{D}$. This will be proved in several steps. We begin with a theorem usually referred to as *Harnack's inequality*.

Let U denote the class of functions defined in (2), for a fixed D, and all q and f as specified, subject furthermore to $\|q\|_D \leq Q$, where Q is a fixed constant. For $\varphi \in \mathscr{E}$ and $A \in \mathscr{E}$, $\|\varphi\|_A = \sup_{x \in A} |\varphi(x)|$; when A is the domain of definition of φ, it may be omitted from the notation.

Theorem 1. *Each u in U is either identically $+\infty$ in D, or everywhere finite in D. For each compact subset K of D, there is a positive constant A depending only on D, K and Q, such that for all finite u in U, and any two points x and x' in K, we have*

$$u(x') \leq Au(x).$$

Proof. Fix a $\delta > 0$ so small that

$$\tfrac{1}{2} \leq E^{\cdot}\{e^{\pm QT_{2\delta}}\} \leq 2 \tag{3}$$

where T_r is defined in (10) of §4.2. This is possible by Exercise 11 of §4.2. We now proceed to prove that if $u(x_0) < \infty$, and $0 < r < \delta \wedge (\rho_0/2)$, where ρ_0 is the distance from x_0 to ∂D, then for all $x \in B(x_0, r)$ we have

$$u(x) \leq 2^{d+2}u(x_0). \tag{4}$$

For $0 < s \leq 2r$, we have $T_s < \tau_D$ under P^{x_0}, since $B(x_0, 2r) \Subset D$. Hence by the strong Markov property

$$\infty > u(x_0) = E^{x_0}\{e_q(T_s)u(X(T_s))\} \geq E^{x_0}\{e^{-QT_s}u(X(T_s))\}. \tag{5}$$

The next crucial step is the stochastic independence of T_s and $X(T_s)$ under any P^x, proved in (XI) of §4.2. Using this and (3) we obtain

$$u(x_0) \geq \tfrac{1}{2}E^{x_0}\{u(X(T_s))\} = \frac{1}{2\sigma(s)} \int_{S(x_0,s)} u(y)\sigma(dy). \tag{6}$$

The step leading from (3) to (7) in §4.3 then yields

$$u(x_0) \geq \frac{1}{2v(2r)} \int_{B(x_0,2r)} u(y)\,dy. \tag{7}$$

For all $x \in B(x_0, r)$, we have $B(x, r) \subset B(x_0, 2r) \Subset D$. Hence we obtain similarly:

$$u(x) = E^x\{e_q(T_s)u(X(T_s))\}$$
$$\leq E^x\{e^{QT_s}u(X(T_s))\} \leq 2E^x\{u(X(T_s))\}. \tag{8}$$

This leads to the first inequality below, and then it follows from (7) that

$$u(x) \leq \frac{2}{v(r)} \int_{B(x,r)} u(y)\,dy \leq \frac{2}{v(r)} \int_{B(x_0,2r)} u(y)\,dy$$

$$\leq \frac{4v(2r)}{v(r)} u(x_0). \tag{9}$$

We have thus proved (4), in particular that $u(x) < \infty$. As a consequence, the set of x in D for which $u(x) < \infty$ is open in D. To show that it is also closed relative to D, let $x_n \to x_\infty \in D$ where $u(x_n) < \infty$ for all n. Then for a sufficiently large value of n, we have $\|x_\infty - x_n\| < \delta \wedge (\rho(x_n, \partial D)/2)$. Hence the inequality (4) is applicable with x and x_0 replaced by x_∞ and x_n, yielding $u(x_\infty) < \infty$. Since D is connected, the first assertion of the theorem is proved.

Now let D_0 be a subdomain with $\bar{D}_0 \subset D$. Let $0 < r < \delta \wedge (\rho(D_0, \partial D)/2)$. Using the connectedness of D_0 and the compactness of $\bar{D}_0$, we can prove the existence of an integer N with the following property. For any two points x and x' in D_0, there exist n points $x_1, \ldots, x_n$ in D_0 with $2 \leq n \leq N + 1$, such that $x = x_1, x' = x_n, \|x_{j+1} - x_j\| < r$ and $\rho(x_{j+1}, \partial D) > 2r$ for $1 \leq j \leq n - 1$. A detailed proof of this assertion is not quite trivial, nor easily found in books, hence is left as Exercise 3 with sketch of solution. Applying (4) successively to x_j and x_{j+1}, $1 \leq j \leq n - 1$ ($\leq N$), we obtain Theorem 1 with $A = 2^{(d+2)N}$. If K is any compact subset of D, then there exists D_0 as described above such that $K \subset D_0$. Therefore the result is *a fortiori* true also for K. ☐

Theorem 1 is stated in its precise form for comparison with Harnack's inequality in the theory of partial differential equations. We need it below only for a fixed u. The next proposition is Exercise 11 of §4.2 and its proof is contained in that of (X) of §4.2. It turns the trick of transforming Theorem 1 into Theorem 3.

Proposition 2. *If $m(D)$ tends to zero, then $E^x\{e^{Q\tau_D}\}$ converges to one, uniformly for $x \in R^d$.*

Theorem 3. *Let u be as in (2), but suppose in addition that f is bounded on ∂D. If $u \not\equiv \infty$ in D, then u is bounded in $\bar{D}$.*

Proof. Let K be a compact subset of D, and $E = D - K$. It follows from Proposition 2 that given any $\varepsilon > 0$, we can choose K to make $m(E)$ so small that

$$\sup_{x \in R^d} E^x\{e^{Q\tau_E}\} \leq 1 + \varepsilon. \tag{10}$$

Put for $x \in \bar{E}$:

$$u_1(x) = E^x\{e_q(\tau_D)f(X(\tau_D)); \tau_E < \tau_D\},$$
$$u_2(x) = E^x\{e_q(\tau_D)f(X(\tau_D)); \tau_E = \tau_D\}.$$

By the strong Markov property, since $\tau_D = \tau_E + \tau_D \circ \theta_{\tau_E}$ on the set $\{\tau_E < \tau_D\}$:

$$u_1(x) = E^x\{e_q(\tau_E)u(X(\tau_E)); \tau_E < \tau_D\}.$$

On $\{\tau_E < \tau_D\}$, $X(\tau_E) \in K$. Since $u \not\equiv \infty$ in D, u is bounded on K by Theorem 1. Together with (10) this implies

$$u_1(x) \le E^x\{e^{Q\tau_E}\}\|u\|_K \le (1 + \varepsilon)\|u\|_K.$$

Since f is bounded we have by (10)

$$u_2(x) \le E^x\{e^{Q\tau_E}\}\|f\| \le (1 + \varepsilon)\|f\|.$$

Thus for all $x \in \bar{E}$ we have

$$u(x) = u_1(x) + u_2(x) \le (1 + \varepsilon)(\|u\|_K + \|f\|). \tag{11}$$

Since $\bar{D} = \bar{E} \cup K$, u is bounded in $\bar{D}$.   □

We shall denote the $u(x)$ in (2) more specifically by $u(D, q, f; x)$.

Corollary. *Let* $f \in b\mathscr{E}$. *If* $u(D, q, |f|; x) \not\equiv \infty$ *in* D, *then* $u(D, q, f; x)$ *is bounded in* $\bar{D}$.

In analogy with the notation used in §4.6, let $\mathscr{B}(D)$ denote the class of functions defined in D which are bounded and Lebesgue measurable; $\mathbb{H}(D)$ the class of functions defined in D which are bounded and satisfy (8) of §4.6 for each compact $K \subset D$. Thus $b\mathbb{C}^{(1)}(D) \subset \mathbb{H}(D) \subset b\mathbb{C}^{(0)}(D)$. Let us state the following analytic lemma.

Proposition 4. *If* $g \in \mathscr{B}(D)$, *then* $G_D g \in \mathbb{C}^{(1)}(D)$. *For* $d = 1$ *if* $g \in b\mathbb{C}^{(0)}(D)$, *for* $d \ge 2$ *if* $g \in \mathbb{H}(D)$, *then* $G_D g \in \mathbb{C}^{(2)}(D)$.

For $d = 1$ this is elementary. For $d \ge 2$, the results follow from (10) and (19) of §4.6, via Theorems 2 and 2′ there, with the observation that $G_D g = G_D(1_D g)$ by (11) of §4.6. Note however that we need the versions of these theorems localized to D, which are implicit in their proofs referred to there. For a curious deduction see Exercise 4 below.

Theorem 5. *For* $d = 1$ *let* $q \in b\mathbb{C}^{(0)}(D)$; *for* $d \ge 2$ *let* $q \in \mathbb{H}(D)$. *Under the conditions of the Corollary to Theorem 3, the function* $u(D, q, f; \cdot)$ *is a solution of the following equation in* D:

$$\left(\frac{\Delta}{2} + q\right)\varphi = 0. \tag{12}$$

Proof. Since u is bounded in $\bar{D}$, we have for $x \in D$:

$$E^x\left\{\int_0^{\tau_D} |q(X_t)u(X_t)|\, dt\right\} \le \|qu\|_D E^x\{\tau_D\} < \infty. \tag{13}$$

[Note that in the integral above the values of q and u on ∂D are irrelevant.] This shows that the function $1_{\{t < \tau_D\}}|q(X_t)u(X_t)|$ of (t, ω) is dominated in the integration with respect to the product measure $m_1 \times P^x$ over $[0, \infty) \times \Omega$, where m_1 denotes the Lebesgue measure on $[0, \infty)$. Therefore, we can use Fubini's theorem to transform the following integral as shown below:

$$E^x\left\{\int_0^{\tau_D} q(X_t)u(X_t)\, dt\right\} = E^x\left\{\int_0^{\infty} 1_{\{t < \tau_D\}} q(X_t) E^{X_t}[e_q(\tau_D)f(X(\tau_D))]\, dt\right\}$$

$$= E^x\left\{f(X(\tau_D)) \int_0^{\infty} 1_{\{t < \tau_D\}} q(X_t) \exp\left[\int_t^{\tau_D} q(X_s)\, ds\right] dt\right\}.$$

The third member above is obtained from the second by first reversing the order of the two integrations, then applying the Markov property at each t under E^x, noting that $\tau_D = t + \tau_D \circ \theta_t$ on $\{t < \tau_D\}$. We can now perform the trivial integration with respect to t to obtain

$$E^x\left\{f(X(\tau_D))\left[\exp \int_0^{\tau_D} q(X_s)\, ds - 1\right]\right\} = u(x) - E^x\{f(X(\tau_D))\}.$$

This result may be recorded in previous notation as follows:

$$G_D(qu) = u - H_D f. \tag{14}$$

Recall that $H_D f$ is harmonic in D hence belongs to $\mathbb{C}^{(\infty)}$ there. Since $qu \in \mathscr{B}(D)$, it follows from Proposition 4 that $u \in \mathbb{C}^{(1)}(D)$. Then $qu \in b\mathbb{C}^{(0)}(D)$ for $d = 1$, $qu \in \mathbb{H}(D)$ for $d \ge 2$, hence $u \in \mathbb{C}^{(2)}(D)$. Taking the Laplacian in (14) and using (12) and (19) of §4.6, we conclude that $-2qu = \Delta u$ in D as asserted. $\square$

The equation (12) is the celebrated Schrödinger's equation in quantum physics. For $q \equiv 0$ it reduces to Laplace's equation.

The next result includes Theorem 2 of §4.4 as a particular case.

Theorem 6. *Under the hypotheses of Theorem 5, if $z \in \partial D$ and z is regular for D^c, and if f is continuous at z, then we have*

$$\lim_{\bar{D} \ni x \to z} u(x) = f(z). \tag{15}$$

Proof. We may suppose $f \geq 0$ by using $f = f^+ - f^-$. Given $\varepsilon > 0$, there exists $r > 0$ such that

$$1 - \varepsilon \leq E^{\cdot}\{e^{\pm 2QT_r}\} \leq 1 + \varepsilon; \tag{16}$$

$$\alpha(\varepsilon) = \sup_{y \in B(z, 2r) \cap (\partial D)} |f(y) - f(z)| \leq \varepsilon. \tag{17}$$

Let $x \in \bar{D} \cap B(z, r)$. Put

$$u_1(x) = E^x\{T_r < \tau_D; e_q(\tau_D)f(X(\tau_D))\} = E^x\{T_r < \tau_D; e_q(T_r)u(X(T_r))\},$$
$$u_2(x) = E^x\{\tau_D \leq T_r; e_q(\tau_D)f(X(\tau_D))\}.$$

We have $X(T_r) \in D$ on $\{T_r < \tau_D\}$, hence by Theorem 3 followed by Schwarz's inequality:

$$u_1(x) \leq E^x\{T_r < \tau_D; e^{QT_r}\}\|u\|_D$$
$$\leq P^x\{T_r < \tau_D\}^{1/2}E^x\{e^{2QT_r}\}^{1/2}\|u\|_D.$$

As $x \to z$, this converges to zero by (8) of §4.4, and (16) above. Next we have for $x \in B(z, r)$:

$$|u_2(x) - f(z)| \leq E^x\{\tau_D \leq T_r; e_q(\tau_D)\}\alpha(\varepsilon)$$
$$+ E^x\{\tau_D \leq T_r; |e_q(\tau_D) - 1|\}|f(z)|$$
$$+ P^x\{T_r < \tau_D\}|f(z)|.$$

The first term on the right is bounded by $E^x\{e^{QT_r}\}\varepsilon \leq (1 + \varepsilon)\varepsilon$ by (16) and (17); the second by $|1 - E^x\{e^{\pm QT_r}\}||f(z)| \leq \varepsilon|f(z)|$ by (16); and the third converges to zero as $x \to z$ by (8) of §4.4. The conclusion (15) follows from these estimates. $\square$

Putting together Theorems 3, 5 and 6, we have proved that for every $f \in \mathbb{C}(\partial D)$, a $\mathbb{C}^{(2)}(D)$-solution of the equation (12) is given by $u(D, q, f; \cdot)$ provided that $u(D, q, 1; \cdot) \not\equiv \infty$ in D. Moreover if D is regular, then this solution belongs to $C^{(0)}(\bar{D})$. Thus we have solved the Dirichlet boundary value problem for the Schrödinger equation by the explicit formula given in (2). It turns out that under the conditions stated above, this is the *unique* solution. For the proof and other closely related results we refer the reader to the very recent paper by Chung and Rao [3]. A special case is given in Exercise 7 below.

Let us remark that contrary to the Laplace case, the uniqueness of solution in the Schrödinger case is in general false. The simplest example is given in R^1 by the equation $u'' + u = 0$ in $D = (0, \pi)$. The particular solution $u(x) = \sin x$ vanishes on ∂D! In general, unicity depends on the size of the domain D as well as the function q. Such questions are related to the eigenvalue problem associated with the Schrödinger operator. Here we see that the quantity $u(D, q, 1, x)$ serves as a *gauge* in the sense that its finiteness for some x in D ensures the unique solvability of all continuous boundary value problems.

Exercises

1. Prove that the function u in (2) is Borel measurable.

2. If $f \in \mathscr{E}_+$ in (2), then either $u \equiv 0$ in D or $u > 0$ in D. [Here D need not be bounded.]

3. (a) Let D be a domain. Then there exist domains D_n strictly contained in D and increasing to D. [Hint: let U_n be the union of all balls at distance $> 1/n$ from ∂D. Fix an x_0 in D and let D_n be the connected component of U_n which contains x_0. Show that $\bigcup_n D_n$ is both open and closed relative to D.]

   (b) Let D_0 be a bounded domain strictly contained in D. Let $0 < r < \frac{1}{2}\rho(D_0, \partial D)$, and $\bar{D}_0 \subset \bigcup_{i=1}^N B(x_i, r/2)$ where all $x_i \in \bar{D}_0$. Define a *connection* "$\to$" on the set of centers $S = \{x_i, 1 \le i \le N\}$ as follows: $x_i \to x_j$ if $\|x_i - x_j\| < r$. Use the connectedness of D_0 to show that for any two elements x_a and x_b of S, there exist distinct elements x_{i_j}, $1 \le j \le l$, such that $x_{i_1} = x_a, x_{i_l} = x_b$, and $x_{i_j} \to x_{i_{j+1}}$ for $1 \le j \le l - 1$. In the language of graph theory, the set S with the connection $\to$ forms a *connected graph*. [This formulation is due to M. Steele.]

   (c) Show that the number N whose existence is asserted at the end of the proof of Theorem 1 may be taken to be the number N in (b) plus one.

In the following problems D is a bounded domain in R^d, $d \ge 1$; $q \in b\mathscr{E}$.]

4. (a) Let D_1 be a subdomain $\Subset D$. If $\lfloor g \subset D - D_1$, then $G_D g$ is harmonic in D_1.

   (b) Let $D_1 \Subset D_2 \Subset D$. If $g \in \mathbb{H}(D)$ then there exists g_1 such that $g_1 \in \mathbb{H}_c(R^d)$ and $g_1 = g$ in $\bar{D}_1$, $g_1 = 0$ in $R^d - D_2$. [Hint: multiply g by a function in $C^{(\infty)}$ as in Lemma 7 of §4.6.]

   (c) Prove Proposition 4 by using Theorems 2 and 2' of §4.6. [This may be putting the horse behind the cart as alluded to in the text, but it is a good exercise!]

5. If $u(D, q, 1; \cdot) \not\equiv \infty$ in D, then it is bounded away from zero in D. Moreover there exists a constant $C > 0$ such that

$$u(D, q, 1; x) \ge C u(E, q, 1; x)$$

for all subdomains $E \Subset D$, and all $x \in \bar{D}$.

6. Prove that $u(D, q, 1; \cdot) \not\equiv \infty$ if and only if for all x in D we have

$$\int_0^\infty E^x\{e_q(t); t < \tau_D\} \, dt < \infty.$$

[Hint: for some $t_0 > 0$ we have two constants $C_1 > 0$ and $C_2 > 0$ such that $C_1 \le E^x\{e_q(\tau_D); \ 0 < \tau_D \le 1\} \le C_2$ for all $x \in D$; now estimate $E^x\{e_q(\tau_D); \ n < \tau_D \le n + 1\}$ and add.]

7. Suppose D is regular and $E^x\{e^{\|q\|\tau_D}\} < \infty$ for some $x \in D$. Then for any $f \in \mathbb{C}(\partial D)$, $u(D, q, f, \cdot)$ is the unique solution of (12) with boundary value f. [Hint: let φ be a solution which vanishes on ∂D. Show that $\varphi = G_D(q\varphi)$. Prove by induction on n that

$$\varphi(x) = \frac{1}{n!} E^x \left\{ \int_0^{\tau_D} q(X_t) \left(\int_0^t q(X_s)\, ds \right)^n \varphi(X_t)\, dt \right\}$$

for all $n \ge 0$. Now estimate $|\varphi(x)|$.]

8. In R^1 let $D = (a, b)$, and $q \in \mathbb{C}(\bar{D})$. Put

$$u_z(x) = E^x\{e_q(\tau_D); \ X(\tau_D) = z\}$$

for $z = a$ and b. Prove that if $u_a \not\equiv \infty$ in D, then both u_a and u_b are bounded in $\bar{D}$. [Hint: to prove $u_b \not\equiv \infty$, use the following result from the elementary theory of differential equations. Either the boundary value problem for the equation $\varphi'' + q\varphi = 0$ in D has a nonzero solution with $\varphi(a) = \varphi(b) = 0$; or it has a unique solution with any given values $\varphi(a)$ and $\varphi(b)$. This is due to M. Hogan.]

An extension of Exercise 8 to higher dimensions has been proved by Ruth Williams. If ∂D is sufficiently smooth, A is an "open" subset of ∂D; and $u(D, q, 1_A; \cdot) \not\equiv \infty$ in D, then $u(D, q, 1; \cdot) \not\equiv \infty$ in D.

Notes on Chapter 4

§4.1 The theory of spatially homogeneous Markov processes is an extension of that of random walks to the continuous parameter case. This is an old theory due largely to Paul Lévy [1]. Owing to its special character classical methods of analysis such as the Fourier transform are applicable; see Gihman and Skorohod [1] for a more recent treatment.

For the theory of dual processes see Blumenthal and Getoor [1], which improved on Hunt's original formulation. Much of the classical Newtonian theory is contained in the last few pages of the book in a condensed manner, but it is a remarkable synthesis not fully appreciated by the non-probabilists.

§4.2. For lack of space we have to de-emphasize the case of dimension $d = 1$ or 2 in our treatment of Brownian motion. So far as feasible we use the general methods of Hunt processes and desist from unnecessary short-cuts. More coverage is available in the cognate books by K. M. Rao [1] and Port and Stone [1]. The former exposition takes a more general probabilistic approach while the latter has more details on several topics discussed here.

§4.3 and §4.4. The force of the probabilistic method is amply illustrated in the solution of the Dirichlet problem. The reader who learns this natural approach first may indeed wonder at the tour de force of the classical treatments, in which some of the basic definitions such as the regularity of boundary would appear to be rather contrived.

As an introduction to the classical viewpoint the old book by Kellogg [1] is still valuable, particularly for its discussion of the physical background. A simpler version may be found in Wermer [1]. Ahlfors [1] contains an elementary discussion of harmonic functions and the Dirichlet problem in R^2, and the connections with analytic functions. Brelot [1] contains many modern developments as well as an elegant (French style) exposition of the Newtonian theory.

The proof of Theorem 8 by means of Lemma 9 may be new. The slow pace adopted here serves as an example of the caution needed in certain arguments. This is probably one of the reasons why even probabilists often bypass such proofs.

§4.5. Another method of treating superharmonic functions is through approximation with smooth ones, based on results such as Theorem 12; see the books by Rao and Port-Stone. This approach leads to their deeper analysis as Schwartz distributions. We choose Doob's method to give further credence to the viability of paths. This method is longer but ties several items together. The connections between (sub)harmonic functions and (sub)martingales were first explored in Doob [2], mainly for the logarithmic potential. In regard to Theorems 2 and 3, a detailed study of Brownian motion killed outside a domain requires the use of Green's function, namely the density of the kernel Q_t defined in (15), due to Hunt [1]. Here we regard the case as a worthy illustration of the general methodology (∂ and all).

Doob proved Theorem 9 in [2] using H. Cartan's results on Newtionian capacity. A non-probabilistic proof of the Corollary to Theorem 10 can be found in Wermer [1]. The general proposition that "semipolar implies polar" is Hunt's Hypothesis (H) and is one of the deepest results in potential theory. Several equivalent propositions are discussed in Blumenthal and Getoor [1]. A proof in a more general case than the Brownian motion will be given in §5.2.

§4.6. The role of the infinitesimal generator is being played down here. For the one-dimensional case it is quite useful, see e.g., Ito [1] for some applications. In higher dimensions the domain of the operator is hard to describe and its full use is neither necessary nor sufficient for most purposes. It may be said that the substitution of integral operators (semigroup, resolvent, balayage) for differential ones constitutes an essential advance of the modern theory of Markov processes. Gauss and Koebe made the first fundamental step in identifying a harmonic function by its averaging property (Theorem 2 in §4.3). This is indeed a lucky event for probability theory.

§4.7. This section is added as an afterthought to show that "there is still sap from the old tree". For a more complete discussion see Chung and Rao [3] where D is not assumed to be bounded but $m(D) < \infty$. The one-dimensional case is treated in Chung and Varadhan [1]. The functional $e_q(t)$ was introduced by Feynman with a purely imaginary q in his "path integrals"; by Kac [1] with a nonpositive q. Its application to the Schrödinger equation is discussed in Dynkin [1] with $q \leq 0$, Khas'minskii [1] with $q \geq 0$. The general case of a bounded q requires a new approach partly due to the lack of a maximum principle.

Let us alert the reader to the necessity of a meticulous verification of domination, such as given in (13), in the sort of calculations in Theorem 5. Serious mistakes have resulted from negligence on this score. For instance, it is *not* sufficient in this case to verify that $u(x) < \infty$, as one might be misled to think after a preliminary (illicit) integration with respect t.

Comparison of the methods used here with the classical approach in elliptic partial differential equations should prove instructive. For instance, it can be shown that the finiteness of $u(D, q, 1; \cdot)$ in D is equivalent to the existence of a strictly positive solution belonging to $\mathbb{C}^{(2)}(D) \cap \mathbb{C}^{(0)}(\bar{D})$. This is also equivalent to the proposition that all eigenvalues λ of the Schrödinger operator, written in the form $(\Delta/2 + q)\varphi = \lambda\varphi$, are strictly negative; see a forthcoming paper by Chung and Li [1]. Further results are on the way.

Chapter 5

Potential Developments

5.1. Quitting Time and Equilibrium Measure

In the next two sections our aim is to establish a number of notable results in classical potential theory by the methods developed in the earlier chapters. In contrast with the preceding sections of last chapter, the horizon will be widened to reach far beyond Brownian motion. We shall deal with Hunt processes satisfying certain general hypotheses and the results will apply to classes of potential kernels including the M. Riesz potentials (see Exercises below) as well as the logarithmic and Newtonian. There are usually different sets of overlapping conditions to yield a particular result. The theory of dual processes in Blumenthal and Getoor [1] gives a framework which has a considerable range, but it is a long and sometimes technically complicated passage. Here instead we offer a more direct and relatively new approach to a number of selected topics, with a view to further development. The case of Brownian motion will be discussed toward the end.

   We begin in this section with the analysis of the sample path when and where it quits a transient set. This will lead to a basic result known as the *equilibrium principle* in potential theory.

   Let $\{X_t, t \geq 0\}$ be a Hunt process on $(E, \mathscr{E})$, as in §3.1. [We banish ∂ from sight if not from thought.] It turns out that initially we need only a Markov process with paths which have left limits in $(0, \infty)$, and for which hitting times have the measurability properties valid for a Hunt process. The other hypotheses for a Hunt process will not be explicitly used until further notice. Let ξ be a σ-finite measure on $\mathscr{E}$, and suppose that the potential kernel U has a density u with respect to ξ, as follows:

$$U(x, A) = \int_A u(x, y)\xi(dy), \qquad x \in \mathbf{E}, A \in \mathscr{E}. \tag{1}$$

Of course $u \geq 0$ and $u \in \mathscr{E} \times \mathscr{E}$; but no further condition will be imposed until later. Let $A \in \mathscr{E}$ and

$$\varDelta_A = \{\omega \mid \exists t > 0 : X_t(\omega) \in A\};$$

see §3.3 for the measurability of Δ_A. We define

$$\gamma_A(\omega) = \begin{cases} \sup\{t > 0 \,|\, X_t(\omega) \in A\}, & \text{if } \omega \in \Delta_A, \\ 0, & \text{if } \omega \in \Omega - \Delta_A; \end{cases}$$

contrast this with

$$T_A(\omega) = \begin{cases} \inf\{t > 0 \,|\, X_t \in A\}, & \text{if } \omega \in \Delta_A, \\ \infty, & \text{if } \omega \in \Omega - \Delta_A. \end{cases}$$

T_A is the hitting time, γ_A is the *quitting time* (or *last exit time*) of A (denoted by L_A in (21) of §3.6). Their dual relationship is obvious from the above. Recall that A is transient if and only if

$$\forall x \in \mathbf{E}: P^x\{\gamma_A < \infty\} = 1. \tag{3}$$

We have for each $t > 0$:

$$\{\gamma_A > t\} = \{T_A \cdot \theta_t < \infty\}; \tag{4}$$

from which it follows that $\gamma_A \in \mathscr{F}_\infty$. The next result is trivial but crucial:

$$\{\gamma_A > 0\} = \{T_A < \infty\}. \tag{5}$$

For $x \in \mathbf{E}$, $B \in \mathscr{E}$, define the *quitting kernel* L_A as follows:

$$L_A(x, B) = P^x\{\gamma_A > 0; X(\gamma_A-) \in B\}. \tag{6}$$

and write as usual for $f \in \mathscr{E}_+$:

$$L_A f(x) = \int L_A(x, dy) f(y) = E^x\{\gamma_A > 0; f(X(\gamma_A-))\}.$$

From here on we fix the transient set A and omit it from the notation until further notice. Let $f \in b\mathbb{C}$ (bounded continuous), $\varepsilon > 0$, and consider the following approximation of $f(X(\gamma-))$:

$$\int_0^\infty \frac{1}{\varepsilon} E^x\{f(X_t); t < \gamma \le t + \varepsilon\} \, dt. \tag{7}$$

Put

$$\psi_\varepsilon(x) = \frac{1}{\varepsilon} P^x\{0 < \gamma \le \varepsilon\}; \tag{8}$$

ψ_ε is universally measurable. Now a basic property of γ may be expressed by means of the shift as follows:

$$\gamma \circ \theta_t = (\gamma - t)^+. \tag{9}$$

For if $\gamma_A \le t$, then $T_A \circ \theta_t = \infty$, namely $\gamma_A \circ \theta_t = 0$; if $\gamma_A > t$, then $\gamma_A = t + \gamma_A \circ \theta_t$. As a consequence of (9), we have for each $t \ge 0$:

$$\{t < \gamma \le t + \varepsilon\} = \{0 < \gamma \circ \theta_t \le \varepsilon\}. \tag{10}$$

Using this we apply the simple Markov property at time t to see that the integrand in (7) is equal to

$$E^x \left\{ f(X_t) \frac{1}{\varepsilon} E^{X_t}[0 < \gamma \le \varepsilon] \right\} = E^x \{ f(X_t)\psi_\varepsilon(X_t) \}.$$

Hence the quantity in (7) is equal to $U(f \cdot \psi_\varepsilon)(x)$. Let us verify that this is uniformly bounded with respect to ε. Since f is bounded it is sufficient to check the following:

$$\int_0^\infty \frac{1}{\varepsilon} P^x \{ t < \gamma \le t + \varepsilon \} \, dt = \int_0^\infty \frac{1}{\varepsilon} \int_\Omega 1_{\{t < \gamma \le t + \varepsilon\}}(\omega) P^x(d\omega) \, dt$$

$$= \int_{\{\gamma > 0\}} \left(\frac{1}{\varepsilon} \int_{(\gamma - \varepsilon)^+}^\gamma 1 \, dt \right) P^x(d\omega) \le P^x \{ \gamma > 0 \} \le 1.$$

Here we have used Fubini's theorem to reverse the order of integration of the double integral above with respect to $P^x \times m$, where m is the Lebesgue measure on $[0, \infty)$. Exactly the same evaluation of the quantity in (7) yields

$$\int_{\{\gamma > 0\}} \frac{1}{\varepsilon} \int_{(\gamma - \varepsilon)^+}^\gamma f(X_t) \, dt \, P^x(d\omega). \tag{11}$$

Since $f \in b\mathbb{C}$, and X has a left limit at γ ($< \infty$), as $\varepsilon \downarrow 0$ the limit of the quantity in (11) is equal to

$$\int_{\{\gamma > 0\}} f(X(\gamma -))P^x(d\omega) = E^x \{ \gamma > 0; f(X(\gamma -)) \}.$$

We have therefore proved that

$$\int L(x, dy)f(y) = \lim_{\varepsilon \downarrow 0} U(f\psi_\varepsilon)(x) = \lim_{\varepsilon \downarrow 0} \int u(x, y)f(y)M_\varepsilon(dy) \tag{12}$$

where M_ε is the measure given by

$$M_\varepsilon(dy) = \psi_\varepsilon(y)\xi(dy).$$

Now we make the following assumptions on the function u, which will be referred to as (R).

(i)   For each $x \in \mathbf{E}$, $y \to u(x, y)^{-1}$ is finite continuous;

$$(R)$$

(ii) $u(x, y) = \infty$ if and only if $x = y$; we put $u(x, x)^{-1} = 0$.

It is clear that condition (i) is equivalent to the following: $u(x, y) > 0$ for all x and y in $\mathbf{E}$; and for each $x \in \mathbf{E}$, $y \to u(x, y)$ is extended continuous in $\mathbf{E}$. Here are some preliminary consequences of the conditions. Since $\int u(x, y) M_\varepsilon(dy) < \infty$ by (12), it follows from (ii) that $M_\varepsilon(\{x\}) = 0$ for every x; namely M_ε is diffuse. Next since $\inf_{y \in K} u(x, y) > 0$ for each compact K by (i), we have $M_\varepsilon(K) < \infty$. Thus M_ε is a Radon measure. Now let $\varphi \in \mathbb{C}_c$; then the function $y \to \varphi(y) u(x, y)^{-1}$ belongs to $\mathbb{C}_c$ for each x, by (i). Substituting this for f in (12), we obtain

$$\int \frac{L(x, dy)}{u(x, y)} \varphi(y) = \lim_{\varepsilon \downarrow 0} \int \varphi(y) M_\varepsilon(dy) \qquad (13)$$

because $u(x, y) u(x, y)^{-1} = 1$ for $y \in \mathbf{E} - \{x\}$, and the point set $\{x\}$ may be ignored since M_ε is diffuse. Since for each x, $L(x, \cdot)$ is a finite measure and $u(x, y)^{-1}$ is bounded on each compact, $L(x, dy) u(x, y)^{-1}$ is a Radon measure. It is well known that two Radon measures are identical if they agree on all φ in $\mathbb{C}_c$ (Exercise 1). Hence the relation (13) implies that there exists a single Radon measure μ on $\mathscr{E}$ such that

$$\int_B \frac{L(x, dy)}{u(x, y)} = \mu(B), \qquad \forall x \in \mathbf{E}, B \in \mathscr{E}. \qquad (14)$$

Let us pause to marvel at the fact that the integral above does not depend on x. This suggests an ergodic phenomenon which we shall discuss in §5.2.

Since M_ε is a Radon measure for each $\varepsilon > 0$, it follows also from (13) that μ is the unique vague limit of M_ε as $\varepsilon \downarrow 0$; but this observation is not needed here. We are now going to turn (14) around:

$$L(x, B) = \int_B u(x, y) \mu(dy), \qquad x \in \mathbf{E}, B \in \mathscr{E}. \qquad (15)$$

When $B = \{x\}$ in (14) the left member is equal to zero by condition (ii). Hence μ is diffuse. Next putting $B = \{y\}$ with $y \neq x$ in (14) we obtain

$$L(x, \{y\}) = u(x, y) \mu(\{y\}) = 0. \qquad (16)$$

For an arbitrary $B \in \mathscr{E}$, we have

$$L(x, B) = \int_{B \setminus \{x\}} u(x, y) \frac{L(x, dy)}{u(x, y)} + L(x, B \cap \{x\})$$

$$= \int_{B \setminus \{x\}} u(x, y) \mu(dy) + L(x, B \cap \{x\})$$

$$= \int_B u(x, y) \mu(dy) + L(x, B \cap \{x\}) \qquad (17)$$

since μ is diffuse. Therefore (15) is true if and only if

$$\forall x: L(x, \{x\}) = P^x\{\gamma > 0; X(\gamma-) = x\} = 0. \tag{18}$$

In the proof of (18) we consider two cases according as the point x is holding or not. Recall that x is a holding point if and only if almost every path starting at x remains at x for a strictly positive time. It follows by the zero-one law that if x is not holding point, then almost every path starting at x must be in $\mathbf{E} - \{x\}$ for some rational value of t in $(0, \delta)$ for any $\delta > 0$, by right continuity of the path (this is the first explicit use of right continuity). Define a sequence of rational-valued optional times $\{S_n, n \geq 1\}$ as follows:

$$S_n = \min\left\{\frac{m}{2^n}\,\middle|\,m \geq 1; X\left(\frac{m}{2^n}\right) \neq x\right\}.$$

If x is not holding, then $P^x\{\lim_n S_n = 0\} = 1$. We have

$$P^x\{\gamma > S_n; X(\gamma-) = x\} = E^x\{P^{X(S_n)}[\gamma > 0; X(\gamma-) = x]\}. \tag{19}$$

(Simple Markov property is sufficient here since S_n is countably valued.) Since $X(S_n) \neq x$, the right member of (19) equals zero by (16) with x and y interchanged. Letting $n \to \infty$ in (19) we obtain (18).

If x is a holding point, then it is clear that

$$0 < U(x, \{x\}) = u(x, x)\xi(\{x\}).$$

It follows firstly that $\xi(\{x\}) > 0$ and secondly that $U(x, \{x\}) = \infty$. Together with the hypothesis that x is holding the latter condition implies that $\{x\}$ is a recurrent set under P^x. This is a familiar result in the theory of Markov chains (where the state space is a countable set). Moreover, another basic result in the latter theory asserts that it is almost impossible for the path to go from a recurrent set to a transient set. Both results can be adapted to the case in question, the details of which are left in Exercise 8 below (strong Markov property is needed). In conclusion, we have proved that if x is holding then $P^x\{T_A < \infty\} = 0$, which implies (18). We summarize the results above with an important addition as follows.

Theorem 1. *Let X be a Hunt process with the potential kernel in (1) satisfying conditions (i) and (ii). Then for each transient set A, there exists a Radon measure μ_A such that for any $x \in \mathbf{E}$ and $B \in \mathscr{E}$:*

$$E^x\{\gamma_A > 0; X(\gamma_A-) \in B\} = \int_B u(x, y)\mu_A(dy). \tag{20}$$

If almost all paths of the process are continuous, then μ_A has support in ∂A. In general if A is open then μ_A has support in $\overline{A}$.

Proof. If the paths are continuous, then clearly we have on $\Delta_A : X(\gamma_A -) = X(\gamma_A) \in \partial A$. In general if A is open, then on Δ_A it is impossible for $X(\gamma_A) \in A$ by right continuity of the paths. Hence there is a sequence of values of t strictly increasing to γ_A at which $X(t) \in A$; consequently the left limit $X(\gamma_A -) \in \bar{A}$. By (6), $L(x, \cdot)$ has support in ∂A in the first case, and in $\bar{A}$ in the second case. Hence so does μ_A by (14). $\qquad \square$

It is essential to see why for a compact A the argument above does not show that μ_A has support in A. For it is possible on Δ_A that $X(\gamma_A) \in A$ while $X(\gamma_A -) \notin \bar{A}$; namely the path may jump from anywhere to ∂A and then quit A forever.

Corollary. *We have*

$$P^x\{T_A < \infty\} = \int u(x, y)\mu_A(dy). \tag{21}$$

This follows from (20) when $B = \mathbf{E}$, in view of (5).

The measure μ_A is called the *equilibrium measure* for A. Its exact determination in electrostatics is known as Robin's problem. Formula (14) above gives the stochastic solution to this problem in R^d, $d \geq 3$. In order to amend Theorem 1 and its corollary so that μ_A will have support in $\bar{A}$ for an arbitrary Borel set A, there are several possibilities. The following expedient is due to John B. Walsh. Consider the left-hitting time T_A^- defined in (23) of §3.3, and the corresponding *left quitting time* γ_A^-;

$$\gamma_A^-(\omega) = \sup\{t > 0 \mid X_{t-}(\omega) \in A\} \tag{22}$$

where $\sup \varnothing = 0$. Since for a Hunt process left limits exist in $(0, \infty)$, and $t \to X_{t-}$ is left continuous, we have $X(\gamma_A^- -) \in \bar{A}$, regardless if the sup in (22) is attained or not. This is the key to the next result.

Theorem 2. *Under the hypotheses of Theorem 1 there exists a Radon measure μ_A^- with support in $\bar{A}$ such that for every $x \in \mathbf{E}$ and $B \in \mathscr{E}$:*

$$E^x\{\gamma_A^- > 0; X(\gamma_A^- -) \in B\} = \int u(x, y)\mu_A^-(dy). \tag{23}$$

Proof. Let us beware that "left" notions are not necessarily the same as "right" ones! The definition of transience is based on $X_t = X_{t+}$, not X_{t-}, and it is not obvious that the transient set A will remain "transient" for the left limits of the process. The latter property means $P^x\{\gamma_A^- < \infty\} = 1$ for every x. That this is indeed true is seen as follows. We have by Theorem 9 of §3.3, $T_A^- \geq T_A$ a.s. for any Borel set A. On the other hand, the left analogue of (4) is true: $\{\gamma_A^- > t\} = \{T_A^- \circ \theta_t < \infty\}$. It follows that

$$\{\gamma_A^- > t\} \subset \{T_A \circ \theta_t < \infty\} = \{\gamma_A > t\}.$$

Therefore, $\{\gamma_A^- = \infty\} \subset \{\gamma_A = \infty\}$, namely A is left-transient if it is (right)-transient. [Is the converse true?]

The rest of the proof of Theorem 2 is exactly the same as that of Theorem 1, and the question of support of μ_A^- is settled by the remark preceding the theorem.   □

Since left notions are in general rather different from their right analogues, Theorem 2 would require re-thinking of several basic notions such as "left-regular" and "left-polar" in the developments to follow. Fortunately under Hypotheses (B) and (L), we know by Theorem 3 of §3.8 that $T_A = T_A^-$ a.s. It follows that $\gamma_A = \gamma_A^-$ a.s. as well (Exercise 6) and so under these hypotheses Theorem 2 contains Theorem 1 as a particular case. We state this as a corollary. We shall denote the support of a measure μ by $\lfloor \mu$.

Corollary. Under Hypothesis (B), (20) and (21) hold with $\lfloor \mu_A \subset \bar{A}$.

Why is Hypothesis (L) not mentioned? Because it is implied by the conditions of the potential kernel, provided we use ξ as the reference measure (Exercise 2).

It is known that Hypothesis (B) holds under certain duality assumptions (see Meyer [3]). In order to state a set of conditions in the context of this section under which Hypothesis (B) holds, we need the next proposition.

Proposition 3. *Under the conditions of Theorem 1, for each y the function $x \to u(x, y)$ is superaveraging. If it is lower semi-continuous then it is excessive.*

Proof. For each $f \in b\mathscr{E}_+$, Uf is excessive by Proposition 2 of §2.1. Hence for each $t > 0$:

$$P_t Uf(x) = \int P_t u(x, y) f(y) \xi(dy) \leq \int u(x, y) f(y) \xi(dy)$$
$$= Uf(x), \tag{24}$$

where

$$P_t u(x, y) = \int P_t(x, dz) u(z, y). \tag{25}$$

Since (24) is true for all $f \in b\mathscr{E}_+$, it follows that for each x there exists N_x with $\xi(N_x) = 0$ such that if $y \notin N_x$:

$$P_t u(x, y) \leq u(x, y).$$

Now the measure ξ charges every nonempty open set (why?). Hence for an arbitrary y we have $y_n \notin N_x$, $y_n \to y$, so that $u(z, y_n) \to u(z, y)$ for every z by condition (i). Therefore by Fatou:

$$P_t u(x, y) \leq \varliminf_n P_t u(x, y_n) \leq \varliminf_n u(x, y_n) = u(x, y).$$

This proves the first assertion of the proposition. The second follows from Proposition 3 of §3.2 and the remark following it.   □

Since $u(\cdot, y)$ is superaveraging, we denote its regularization by $\underline{u}(\cdot, y)$:

$$\underline{u}(x, y) = \lim_{t \downarrow 0} P_t u(x, y).$$

For each y, $x \to \underline{u}(x, y)$ is excessive by Proposition 5 of §3.2. Observe that the function $\underline{u}$ may not satisfy the conditions (i) and (ii), in particular $\underline{u}(x, x)$ may not be infinite. The following results is proved in Chung and Rao [1] but the proof is too difficult to be given here. A simpler proof would be very interesting indeed.

Theorem 4. *Under the conditions of Theorem 1, if we assume also that*

    (a)   *each compact is transient,*
    (b)   *for each x, $\underline{u}(x, x) = +\infty$,*

then Hypothesis (B) is true.

In view of Proposition 3, condition (b) above hold if $u(\cdot, y)$ is lower semi-continuous for each y. In this case condition (a) is satisfied if the process is transient according to the definition given in §3.2, by part of Theorem 2 of §3.7. We shall return to these conditions in §5.2.

It is clear in the course of this book that the systematic use of hitting times (balayage) constitutes a major tool in the theory of Hunt processes. By comparison, the notion of quitting times was of recent origin and its potentials remain to be explored. The next result serves as an illustration of the method. Further work along this line should prove rewarding.

Theorem 5. *Assume Hypothesis (B) as well as (R). Let A_n be transient sets such that $A_n \downarrow A$ and $\bigcap_n \bar{A}_n = A$. Then we have for each $x \in A^c \cup A^r$ and each $f \in b\mathbb{C}$:*

$$\lim_n L_{A_n}(x, f) = L_A(x, f). \tag{26}$$

In other words the sequence of measures $L_{A_n}(x, \cdot)$ converges tightly to $L_A(x, \cdot)$.

Proof. We begin with the following basic relation. For each $x \in A^c \cup A^r$:

$$\bigcap_n \{T_{A_n} < \infty\} = \{T_A < \infty\}, \qquad P^x\text{-a.s.} \tag{27}$$

This has been proved before in this book (where?), though perhaps not exactly in this form. The reader should ponder why the conditions on x and on the transience of A_n are needed, as well as the quasi left continuity of the process. Now we write (27) in terms of quitting times as follows:

$$\bigcap_n \{\gamma_{A_n} > 0\} = \{\gamma_A > 0\}. \tag{28}$$

We will omit below the cliche "almost surely" when it is obvious. Clearly $\gamma_{A_n} \downarrow$ and $\gamma_{A_n} \geq \gamma_A$. Let $\beta = \lim_n \gamma_{A_n}$. Then on $\{\gamma_A > 0\}$, we have $X(\gamma_{A_n}-) = X(\gamma_{A_n}^-) \in \bar{A}_n$ as shown above on account of Hypothesis (B). It follows by right continuity that if $\beta < \gamma_{A_n}$ for all n then

$$X(\beta) = \lim_n X(\gamma_{A_n}-) \in \bigcap_n \bar{A}_n = A. \tag{29}$$

Thus $\beta \leq \gamma_A$ and so $\beta = \gamma_A$. The last equation is trivial if $\beta = \gamma_{A_n}$ for some n. Next, we prove that on $\{\gamma_A > 0\}$ we have

$$\lim_n X(\gamma_{A_n}-) = X(\gamma_A-). \tag{30}$$

This is trivial if X is continuous at γ_A. The general argument below is due to John B. Walsh and is somewhat delicate. If X has a jump at γ_A, this jump time must be one of a countable collection of optional times $\{\alpha_n\}$, by Exercise 5 of §3.1. This means that for almost every ω, $\gamma_A(\omega) = \alpha_n(\omega)$ where n depends on ω. We can apply the strong Markov property at α_n for all n to "cover γ_A" whenever X is discontinuous there. (We cannot apply the property at γ_A!) Two cases will be considered for each α_n, written as α below.

Case 1. $X(\alpha) \notin A$. Applying (27) with $x = X(\alpha)$, we see that since $T_A \circ \theta_\alpha = \infty$ on $\{\alpha = \gamma_A\}$, there exists $N(\omega) < \infty$ such that $T_{A_n} \circ \theta_\alpha = \infty$, hence $\alpha = \gamma_{A_n}$ for $n \geq N(\omega)$. Thus (30) is trivially true because $\gamma_{A_n} = \gamma_A$ for all sufficiently large values of n.

Case 2. $X(\alpha) \in A$. Then on $\{\alpha = \gamma_A\}$ we must have $X(\alpha) \in A \backslash A^r$ because the path does not hit A at any time strictly after α. Since α is a jump time and $A \backslash A^r$ is semipolar by Theorem 6 of §3.5, this possibility is ruled out under Hypothesis (B), by Theorem 1 (iv) of §3.8. This ends the proof of (30).

Now let $x \in A^c \cup A^r$ and $f \in b\mathbb{C}$. Then we have by (28), $\{\gamma_{A_n} > 0\} \downarrow \{\gamma_A > 0\}$ P^x-a.s.; hence by (30) and bounded convergence:

$$\lim_n E^x\{\gamma_{A_n} > 0; f(X(\gamma_{A_n}-))\} = E^x\{\gamma_A > 0; f(X(\gamma_A-))\}. \tag{31}$$

Recalling (6) this is the assertion in (26).   □

Corollary. *If A_1 is compact then*

$$\lim_n \mu_{A_n}(\mathbf{E}) = \mu_A(\mathbf{E}). \tag{32}$$

Proof. We may suppose $A_1 \neq \mathbf{E}$. There exists $x_0 \notin A_1$, and $u(x_0, \cdot)^{-1}$ is bounded continuous on A_1. Hence we have by (26).

$$\lim_n \int_{A_n} \frac{L_{A_n}(x_0, dy)}{u(x_0, y)} = \int_A \frac{L_A(x_0, dy)}{u(x_0, y)}$$

which reduces to (32). $\qquad\square$

The corollary will be applied in the next section to yield important results, under further conditions on the potential kernel.

Exercises

1. Let $(\mathbf{E}, \mathscr{E})$ be as in §1.1. A measure μ on $\mathscr{E}$ is called a Radon measure iff $\mu(K) < \infty$ for each compact subset K on $\mathbf{E}$. Prove that if μ and v are two Radon measures such that $\int f\, d\mu = \int f\, dv$ for all $f \in \mathbb{C}_c$, then $\mu \equiv v$. [Hint: use Lemma 1 of §1.1 to show that $\mu(D) = v(D) < \infty$ for each relatively compact open D; then use Lemma 2 of §1.1 to show that $\mu(B \cap D) = v(B \cap D)$ for all $B \in \mathscr{E}$. Apply this to a sequence of D_k such that $D_k \uparrow \mathbf{E}$. This exercise is given here because it does not seem easy to locate it in textbooks.]

In Exercises 2 and 5, we assume the conditions of Theorem 1.

2. Let A be transient. If $\mu_A(\mathbf{E}) = 0$, then A is polar. Conversely if $P_A 1(x) = 0$ for some x, then $\mu_A(\mathbf{E}) = 0$.

3. Assume $U(x, K) < \infty$ for each x and compact K. Let f be any excessive function. If $f(x) > 0$ for some x then $f(x) > 0$ for all x. [Hint: use Proposition 10 of §3.2.]

4. Assume each compact set is transient. Let f be an excessive function. If $f(x) < \infty$ for some x then the set $\{x \mid f(x) = \infty\}$ is a polar set. [Hint: use Theorem 7 of §3.4 and $u > 0$.]

5. Under the same conditions as Exercise 4 prove that each singleton is polar. [Hint: let D_n be open relatively compact, $D_n \downdownarrows \{x_0\}$; $P_{D_n} 1 = U\mu_n$ with $|\mu_n \subset \bar{D}_n$. Show that (a subsequence of) $\{\mu_n\}$ converges vaguely to $\lambda \delta_{x_0}$, and $\lambda u(x, x_0) \leq \underline{\lim}_n U\mu_n(x) \leq 1$ so that $\lambda = 0$. For $x_1 \notin \bar{D}_1$, $\lim_n U\mu_n(x_1) = 0$ because $u(x_1, \cdot)$ is bounded continuous in $\bar{D}_1$. Now use Exercise 2.]

6. Prove that for a Hunt process, $\gamma_A \geq \gamma_A^-$ a.s. for any $A \in \mathscr{E}$. If $T_A = T_A^-$ a.s. then $\gamma_A = \gamma_A^-$ a.s.

7. Prove that a polar set is left-polar; a thin set is left-thin.

8. The following is true for any Hunt process (even more generally). Let x be a holding but not absorbing point. Define the *sojourn time* $S = \inf\{t > 0 \mid X_t \neq x\}$, and the re-entry time $R = \inf\{t > S \mid X_t = x\}$. Show that

$P^x\{S > t\} = e^{-\lambda t}$ for some $\lambda: 0 < \lambda < \infty$. Let $P^x\{R < \infty\} = \rho$. Show that $U(x, \{x\}) = \lambda^{-1}(1 - \rho)^{-1}$; hence $U(x, \{x\}) = \infty$ if and only if $\{x\}$ is "recurrent under P^x". Let $A \in \mathscr{E}$; prove that if $P^x\{T_A < \infty\} > 0$ then $P^x\{T_A < R\} > 0$. Consequently if $\{x\}$ is recurrent under P^x, then so is A. Note: without further assumption on the process, a holding point x may not be hit under P^y, $y \neq x$; hence the qualification "under P^x" above.

9. Derive a formula for $E^x\{e^{-\lambda\gamma}f(X_\gamma-); \gamma > 0\}$ for $\lambda > 0$, $\gamma = \gamma_A$ as in Theorem 1.

10. Find the equilibrium measure μ_A when A is a ball in the Newtonian case (namely Brownian motion in R^3).

5.2. Some Principles of Potential Theory

In this section we continue the study of a Hunt process with the potential kernel given in (1) of §5.1, under further conditions. Recall that Hypothesis (L) is in force with ξ as the reference measure. The principal new condition is that of symmetry:

$$\forall(x, y): u(x, y) = u(y, x). \tag{S}$$

Two alternative conditions of transience will be used:

$$\forall x, \forall \text{ compact } K: U(x, K) < \infty; \tag{T_1}$$

$$\text{each compact set is transient.} \tag{T_2}$$

Two alternative conditions of regularity (besides (R) in §5.1) will be used:

$$\forall y: x \to u(x, y) \text{ is lower semi-continuous}; \tag{U_1}$$

$$\forall y: x \to u(x, y) \text{ is excessive.} \tag{U_2}$$

There are various connections between the conditions. Recall that (T_1) and (U_1) imply (T_2) by Theorem 2 of §3.7; (R) and (U_1) imply (U_2) by Proposition 3 of §5.1; (R), (T_2) and either (U_1) or (U_2) imply Hypothesis (B) by Theorem 4 of §5.1; (R) and (S) imply (U_1) trivially.

Readers who do not wish to keep track of the various conditions may assume (R), (S) and (T_1), and be assured that all the results below hold true with the possible exception of those based on the energy principle. However, it is one of the fascinating features of potential theory that the basic results are interwoven in the manner to be illustrated below. This has led to the development of *axiometic potential theory* by Brelot's school.

The *Equilibrium Principle* of potential theory may be stated as follows. For each compact K, there exists a finite measure μ_K with support in K such that

$$P_K 1 = U\mu_K. \tag{E}$$

The principle holds under (R), (T_2) and either (U_1) or (U_2), because then Hypothesis (B) holds and (E) holds by the Corollary to Theorem 2 of §5.1.

We are interested in the mutual relationship between several major principles such as (E). Thus we may assume the validity of (E) itself in some of the results below. Let us first establish a basic relation known as Hunt's "switching formula" in the duality theory. The proof is due to K. M. Rao. From here on in this section the letters K, D, and B are reserved respectively for a compact, relatively compact open, and (nearly) Borel set.

Theorem 1. *Under (S), (T_1) and (U_2), we have for each B:*

$$P_B u(x, y) = P_B u(y, x). \tag{1}$$

Proof. We begin by noting that we have by (S)

$$P_B u(x, y) = \int u(y, z) P_B(x, dz); \tag{2}$$

hence for each x, $y \to P_B u(x, y)$ is excessive by simple analysis (cf., Exercise 8 of §4.1). On the other hand for each y, $x \to P_B u(x, y)$ is excessive by (U_2) and a general property of excessive function (Theorem 4 of §3.4).

Let $K \subset D$, we have

$$U 1_K = P_D U 1_K < \infty$$

by Theorem 3 of §3.4 and (T_1); namely

$$\int_K u(x, y) \xi(dy) = \int_K P_D u(x, y) \xi(dy) < \infty. \tag{3}$$

Since $u(x, y) \geq P_D u(x, y)$ by Theorem 4 of §3.4, and K is arbitrary for a fixed D, it follows from (3) that

$$u(x, y) = P_D u(x, y) \tag{4}$$

for each x and ξ-a.e. y in D. Both members of (4) are excessive in y for each fixed x, hence (4) holds for each x in $\mathbf{E}$ and $y \in D$, by the Corollary to Proposition 4 of §3.5. [This property of u is of general importance.]

Next we have

$$P_K u(x, y) = \int P_K(x, dz) u(y, z) = \int P_K(x, dz) P_D u(y, z) \tag{5}$$

because $P_K(x, \cdot)$ is supported by K, and $u(y, z) = P_D u(y, z)$ for $z \in K \subset D$ by (4) as just proved. It follows by Fubini and (S) that the quantity in (5) is equal to

$$\int P_D(y, dw) \int P_K(x, dz) u(z, w) = \int P_D(y, dw) P_K u(x, w)$$
$$\leq \int P_D(y, dw) u(x, w)$$
$$= P_D u(y, x).$$

Thus we have proved that for all x and y, and $K \subset D$:

$$P_K u(x, y) \leq P_D u(y, x). \tag{6}$$

Integrating this with respect to $\xi(dx)$ over an arbitrary compact set C, we have

$$\int_C \xi(dx) P_K u(x, y) \leq \int_C P_D u(y, x) \xi(dx) = P_D U 1_C(y). \tag{7}$$

Taking a sequence of such sets $D_n \downarrow\downarrow K$, we have $P_{D_n} U 1_C(y) \downarrow P_K U 1_C(y)$, provided $y \notin K \backslash K^r$, by the Corollary to Theorem 5 of §2.4. Using this sequence in (7), we obtain

$$\int_C \xi(dx) P_K u(x, y) \leq P_K U 1_C(y) = \int_C P_K u(y, x) \xi(dx), \qquad y \notin K \backslash K^r. \tag{8}$$

It is essential to notice that the last member above is finite by (T_1) because it does not exceed $U(y, C)$. Since C is arbitrary we deduce from (8) that

$$P_K u(x, y) \leq P_K u(y, x) \tag{9}$$

for each $y \notin K \backslash K^r$, and ξ-a.e. x. Both members of (9) are excessive in each variable when the other variable is fixed and $\xi(K \backslash K^r) = 0$ because $K \backslash K^r$ is semipolar. Therefore (9) holds in fact for all x and y: since it is symmetric in x and y, we conclude that

$$\forall(x, y): P_K u(x, y) = P_K u(y, x). \tag{10}$$

Now given B and x, there exist compacts $K_n \subset B$, such that $T_{K_n} \downarrow T_B$, P^x-a.s. by Theorem 8(b) of §3.3. Since for each y, $u(X_t, y)$ is right continuous in t under (U_2) by Theorem 6 of §3.4, it follows by Fatou and (10) that

$$P_B u(x, y) \leq \varliminf_n P_{K_n} u(x, y) = \varliminf_n P_{K_n} u(y, x) \leq P_B u(y, x). \tag{11}$$

Interchanging x and y we obtain (1).   □

We are now ready to establish another major principle, known as the Maria-Frostman *Maximum Principle* as follows. For any σ-finite measure μ supported by the compact K, we have

$$\sup_{x \in E} U\mu(x) = \sup_{x \in K} U\mu(x) \tag{M}$$

This principle must be intuitively obvious to a physicist, since it says that the potential induced by a charge is greatest where the charge lies. Yet its proof seems to depend on some sort of duality assumption on the distribution of the charge. This tends to show that physical processes carry an inherent duality suggested by the reversal of time.

Theorem 2. *The maximum principle* (M) *holds under* (S), (T_1) *and* (U_2).

Proof. There is nothing to show if the right member of (M) is infinite, so we may suppose it finite and equal to M. For $\varepsilon > 0$ define the set

$$B = \{x \in E \mid U\mu(x) \le M + \varepsilon\}. \tag{12}$$

Since $U\mu$ is excessive under (U_2), it is finely continuous (Corollary 1 to Theorem 1 of §3.5) and so B is finely closed. The fine continuity of $U\mu$ also implies that $K \subset B^r$ (why?). Therefore we have by Theorem 1 and Fubini:

$$P_B U\mu(x) = \int P_B u(x, y)\mu(dy) = \int P_B u(y, x)\mu(dy)$$
$$= \int u(y, x)\mu(dy) = U\mu(x), \tag{13}$$

because $\lfloor \mu \subset K$ and for each $y \in K$, $P_B u(y, x) = u(y, x)$ trivially since $y \in B^r$. On the other hand, $P_B(x, \cdot)$ has support in B by Theorem 2 of §3.4, since B is finely closed. Hence we have

$$P_B U\mu(x) = \int P_B(x, dy)U\mu(y) \le \sup_{y \in B} U\mu(y) \le M + \varepsilon. \tag{14}$$

Putting (13) and (14) together we conclude that $U\mu \le M$ since ε is arbitrary. $\square$

The argument leading to (13) is significant and recorded below.

Corollary. *For any σ-finite measure μ such that* $\lfloor \mu \subset B^r$, *we have*

$$U\mu = P_B U\mu. \tag{15}$$

In conjunction with (E), an interesting consequence of (15) is the following particular case of Hypothesis (B):

$$P_K 1 = P_G P_K 1. \tag{16}$$

In some situations this can take the place of Hypothesis (B).

The next principle will be named the *Polarity Principle*. For Brownian notion it is proved in Theorem 10 of §4.5. Its general importance was recognized by Hunt. The depth of this principle is evident from the following formulation, in which we follow a proof of K. M. Rao's with some modification.

Theorem 3. *Assume* (T_2), *Hypothesis* (B), *and both the equilibrium principle* (E) *and the maximum principle* (M). *Then the polarity principle holds as follows*:

$$\text{Every semipolar set is polar.} \tag{P}$$

Remark. For a thin set this says that if it cannot be hit immediately from some point, then it cannot be hit at all from any point!

Proof. By Theorem 8(b) of §3.3 and Theorem 6 of §3.5, it is sufficient to prove that each compact thin set is polar. Let K be such a set, and define for each $n \geq 1$:

$$A_n = K \cap \left\{ x \in \mathbf{E} \,\middle|\, P_K 1(x) \leq 1 - \frac{1}{n} \right\}.$$

Since $P_K 1$ is finely continuous, each A_n is finely closed. Let L be a compact subset of A_n. Then $P_L 1(x) \leq 1 - 1/n$ for $x \in L$; hence for all x by (M). This implies $E^x\{e^{-T_L}\} \leq 1 - 1/n$ for all x; hence L is polar by Exercise 6 of §3.8, where Hypothesis (B) is used. Thus A_n is polar and so is $A = \bigcup_{n=1}^{\infty} A_n$. Put $C = K - A$. Then C is finely closed (why?); and we have

$$\forall x \in \mathbf{E}: P_K 1(x) = P_C 1(x); \qquad \forall x \in C: P_C 1(x) = 1. \tag{17}$$

Put $T_1 = T_K$; if α is a countable ordinal which has the predecessor $\alpha - 1$, put

$$T_\alpha = T_{\alpha-1} + T_K \circ \theta_{T_{\alpha-1}};$$

if α is a limit countable ordinal, put

$$T_\alpha = \sup_{\beta < \alpha} T_\beta$$

where β ranges over all ordinals preceding α. The following assertions are almost surely true for each α. Since A is polar, we have $X(T_\alpha) \in C$ on

$\{T_\alpha < \infty\}$. For a limit ordinal α this is a consequence of quasi left continuity. In view of (17), we have for each α:

$$\{T_\alpha < \infty\} = \bigcap_{n=1}^{\infty} \{T_{\alpha+n} < \infty\}. \tag{18}$$

On the other hand, since K is thin, strong Markov property implies that $T_\alpha < T_{\alpha+1}$ on $\{T_\alpha < \infty\}$. It follows as in the proof of Theorem 6 of §3.4 that there exists a first countable ordinal α^* (not depending on ω) for which $T_{\alpha^*} = \infty$. This α^* must be a limit ordinal by (18). Therefore, on the set of ω where K is ever hit, it is hit at a sequence of times which are all finite and increase to infinity. This contradicts the transience of K. Hence K cannot be hit at all, so it is polar.   □

Corollary. *For any $B \in \mathscr{E}^*$ the set $B \backslash B^r$ is polar.*

In particular for any open D, the set of points on ∂D which are not regular for D^c is a polar set, because it is a subset of $D^c \backslash (D^c)^r$. This is the form of the result known in classical potential theory as Kellogg-Evans theorem. The general form of (P) goes back to Brelot, Choquet and H. Cartan. Its relevance to the Dirichlet problem has been discussed in Proposition 11 of §4.5.

Now we return to the setting of §5.1, and adduce some important consequences about the equilibrium measure under the new assumptions (S), (T_2) and (M). For each transient set A, we define

$$C(A) = \mu_A(\mathbf{E}) \tag{19}$$

to be the *capacity* of A. Under (T_2), $C(K)$ is defined for each compact K and is a finite number since μ_K is a Radon measure. For any two σ-finite measures λ_1 and λ_2, we put

$$\langle \lambda_1, \lambda_2 \rangle = \iint \lambda_1(dx) u(x, y) \lambda_2(dy) = \int U\lambda_2 \, d\lambda_1. \tag{20}$$

The condition of σ-finiteness is made to ensure the applicability of Fubini's theorem. Under (S) it then follows that

$$\langle \lambda_1, \lambda_2 \rangle = \langle \lambda_2, \lambda_1 \rangle \tag{21}$$

whether finite or infinite. This symmetry plays a key role in what follows. The next result is a characterization of the capacity well known in classical potential theory.

Theorem 4. *Let v be any σ-finite measure with $\lfloor v \subset K$, such that*

$$\forall x \in K: Uv(x) \leq 1. \tag{22}$$

Then we have

$$v(K) \le C(K). \tag{23}$$

Proof. Let $K \subset D$; under (T_2) both K and D are transient and so μ_D as well as μ_K exist. Under (S) we have as a case of (20):

$$\int U\mu_D \, dv = \int Uv \, d\mu_D. \tag{24}$$

The left member above equals $v(K)$ because $U\mu_D = P_D 1 = 1$ on K. The right member does not exceed $\mu_D(\mathbf{E})$ because $Uv \le 1$ by (22) and (M). Now let $D_n \downarrow\downarrow K$; then we have proved that $v(K) \le \mu_{D_n}(\mathbf{E}) = C(D_n)$ for each n, and consequently (23) follows by the Corollary to Theorem 5 of §5.1 $\square$

Note that if the inequality in (22) holds for all $x \in \mathbf{E}$, then the intervention of (M) is not needed in the preceding proof. This is the case for Corollary 1 below.

Corollary 1. *If $K_1 \subset K_2$, then $C(K_1) \le C(K_2)$.*

Corollary 2. *If v is a σ-finite measure on $\mathscr{E}$ such that $Uv < \infty$, v-a.e, then v does not charge any polar set.*

Proof. Let $A_n = \{x \mid Uv(x) \le n\}$. If v charges a polar set it must charge a polar subset of A_n for some n, and so also a compact polar subset of A_n, because v is necessarily inner regular (Exercise 1). Let v_1 be the restriction of v to such a set K, and $v_2 = v_1/n$. Then $Uv_2 \le 1$ on K, hence by the theorem $v_2(K) \le C(K)$. But $C(K) = 0$ by Exercise 2 of §5.1. Thus $v_2 \equiv 0$ and v does not charge K after all. $\square$

Let λ and v be two finite measures on $\mathscr{E}$; then $\lambda - v$ is well defined by $(\lambda - v)(B) = \lambda(B) - v(B)$; integration with respect to $\lambda - v$ is also well defined by $\int f \, d(\lambda - v) = \int f \, d\lambda - \int f \, dv$ when both $\int f \, d\lambda$ and $\int f \, dv$ are finite. We extend the definition (20) as follows. The *energy* of $\lambda - v$ is defined to be

$$\langle \lambda - v, \lambda - v \rangle = \langle \lambda, \lambda \rangle - \langle \lambda, v \rangle - \langle v, \lambda \rangle + \langle v, v \rangle \tag{25}$$

provided that all four terms on the right are finite. This amounts to the condition that $\langle \lambda + v, \lambda + v \rangle < \infty$. Let us remark that if $\lambda_1 - \lambda_2 = \lambda_3 - \lambda_4$ and both $\langle \lambda_1 - \lambda_2, \lambda_1 - \lambda_2 \rangle$ and $\langle \lambda_3 - \lambda_4, \lambda_3 - \lambda_4 \rangle$ are defined then they are equal. If the energies of $\lambda_1 - \lambda_2$ and of $\lambda_3 - \lambda_4$ are defined then so is the energy of $(\lambda_1 - \lambda_2) - (\lambda_3 - \lambda_4) = (\lambda_1 + \lambda_4) - (\lambda_2 + \lambda_3)$. In particular we can define the energy of a signed measure using its Hahn-Jordan decomposition, but there is no need for that. We are ready to announce another principle.

Energy Principle. For any two finite measures λ and ν on $\mathscr{E}$, we have $\langle \lambda - \nu, \lambda - \nu \rangle \geq 0$ provided it is defined; if it is equal to zero then $\lambda \equiv \nu$.

Theorem 5. *Suppose that the Hunt process has a transition density function* p_t *with respect to* ξ *which is symmetric, namely for* $t > 0$:

$$P_t(x, dy) = p_t(x, y)\xi(dy); \qquad p_t(x, y) = p_t(y, x) \tag{26}$$

for all (x, y); *where* $p_t \geq 0$, $p_t \in \mathscr{E} \times \mathscr{E}$. *Then the energy principle holds.*

Proof. We have

$$u(x, y) = 2 \int_0^\infty p_{2t}(x, y)\, dt$$

and

$$p_{2t}(x, y) = \int p_t(x, z) p_t(z, y) \xi(dz).$$

Writing μ for $\lambda - \nu$ purely for abbreviation (not as "signed measure"), we have if $\langle \mu, \mu \rangle < \infty$:

$$\langle \mu, \mu \rangle = \iint \mu(dx) u(x, y) \mu(dy)$$

$$= 2 \int_0^\infty dt \int \left[\iint \mu(dx) p_t(x, z) p_t(z, y) \mu(dy) \right] \xi(dz)$$

$$= 2 \int_0^\infty dt \int \left(\int \mu(dx) p_t(x, z) \right)^2 \xi(dz),$$

by the symmetry of p_t. Hence $\langle \mu, \mu \rangle \geq 0$. If $\langle \mu, \mu \rangle = 0$ then there exists a set N of t with $m(N) = 0$ and for each $t \notin N$ a set Z_t of z with $\xi(Z_t) = 0$ such that

$$\int \mu(dx) p_t(x, z) = 0 \quad \text{if } t \notin N, \, z \notin Z_t. \tag{27}$$

Let $f \in b\mathbb{C}$; it follows from (27) that

$$\int \mu(dx) P_t f(x) = 0, \quad \text{if } t \notin N. \tag{28}$$

Since X is right continuous, $\lim_{t \downarrow 0} P_t f(x) = f(x)$ for each x. Taking a sequence of $t \notin N$, $t \downarrow 0$ in (28) we obtain

$$\forall f \in b\mathbb{C}: \int f\, d\lambda - \int f\, d\nu = 0,$$

where both integrals are finite. Hence $\lambda \equiv \nu$ (see Exercise 1 of §5.1). $\qquad \square$

From here on all the results so far proved will be used wherever appropriate. Let K be a fixed compact set and $\Phi(K)$ denote the class of finite measures v such that $\underline{|v} \subset K$ and $\langle v, v \rangle < \infty$. It follows from Corollary 2 to Theorem 4 that such a v does not charge any polar set, since $\int Uv \, dv < \infty$. For the equilibrium measure μ_K, we have

$$\langle \mu_K, \mu_K \rangle = \int U\mu_K \, d\mu_K = \int P_K 1 \, d\mu_K \leq \mu_K(K) = C(K) < \infty. \qquad (29)$$

Hence $\mu_K \in \Phi(K)$. But then μ_K does not charge the set A below:

$$A = K \cap \{x \in \mathbf{E} \,|\, P_K 1(x) < 1\} \qquad (30)$$

because A is polar as in the proof of Theorem 3. Consequently the inequality in (29) is an equality:

$$\langle \mu_K, \mu_K \rangle = C(K). \qquad (31)$$

We know (Exercise 2 of §5.1) that $C(K) = 0$ if and only if K is polar. If $C(K) > 0$ let $v \in \Phi(K)$ with $v(K) = C(K)$. Let us compute

$$\langle v - \mu_K, v - \mu_K \rangle = \langle v, v \rangle - \langle v, \mu_K \rangle - \langle \mu_K, v \rangle + \langle \mu_K, \mu_K \rangle. \qquad (32)$$

Since v does not charge A we have

$$\langle v, \mu_K \rangle = \langle \mu_K, v \rangle = \int U\mu_K \, dv = \int P_K 1 \, dv = v(K). \qquad (33)$$

It follows from (31) and (33) that the energy of $v - \mu_K$ is equal to $\langle v, v \rangle - C(K)$. We have therefore proved the following result under the energy principle.

Theorem 6. *Let* $C(K) > 0$. *For any* $v \in \Phi(K)$ *with* $v(K) = C(K)$, *we have*

$$C(K) \leq \langle v, v \rangle \qquad (34)$$

where equality holds if and only if $v = \mu_K$.

In other words, the equilibrium measure for K is the unique measure among all measures on K having the same total mass which minimizes the energy. This is the way it was found by Gauss and Frostman for Newtonian and M. Riesz potentials (see Exercises 7–12 below). Here we have turned the table around by first establishing the existence of the equilibrium measure in §5.1 under fairly general conditions, then verifying its minimization property as shown above. Although more restrictive assumptions, in particular the symmetry of transition density, are imposed to ensure Theorem 6, these conditions still cover the classical cases with room to spare.

Another uniqueness theorem for the equilibrium measure based on duality (Theorem 3 of §4.1) is given in Exercise 2.

The word "equilibrium" connotes in physics a "steady state", namely a stationary distribution of some random process. Such an identification seems missing in the literature and is supplied below. Although it is a kind of "Columbus' egg" stood on its head, there may be possibilities for exploration: an extension to the asymmetric case is discussed in Chung and Rao [2].

We assume that the compact K is not a polar set and is regular: $K = K^r$. The latter condition amounts to the absence of the polar set A in (30). In this case the quitting kernel L_K in §5.1 is a strict probability kernel, namely:

$$\forall x \in K: L_K(x, K) = \int_K u(x, y)\mu_K(dy) = 1. \tag{35}$$

Hence we can construct a discrete parameter homogeneous Markov process on the state space K, with L_K as its transition probability function. Under (R) and (S), it is easy to see that the family of functions $\{u(x, \cdot), x \in K\}$, restricted to K, is uniformly integrable with respect to μ_K. For if $x_n \in K$ and $x_n \to x$, then $\lim_n u(x_n, y) = u(x, y)$ for each y; at the same time $L(x_n, K) = L(x, K)$ for all n. The asserted uniform integrability then follows from an elementary result (Theorem 4.5.4 of *Course*). This is stronger than the following condition, known as Doblin's Hypothesis (D); see Doob [1], p. 192. There exists $\varepsilon > 0$ such that if $A \subset K$, $A \in \mathscr{E}$ with $\mu_K(A) \leq \varepsilon$, then $L_K(x, A) \leq 1 - \varepsilon$ for all $x \in K$. It is an easy case of Doblin's general results that there exists a probability measure π on A such that for every $A \subset K$, $A \in \mathscr{E}$ and every $x \in K$:

$$\lim_n \frac{1}{n} \sum_{j=1}^{n} L_K^{(j)}(x, A) = \pi(A); \tag{36}$$

where the $L_K^{(j)}$'s are the iterates of $L_K = L_K^{(1)}$. The measure π is an invariant (stationary) measure for L, namely $\pi L_K = \pi$. Now under (S) there is an obvious invariant measure obtained by normalizing the equilibrium measure:

$$\mu_K^\circ = \frac{1}{\mu_K(E)} \mu_K = \frac{1}{C(K)} \mu_K. \tag{37}$$

It follows that $\pi = \mu_K^\circ$ by the remarks below.

The results above can be improved in several respects, for an arbitrary probability kernel on a compact space K, of the form $L(x, dy) = u(x, y)\mu(dy)$ where $\mu(K) < \infty$. If $u > 0$ and satisfies (U_1) (without (R) or (S)), then there is a unique invariant probability measure π and the Cesáro mean in (36) may be replaced by a simple limit; see Chung and Rao [2]. Thus the hypotheses of symmetry imposed in the preceding discussion are merely expedient to ensure the polarity and energy principles. A theory of energy for nonsymmetric kernels would be a major advance in this circle of ideas.

Let us now return briefly to the case of Brownian motion in R^3 which corresponds to the Newtonian potential. This is also referred to as the electrostatic potential induced by charges on conductors. In this case the equilibrium principle may be stated as follows. For each bounded Borel set A, there exists a finite measure μ_A supported by ∂A such that

$$P_A 1(x) = \frac{1}{2\pi} \int_{\partial A} \frac{\mu_A(dy)}{\|x - y\|}. \tag{38}$$

This representation is very fruitful. First, it yields the following formula for the capacity of A:

$$C(A) = 2\pi \lim_{x \to \infty} \|x\| P_A 1(x). \tag{39}$$

It follows at once that if $A_1 \subset A_2$, then $C(A_1) \leq C(A_2)$. Next, if A_n is a sequence of compacts decreasing to A, then the convergence in (39) is uniform with respect to n by an easy estimate based on (38). Consequently we can interchange limits in x and n to obtain

$$\lim_n C(A_n) = C(A). \tag{40}$$

Finally, for any two Borel sets A and B we have the inequality:

$$P_A 1 + P_B 1 \geq P_{A \cup B} 1 + P_{A \cap B} 1, \tag{41}$$

which is proved as follows. For each x, $P_{A \cup B} 1(x) - P_B 1(x)$ is the probability that the path from x hits $A \cup B$ but not B. Such a path must hit A but not $A \cap B$, the probability of which is $P_A 1(x) - P_{A \cap B} 1(x)$, hence the inequality. Applying (39) to (41) we obtain

$$C(A) + C(B) \geq C(A \cup B) + C(A \cap B). \tag{42}$$

This is the strong subadditivity of the capacity functions discussed in Proposition 3 of §3.3. Together with the other properties verified above, the development in §3.3 shows that we can extend the function $C(\cdot)$ to all subsets of R^3 to be a Choquet capacity. Cf. Helms [1, §7.5].

As another illustration, let us determine the equilibrium measure for a ball. Take $B = B(o, r)$ and $x = o$ in (14) of §5.1. Since $u(x, y) = (2\pi \|x - y\|)^{-1}$, we have for any Borel set A:

$$\mu_B(A) = 2\pi \int_A \|y\| L(o, dy). \tag{43}$$

Rotational symmetry implies that $L(o, \cdot)$ is the uniform distribution on ∂B. A formal proof may be given along the lines of Proposition (XI) of §4.2. It follows that

$$\mu_B(A) = 2\pi r \frac{\sigma(\partial B \cap A)}{4\pi r^2} = \frac{1}{2r}\sigma(\partial B \cap A);$$

in particular

$$C(B) = C(\partial B) = \mu_B(\partial B) = 2\pi r. \tag{44}$$

We can check this from (41) and (16) of §4.4:

$$\lim_{x\to\infty} 2\pi\|x\|P_B1(x) = \lim_{x\to\infty} 2\pi\|x\|\frac{r}{\|x\|} = 2\pi r.$$

For another application of (39), let A_1 be a Borel set with its closure contained in the interior of A. Then it is obvious that $P_A1(x) = P_{A-A_1}1(x)$ for $x \in A^c$. Hence $C(A) = C(A - A_1)$ by (39). It is interesting to see how this result can be sharpened and extended to the class of processes considered in §5.1, with continuous paths. We have by (19)

$$C(A) = \int_{\mathbf{E}} \frac{L_A(x, dy)}{u(x, y)}$$

for any x. Choose $x \in A^c$. Then the continuity of paths implies that under P^x we have $\gamma_A = \gamma_{A-A_1}$ almost surely, because the path from outside A cannot hit A without hitting $A - A_1$, and cannot quit $A - A_1$ (forever) without quitting A. Therefore, $L_A(x, \cdot) = L_{A-A_1}(x, \cdot)$ which says that the equilibrium measures for A and $A - A_1$ are the same. The fact that $\mu_{\bar{A}} = \mu_{\partial A}$ may be regarded as a limiting case.

In the language of electrostatics, if the interior of a solid conductor is partially hollowed out, distribution of the induced charge on the outside surface remains unchanged when it is grounded (in equilibrium). An analytic proof of this physical observation may be based on the energy minimizing characterization of the equilibrium measure (Theorem 6). To some of us such a proof may seem more devious than the preceding one, but this is a matter of subjective judgment and previous conditioning. Be it as it may, here is an appropriate place to end these notes (for the moment), leaving the reader with thoughts on the emprical origin of mathematical theory, the grand old tradition of analysis, and the relatively new departure founded on the theory of probability.

Exercises

1. Let $(\mathbf{E}, \mathcal{E})$ be as specified in §3.1. Show that any σ-finite measure ν on $\mathcal{E}$ is inner regular; namely: $\nu(B) = \sup_{K \subset B} \nu(K)$. [Hint: for a finite measure this follows from Exercise 12 of §2.1 of *Course*.]

2. Show that under the hypothesis of Theorem 5 the duality relation (18) of §4.1 holds with $\hat{u}^{\alpha} = u^{\alpha}$. Assume also (P). Prove the uniqueness of the equilibrium measure μ_K in the following form: if $Uv = U\mu_K$ on K and $\lfloor v \subset K^r$, then $v = \mu_K$. [Hint: use the Corollary to Theorem 2; and note that $\lfloor \mu_K \subset K^r$ by (P).]

3. Under (S) and (E), prove that if μ and v are σ-finite measures such that $U\mu \leq Uv$, then $\mu(\mathbf{E}) \leq v(\mathbf{E})$. [Hint: integrate with respect to μ_K.]

4. Let $\Phi^{\pm}(K)$ denote the class of $\lambda = \lambda_1 - \lambda_2$ where $\lambda_1 \in \Phi(K)$, $\lambda_2 \in \Phi(K)$. Under the energy principle prove that μ_K is the unique member of $\Phi^{\pm}(K)$ which minimizes G below:

$$G(\lambda) = \langle \lambda, \lambda \rangle - 2\lambda(K).$$

This is the quadratic actually used by Gauss.

5. Suppose that $\langle \lambda, \lambda \rangle \geq 0$ for every $\lambda \in \Phi^{\pm}(K)$ as defined in Exercise 4. This is called the *Positivity Principle* and forms part of the energy principle. Show that under (S) we have the Cauchy-Schwarz inequality: $\langle \lambda, v \rangle^2 \leq \langle \lambda, \lambda \rangle \langle v, v \rangle$ for any λ and v in $\Phi^{\pm}(K)$.

6. Generalize (34) as follows. Let K be compact: v a finite measure on K; u defined on $K \times K$ is symmetric and ≥ 0. Define $L(x, dy) = u(x, y)v(dy)$ and assume that for every $x \in K$ we have $L(x, K) \leq 1$. Prove that for any measure λ on K with $\lambda(K) \leq 1$ we have

$$\langle \lambda L, \lambda L \rangle \leq \langle \lambda, \lambda L \rangle.$$

Hence if the positivity principle in Exercise 5 holds, then $\langle \lambda L, \lambda L \rangle \leq \langle \lambda, \lambda \rangle$. Thus the energy of a subprobability measure decreases after each transformation by L. This is due to J. R. Baxter. [Hint: put $\varphi(y) = \int \lambda(dx)u(x, y)$. Then $\int (L\varphi)^2 \, dv \leq \int L(\varphi^2) \, dv \leq \int \varphi^2 \, dv$; hence $\langle \lambda L, \lambda L \rangle = \int (\varphi L \varphi) \, dv \leq \int \varphi^2 \, dv = \langle \varphi, \varphi L \rangle$.]

The next few problems are taken from Frostman [1] and meant to give some idea of the classical approach to the equilibrium problem. They are analytical propositions without reference to a process, where $(\mathbf{E}, \mathscr{E}) = (R^d, \mathscr{B}^d)$, $\xi = m$ (Lebesgue measure).

7. Let $u \geq 0$ and $(x, y) \to u(x, y)$ be lower semi-continuous. Prove that there exists a probability measure v such that

$$\langle v, v \rangle = \inf_{\lambda} \langle \lambda, \lambda \rangle$$

where λ ranges over all probability measures on a compact K. [Hint: take a vaguely convergent subsequence of a sequence $\{\lambda_n\}$ such that $\langle \lambda_n, \lambda_n \rangle$ tends to the infinum.]

8. Let v and λ be as in Exercise 7. Assume in addition that u is symmetric. Prove that $\langle v, v \rangle \leq \langle v, \lambda \rangle$ for any λ. Deduce that if $\langle v, v \rangle = \langle \lambda, \lambda \rangle$ then

$v \equiv \lambda$ under the energy principle. [Hint: consider the energy of $(1 - \varepsilon)v + \varepsilon\lambda$ for $\varepsilon \downarrow 0$.]

9. Suppose $\langle \lambda, \lambda \rangle < \infty$. Then $Uv = \langle v, v \rangle$, λ-a.e. on K. [Hint: let $\langle v, v \rangle = c > b > a > 0$ and suppose that $Uv < a$ on A with $\lambda(A) > 0$. It follows from lower semi-continuity that there exists a ball B with $v(B) > 0$ such that $Uv > b$ on B. Now transfer the mass $v(B)$ and re-distribute it over A proportionately with respect to λ; namely put for all S: $v'(S) = v(S \backslash B) + v(B)\lambda(S \cap A)\lambda(A)^{-1}$. Compute $\langle v, v' \rangle$ to get a contradiction with Exercise 8. Hence $Uv \geq c$, λ-a.e. on K.]

Note. It follows from (31) and (37) that $Uv_K = \langle v_K, v_K \rangle$ on $K \cap K^r$, hence except for a polar set by (P). If λ has finite energy then λ does not charge any polar set, hence the result in Exercise 9 is true by the methods used in §§5.1–5.2. The notion of "capacity" was motivated by such considerations.

10. Let $u(x, y) = \|x - y\|^{-\alpha}$ where $\alpha > 0$ in R^2, and $\alpha > 1$ in R^3. Prove that if $U\mu$ is continuous, then (M) holds. [Hint: the Laplacian of $U\mu$ at a point where maximum is attained must be < 0 by calculus. This is called the "elementary maximum principle" by Frostman and is an illustration of the humble origin of a noble result.]

11. Let u be as in Exercise 10 but $1 < \alpha < 3$ in R^3. Let μ have the compact support K and suppose that $U\mu$ considered as a function on K only is continuous there. Then it is continuous everywhere. This is known as the *Continuity Principle* and is an important tool in the classical theory. [Hint: $U\mu$ is continuous in K^c. It is sufficient to show that $\varphi_\delta(x) = \int_{B(x,\delta)} u(x, y)\mu(dy) \downarrow 0$ uniformly in x as $\delta \downarrow 0$. For $x \in K$ this is true by the hypothesis and Dini's theorem. For $x \notin K$ let y be a point in K nearest x. Then $\|y - z\| \leq 2\|x - z\|$ for all $z \in K$; if $\|x - y\| \leq \delta$ then $\varphi_\delta(x) \leq 2^\alpha \varphi_{2\delta}(y)$.]

12. Establish the celebrated M. Riesz convolution formula below:

$$\int_{R^d} \|x - y\|^{\alpha - d}\|y - z\|^{\beta - d}\,dy = C(d, \alpha, \beta)\|x - y\|^{\alpha + \beta - d}$$

for $d \geq 1$ and $0 < \alpha + \beta < d$, where C is a constant. [Hint: put $x = o$, $\|z\| = 1$ to see that the existence of C is trivial. For its explicit evaluation use the formula

$$k(\alpha)\|x - y\|^{\alpha - d} = \int_0^\infty t^{(\alpha/2) - 1}p_t(x, y)\,dt$$

where p_t is given in (1) of §4.2, and $k(\alpha)$ is easily computed.]

The next three exercises concern Brownian motion in R^3.

13. Let $B = B(o, r)$. Prove that

$$P^o\{\gamma_B \in dt\} = r(2\pi t^3)^{-1/2}e^{-r^2/2t}\,dt.$$

[Hint: $P^o\{\gamma_B > t\} = E^o\{P^{X(t)}[T_B < \infty]\}$; use (16) of §4.4.]

14. As in Exercise 13, prove that

$$E^o\{e^{-\alpha T_B}\} = 2r\sqrt{2\alpha}(1 - e^{-r\sqrt{2\alpha}})^{-1}.$$

[Hint: by Exercise 13 of §4.2, with $\lambda = (\alpha, 0, 0)$, we have

$$E^o\{e^{-\alpha T_B}\}^{-1} = E^o\{\exp[\sqrt{2\alpha}X_1(t)]\} = \frac{1}{\sigma(r)}\int_{\partial B}\exp[\sqrt{2\alpha}x_1]\sigma(dy).$$

Both results above for R^d, $d \geq 3$ are in Getoor [3]].

15. For any bounded Borel set A, and any Borel set S, prove that

$$\int_0^\infty dx\, P^x\{0 < \gamma_A \leq t; X(\gamma_A) \in S\} = t\mu_A(S)$$

where μ_A is the equilibrium measure for A. [Hint: use

$$\int_0^\infty dx(u(x, y) - P_t u(x, y)) = t.$$

Apart from the use of γ_A the result (valid for R^d, $d \geq 3$) is implicit in Spitzer [1].]

NOTES ON CHAPTER 5

The title of this chapter is a *double entendre*. It is not the intention of this book to treat potential theory except as a concomitant of the underlying processes. Nevertheless we shall proceed far enough to show several facets of this old theory in a new light.

§5.1 The content of this section is based on Chung [1]. This was an attempt to utilize the inherent duality of a Markov process as evidenced by its hitting and quitting times. Formula (20) or its better known corollary (21) is a particular case of the representation of an excessive function by the potential of a measure plus a generalized harmonic function. Such a representation is proved in Chung and Rao [1] for the class of potentials which satisfies the conditions of Theorem 1. In classical potential theory this is known as F. Riesz's decomposition of a subharmonic function; see Brelot [1]. We stop short of this fundamental result as it would take us too far into a well-entrenched field.

Let us point out that Robin's problem of determining the equilibrium charge distribution is *solved* by the formula (14). One might employ a kind of Monte Carlo method to simulate the last exit probability there to obtain empirical results.

§5.2. The various principles of Newtonian potential theory are discussed in Rao [1]. The extension of these to M. Riesz potentials was a major advance made by Frostman [1]. Their mutual relations form the base of an axiomatic potential theory founded by Brelot. Wermer [1] gives a brief account of the physical concepts of capacity and energy leading to the equilibrium potential.

In principle, it should be possible to treat questions of duality by the method of time reversal. Although a general theory of reversing has existed for some time, it seems still too difficult for applications. A prime example is the polarity principle (Theorem 3) which is not *prima facie* such a problem. Yet all known probabilistic proofs use some kind of reversing (cf. Theorem 10 of §4.5). It would be extremely interesting to obtain this and perhaps also the maximum principle (Theorem 2) by a manifest reversing argument. Such an approach is also suggested by the concept of reversibility of physical processes from which potential theory sprang.

Bibliography

L. V. Ahlfors
[1] *Complex Analysis*, Second edition, McGraw Hill Co., New York, 1966.

J. Azéma
[1] Théorie générale des processus et retournement du temps, *Ann. École Norm Sup. Ser. 4*, **6** (1973), 459–519.

R. M. Blumenthal and R. K. Getoor
[1] *Markov Processes and Potential Theory*, Academic Press, New York, 1968.

M. Brelot
[1] *Eléments de la Théorie Classique du Potentiel*, Fourth Edition, Centre de Documentation Universitaire, Paris, 1969.

K. L. Chung
[1] *A Course in Probability Theory*, Second Edition, Academic Press, New York, 1974.
[2] *Markov Chains with Stationary Transition Probabilities*, Second Edition, Springer-Verlag 1967.
[3] *Boundary Theory for Markov Chains*, Princeton University Press, Princeton NJ, 1970.
[4] A simple proof of Doob's theorem, Sém. de Prob. V, (1969/70), p. 76. *Lecture Notes in Mathematics No. 191*, Springer-Verlag, Berlin Heidelberg New York.
[5] On the fundamental hypotheses of Hunt Processes, Instituto di alta matematica, *Symposia Mathematica* **IX** (1972), 43–52.
[6] Probabilistic approach to the equilibrium problem in potential theory, *Ann. Inst. Fourier* **23** (1973), 313–322.
[7] Remarks on equilibrium potential and energy, *Ann. Inst. Fourier* **25** (1975), 131–138.
[8] Excursions in Brownian motion, *Arkiv för Mat.* **14** (1976), 155–177.

K. L. Chung and J. L. Doob
[1] Fields, optionality and measurability, *Amer. J. Math.* **87** (1965), 397–424.

K. L. Chung and J. Glover
[1] Left continuous moderate Markov processes, *Z. Wahrscheinlichkeitstheorie* **49** (1979), 237–248.

K. L. Chung and P. Li
[1] Comparison of probability and eigenvalue methods for the Schrödinger equation, *Advances in Math.* (to appear).

K. L. Chung and K. M. Rao
[1] A new setting for potential theory (Part I), *Ann. Inst. Fourier* **30** (1980), 167–198.
[2] Equilibrium and energy, *Probab. Math. Stat.* **1** (1981).
[3] Feynman-Kac functional and the Schrödinger equation, *Seminar in Stochastic Processes* **1** (1981), 1–29.

K. L. Chung and S. R. S. Varadhan
[1] Kac functional and Schrödinger equation, *Studia Math.* **68** (1979), 249–260.

K. L. Chung and J. B. Walsh
[1] To reverse a Markov process, *Acta Math.* **123** (1969), 225–251.

R. Courant
[1] *Differential and Integral Calculus*, Interscience 1964.

C. Dellacherie
[1] Ensembles aléatories I, II, Sém. de Prob. III (1967/68), pp. 97–136, *Lecture Notes in Mathematics No. 88*, Springer-Verlag.
[2] Capacités et Processus Stochastiques, Springer-Verlag Berlin Heidelberg New York 1972.

[3] Mesurabilité des debuts et théorème de section: le lot a la portée de toutes les bourses, Sém. de Prob. XV (1979/80), pp. 351–370 *Lecture Notes in Mathematics No.*, Springer-Verlag, Berlin Heidelberg New York.

C. Dellacherie and P. A. Meyer
[1] *Probabilités et Potentiel*, Hermann, Paris 1975.
[2] *Probabilités et Potentiel: Théorie des Martingales*, Hermann, Paris 1980.

J. L. Doob
[1] *Stochastic Processes*, Wiley and Sons, New York, 1953.
[2] Semimartingales and subharmonic functions, *Trans. Amer. Math. Soc.* **77** (1954), 86–121.
[3] A probability approach to the heat equation, *Trans. Amer. Math. Soc.* **80** (1955), 216–280.

E. B. Dynkin
[1] *Markov Processes*, Springer-Verlag, Berlin Heidelberg New York 1965.

E. B. Dynkin and A. A. Yushekevich
[1] *Theorems and Problems in Markov Processes*, (in Russian), Nauk, Moscow, 1967.

O. Frostman
[1] Potentiel d'équilibre et capacités des ensembles avec quelques applications à la théorie des fonctions, *Medd. Lunds Univ. Math. Sem.* **3** (1935), 1–118.

R. K. Getoor
[1] Markov processes: Ray processes and right processes, *Lecture Notes in Mathematics No. 440*, Springer-Verlag, Berlin Heidelberg New York 1975.
[2] Transience and recurrence of Markov Processes, Sém. de Prob. XIV (1978/79), pp. 397–409, *Lecture Notes in Mathematics No. 784*, Springer-Verlag, Berlin Heidelberg New York.
[3] The Brownian escape process, *Ann. Probability* **7** (1979), 864–867.

I. I. Gihman and A. V. Skorohod
[1] *The Theory of Stochastic Processes II* (translated from the Russian), Springer-Verlag Berlin Heidelberg New York 1975.

L. L. Helms
[1] *Introduction to Potential Theory*, Wiley and Sons, New York, 1969.

G. A. Hunt
[1] Some theorems concerning Brownian motion, *Trans. Amer. Math. Soc.* **81** (1956), 294–319.
[2] Markoff processes and potentials, I, *Illinois J. Math.* **1** (1957), 44–93.
[3] *Martingales et Processus de Markov*, Dunod, Paris, 1966.

K. Ito
[1] *Lectures on Stochastic Processes*, Tate Institute, Bombay 1961.

M. Kac
[1] On some connections between probability theory and differential and integral equations. *Proc. Second Berkeley Symposium on Math. Stat. and Probability*, 189–215. University of California Press, Berkeley 1951.

S. Kakutani
[1] Two-dimensional Brownian motion and harmonic functions, *Proc. Imp. Acad. Tokyo* **22** (1944), 706–714.

J. L. Kelley
[1] *General Topology*, D. Van Nostrand Co., New York 1955.

O. D. Kellogg
[1] *Foundations of Porential Theory*, Springer-Verlag Berlin, 1929.

R. Z. Khas'minskii
[1] On positive solutions of the equation $\mathscr{A}u + Vu = 0$. *Theory of Probability and its Applications* (translated from the Russian) **4** (1959), 309–318.

P. Lévy
[1] *Théorie de l'Addition des Variables Aléatoires*, Second edition, Gauthier-Villars, Paris, 1954.

P. A. Meyer
[1] *Probability and Potentials*, Blaisdell Publishing Co. 1966.
[2] Processus de Markov, *Lecture Notes in Mathematics No. 26*, Springer-Verlag, Berlin Heidelberg New York 1967.
[3] Processus de Markov: la frontière de Martin, *Lecture Notes in Mathematics No. 77*, Springer-Verlag, Berlin Heidelberg New York 1968.

P. A. Meyer, R. T. Smythe and J. B. Walsh
  [1] Birth and death of Markov processes, *Proc. Sixth Berkeley Symposium on Math. Stat. and Probability*, Vol. III, pp. 295–305, University of California Press, 1972.
S. Port and C. Stone
  [1] *Brownian Motion and Classical Potential Theory*, Academic Press, New York 1978.
K. M. Rao
  [1] Brownian motion and classical potential theory, *Aarhus University Lecture Notes* No. **47**, 1977.
H. L. Royden
  [1] *Real Analysis*, Second Edition, Macmillan, New York, 1968.
R. T. Smythe and J. B. Walsh
  [1] The existence of dual processes, *Invensiones Math.* **19** (1973), 113–148.
F. Spitzer
  [1] Electrostatic capacity, heat flow, and Brownian motion, *Z. Wahrscheinlichkeitstheorie* **3** (1964), 110–121.
J. M. Steele
  [1] A counterexample related to a criterion for a function to be continuous, *Proc. Amer. Math. Soc.* **79** (1980), 107–109.
E. C. Titchmarsh
  [1] *The Theory of Functions*, Second edition, Oxford University Press, 1939.
J. B. Walsh
  [1] The cofine topology revisited, *Proceedings of Symposia in Pure mathematics XXXI, Probability*, Amer. Math. Soc. (1977), pp. 131–152.
J. Wermer
  [1] Potential theory, *Lecture Notes in Mathematics No. 408*, Springer-Verlag, Berlin Heidelberg New York 1974.
J. G. Wendel
  [1] Hitting spheres with Brownian motion, *Ann. of Probability* **8** (1980), 164–169.
D. Williams
  [1] *Diffusions, Markov Processes, and Martingales*, Vol. 1, Wiley and Sons, New York 1979.
K. Yosida
  [1] *Functional Analysis*, Springer-Verlag, Berlin Heidelberg New York 1965.

Index

Grundlehren der mathematischen Wissenschaften

A Series of Comprehensive Studies in Mathematics

A Selection